Nonlinear Process Control

MICHAEL A. HENSON
DEPARTMENT OF CHEMICAL ENGINEERING
LOUISIANA STATE UNIVERSITY

DALE E. SEBORG
DEPARTMENT OF CHEMICAL ENGINEERING
UNIVERSITY OF CALIFORNIA, SANTA BARBARA

EDITORS

Prentice Hall PTR
Upper Saddle River, New Jersey 07458

Library of Congress Cataloging-in-Publication Data
Nonlinear process control / edited by Michael A. Henson and Dale E. Seborg
p. cm.
Includes bibliographical references and index.
ISBN 0-13-625179-X
1. Automatic control. 2. Nonlinear control theory. 3. Chemical process control.
I. Henson, Michael A. II. Seborg, Dale E.
TJ213.N5697 1997
629.8--dc21 96-46917
CIP

Editorial/production supervision: *Kathleen M. Caren*
Manufacturing manager: *Alexis R. Heydt*
Acquisitions editor: *Bernard M. Goodwin*
Editorial assistant: *Diane Spina*
Cover director: *Jayne Conte*
Marketing Manager: *Miles Williams*

Prentice-Hall, Inc.
A Simon & Schuster Company
Upper Saddle River, New Jersey 07458

The publisher offers discounts on this book when ordered in bulk quantities.
For more information, contact:

Corporate Sales Department
Prentice Hall PTR
1 Lake Street
Upper Saddle River, NJ 07458

Phone: 800-382-3419, Fax: 201-236-7141
E-mail: corpsales@prenhall.com

Printed in the United States of America
10 9 8 7 6 5 4 3 2 1

ISBN 0-13-625179-X

Prentice-Hall International (UK) Limited, *London*
Prentice-Hall of Australia Pty. Limited, *Sydney*
Prentice-Hall Canada Inc., *Toronto*
Prentice-Hall Hispanoamericana, S.A., *Mexico*
Prentice-Hall of India Private Limited, *New Delhi*
Prentice-Hall of Japan, Inc., *Tokyo*
Simon & Schuster Asia Pte. Ltd., *Singapore*
Editora Prentice-Hall do Brasil, Ltda., *Rio de Janeiro*

Contributors

RANCIS J. DOYLE III
chool of Chemical Engineering
urdue University
Vest Lafayette, IN

THOMAS F. EDGAR
Department of Chemical Engineering
University of Texas
Austin, TX

IICHAEL A. HENSON
epartment of Chemical Engineering
ouisiana State University
aton Rouge, LA

THOMAS J. MCAVOY
Department of Chemical Engineering
University of Maryland
College Park, MD

DWARD S. MEADOWS
epartamento de Eletrônica
ontifícia Universidade Católica
do Rio Grande do Sul
orto Alegre, RS Brazil

KENNETH R. MUSKE
Computational Science Methods Group
Los Alamos National Laboratory
Los Alamos, NM

ABATUNDE A. OGUNNAIKE
entral Science and Engineering
.I. du Pont de Nemours and Co., Inc.
Vilmington, DE

RONALD K. PEARSON
Central Science and Engineering
E.I. du Pont de Nemours and Co., Inc.
Wilmington, DE

AMES B. RAWLINGS
epartment of Chemical Engineering
Jniversity of Wisconsin
Iadison, WI

DALE E. SEBORG
Department of Chemical Engineering
Unversity of California
Santa Barbara, CA

IONG-TE SU
Ioneywell Technology Center
Ioneywell Inc.
Iinneapolis, MN

Contents

Preface

In the past decade, the control of nonlinear systems has received considerable attention in both academia and industry. The recent interest in the design and analysis of nonlinear control systems is due to several factors. First and foremost, linear controllers usually perform poorly when applied to highly nonlinear systems or moderately nonlinear systems that operate over a wide range of conditions. On the other hand, significant progress has been made in the development of model-based controller design strategies for nonlinear systems. These techniques employ the nonlinear model directly in the controller calculation without the need for local linearization about an operating point. Finally, the development of inexpensive and powerful computers have made on-line implementation of these nonlinear model-based controllers feasible.

This research monograph is intended as an introduction to the design, analysis, and application of nonlinear control strategies for process systems. Individual chapters have been prepared by leading academic and industrial researchers. To make the monograph accessible to a larger audience, we have attempted to present a balanced view of the theoretical and practical issues. However, the very nature of nonlinear systems dictates the use of some advanced mathematical tools. Although the control techniques presented are applicable to a broad range of nonlinear systems, the application of these methods to nonlinear process control problems is emphasized. Case studies are presented to illustrate the controller design procedures. Comparisons with linear control techniques also are included to demonstrate the performance improvements that can be achieved by employing a nonlinear control strategy.

The monograph has seven chapters which cover three major topics related to nonlinear process control: nonlinear identification, nonlinear controller design, and nonlinear state estimation. Chapter 1 motivates the need for nonlinear process control systems and contains a discussion of several classic nonlinear control techniques. Because all of the control strategies discussed in the subsequent chapters presume the availability of a nonlinear process model, identification techniques for nonlinear systems are presented in Chapter 2. Several important topics from nonlinear systems theory are presented in Chapter 3. Although both Chapters 2 and 3 could easily warrant an entire book, they are presented primarily as a prelude to the nonlinear control and estimation techniques presented in the subsequent chapters.

The two most important controller design approaches for nonlinear process applications are discussed in Chapters 4 and 5. Feedback linearization techniques which yield exact linearization of the closed-loop system are presented in Chapter 4. Nonlinear model predictive control strategies, which are nonlinear generalizations of linear model predictive control schemes successfully applied in the process industries, are discussed in Chapter 5. Because it may be difficult or even impossible to obtain on-line measurements of all the state variables required for control, the design of nonlinear state observers is presented in Chapter 6. The use of artificial neural networks for nonlinear process identification and control is discussed in Chapter 7.

We have assumed that the reader has completed an introductory course in process control and has a working knowledge of elementary calculus. Some familiarity with linear state-space systems and linear controller design techniques is helpful but not essential. The book may serve as a concise reference for control engineers interested in nonlinear process control theory and applications. The book also can be used as a textbook in a graduate course on process control. Chapters 3–5 could be covered in a course in which nonlinear control is one of several topics.

Acknowledgements

First and foremost, we would like to acknowledge the individuals who contributed chapters to this book. Their hardwork and patience are gratefully appreciated. We also wish to express our gratitude to Michael Kurtz for his assistance with LaTeX and the camera-ready copy. The cooperation of Bernard Goodwin and Sophie Papanikolaou from Prentice Hall also is appreciated. MAH would like to acknowledge support from a Career Development Award (CTS-9501368) from the National Science Foundation. DES would like to acknowledge support from a NSF/GOALI award (CTS-9424094) and the DuPont Company.

Michael A. Henson
Dale E. Seborg

Chapter 1

Introduction

DALE E. SEBORG
Department of Chemical Engineering
University of California
Santa Barbara, CA 93106-5080

MICHAEL A. HENSON
Department of Chemical Engineering
Louisiana State University
Baton Rouge, LA 70803-7303

Many common process control problems exhibit nonlinear behavior, in that the relationship between the controlled and manipulated variables depends on the operating conditions. For example, if the dynamic behavior of a nonlinear process is approximated by a linear model such as a first-order transfer function, the model parameters (*e.g.* steady-state gain, time constant, time delay) depend on the nominal operating condition. If the process is only mildly nonlinear or remains in the vicinity of a nominal steady state, then the effects of the nonlinearities may not be severe. In these situations, conventional feedback control strategies can provide adequate performance.

But many important industrial processes including high purity distillation columns, highly exothermic chemical reactions, pH neutralizations, and batch systems can exhibit highly nonlinear behavior. These processes may be required to operate over a wide range of conditions due to large process upsets or setpoint changes. When conventional PID controllers are used to control highly nonlinear processes, the controllers must be tuned very conservatively in order to provide stable behavior over the entire range of operating conditions. But conservative controller tuning can result in serious degradation of control system performance. There are other situations where conventional PID control is inadequate, for example, when the process gain changes sign (*e.g.* some reactor control problems).

Process control research has largely emphasized the analysis of linear systems (via transfer function and state-space models) and the design of linear controllers. In a similar vein, industrial practice has traditionally relied on linear control laws, the ubiquitous PI and PID control algorithms. But within the last 15 years, model-based control strategies such as model predictive control (MPC) have become the preferred control technique for difficult mutivariable control problems in oil refineries and pretrochemical plants [11]. Because the current generation of MPC systems are largely based on linear dynamic models such as step response and impulse response models, the resulting linear controllers must be conservatively tuned for highly nonlinear process.

In view of the shortcomings of linear controllers for highly nonlinear processes, there are considerable incentives for developing more effective control strategies that incorporate knowledge of the nonlinear characteristics. During the past decade, there have been a resurgence of interest in developing nonlinear control strategies that are appropriate for process control. The major objective of this book is to provide an overview of key issues and new research results in this important area.

1.1 Conventional Nonlinear Control Strategies

In the traditional strategy for nonlinear control problems, the objective is to make the closed-loop system behave more linearly by keeping the loop gain constant. For example, the nonlinearities associated with control valves can be reduced by using valve positioners or cascaded flow control loops. Nonlinear transformations of input or output variables also can make the loop gain more constant. For example, the nonlinear characteristics of orifice plate flow measurement can be compensated by taking the square root of the measurement prior to the control calculation. For composition control of high purity distillation columns, using the logarithm of a composition as the controlled variable can make the control loop more linear.

The gain scheduling technique has been widely used to compensate for nonlinear process characteristics [12]. In this approach the controller settings are adjusted to compensate for known nonlinearities so that the loop gain is kept as constant as possible. For example, if the process gain K_p varies with the throughput of the process, the controller gain K_c should be varied so the product, $K_c K_p$, is kept constant. In the general case, the controller could include the inverse of a known static nonlinearity such as a pH titration curve. But if there are significant time delays, this simple gain scheduling approach may be inferior to standard PID control [15].

Model-based control strategies for nonlinear processes have traditionally been based on local linearization and linear controller design based on the linearized model. If the model is updated on-line, the controller should be effective over a wider range of operating conditions. This philosophy provides the basis for the self-tuning approach to adaptive control [1].

1.2 Recent Developments

In recent years, there has been a resurgence of interest in developing improved control and identification strategies for nonlinear systems. The renewed interest has been motivated by several developments:

1. Advances in nonlinear systems theory which have led to controller design methods that are applicable to broad classes of nonlinear control problems.
2. The development of efficient identification methods for empirical nonlinear models, and their widespread availability in commercial software packages.
3. Continued improvement in the capabilities of computer-control hardware and software, thus making it feasible to incorporate complex nonlinear models in plant control systems.

Next we introduce two nonlinear controller design methods that have received considerable attention: input/output linearization and nonlinear predictive control. Other techniques such as sliding mode control [13] and fuzzy control [14] are beyond the scope of this book.

Recently, a new controller design method that provides exact linearization of nonlinear models has been developed [7]. Unlike conventional linearization via Taylor series expansion, this technique produces a linearized model that is independent of the operating point. An analytical expression for the nonlinear control law can then be derived for broad classes of nonlinear systems. This approach is usually referred to as *input/output linearization* or *feedback linearization* and is based on concepts from nonlinear systems theory. The resulting controller includes the inverse of the dynamic model of the process, providing that such an inverse exists. This general approach has been utilized in several process control design methods such as: generic model control [9], globally linearizing control [8], internal decoupling control [2], reference system synthesis [3], and a nonlinear version of internal model control [6]. Nonlinear systems theory and the input/output linearization approach are discussed in Chapters 3 and 4, respectively.

The success of linear model predictive control systems has motivated the extension of this methodology to nonlinear control problems. This general approach is referred to as *nonlinear predictive control*. The control problem formulation is analogous to linear model predictive control except that a nonlinear dynamic model is used to predict future process behavior. The required control actions are calculated by solving a nonlinear programming problem at each sampling instant [4, 5]. Thus, the potential benefits of the nonlinear approach must justify the greater computational complexity in comparison with standard linear techniques. Nonlinear predictive control is the subject of Chapter 5.

A key issue in model-based nonlinear control is what type of process model should be used. Physically-based models derived from first principles (*e.g.* mass and energy balances) are appealing because of the physical insight they provide and their applicability over relatively wide ranges of operating conditions. However, such models often are not available due the engineering effort and cost it takes to develop and maintain them. Another disadvantage is that they often contain a number of process variables that cannot be measured. One possible solution to this problem is state estimation, the subject of Chapter 6.

An alternative approach is to design the nonlinear controller using an empirical model or semi-empirical model which is developed from experimental data. In recent years, there has been considerable interest in developing nonlinear dynamic models from input/output data. While a variety of modeling techniques are available, the predominant tool for obtaining nonlinear empirical models is *artificial neural networks*. Neural network software packages are now commercially available which should hasten their widespread use in industry, including process control applications. An overview of nonlinear empirical models is presented in Chapter 2 while neural network models are considered in Chapter 7.

Table 1.1: Nominal Operating Conditions for the CSTR.

Variable	Value	Variable	Value
q	100 L/min	$\frac{E}{R}$	8750 K
C_{Af}	1 mol/L	k_0	7.2 x 10^{10} min^{-1}
T_f	350 K	UA	5 x 10^4 J/min·K
V	100 L	T_c	300 K
ρ	1000 g/L	C_A	0.5 mol/L
C_p	0.239 J/g·K	T	350 K
$(-\Delta H)$	5 x 10^4 J/mol		

1.3 Illustrative Example

In order to demonstrate the potential benefits that can result from using nonlinear control techniques, we consider a simulated chemical reactor. A nonlinear controller designed using the input/output linearization approach is compared to a linear MPC controller.

Consider the classical continuous stirred tank reactor (CSTR) for an exothermic, irreversible reaction, $A \rightarrow B$. Assuming constant liquid volume, the following dynamic model can be derived based on a component balance for reactant A and an energy balance [12]:

$$\begin{aligned} \dot{C}_A &= \frac{q}{V}(C_{Af} - C_A) - k_0 \exp\left(-\frac{E}{RT}\right) C_A \qquad (1.1) \\ \dot{T} &= \frac{q}{V}(T_f - T) + \frac{(-\Delta H)}{\rho C_p} k_0 \exp\left(-\frac{E}{RT}\right) C_A + \frac{UA}{V\rho C_p}(T_c - T) \end{aligned}$$

We have used standard notation where C_A is the concentration of A in the reactor, T is the reactor temperature, and T_c is the temperature of the coolant stream. The objective is to control T by manipulating T_c. Table 1.1 contains nominal operating conditions, which correspond to an unstable steady state. The open loop response in Figure 1.1 demonstrates that the reactor exhibits highly nonlinear behavior in this operating regime.

The nonlinear state-space model,

$$\begin{aligned} \dot{x} &= f(x) + g(x)u \qquad (1.2) \\ y &= h(x) \end{aligned}$$

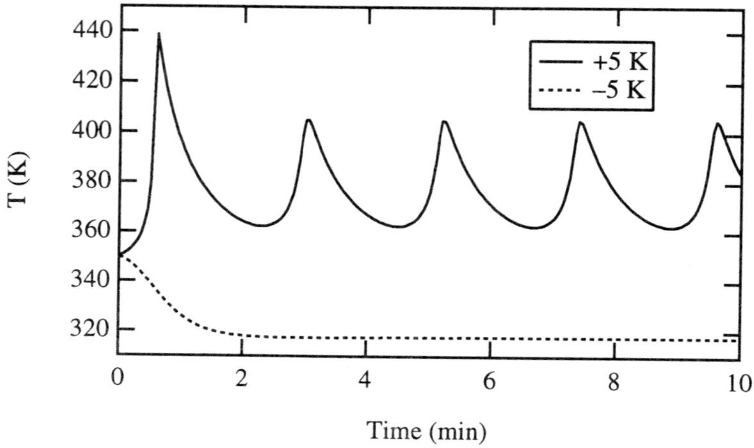

Figure 1.1: Open-loop response for ±5 K changes in T_c.

can be obtained by defining the state vector as $x = [C_A \;\; T]^T$, the manipulated input as $u = T_c$, and the controlled output as $y = T$. This model can be used directly in the input-output linearizing controller design, as discussed in Chapter 4. The linearizing controller is tuned such that the closed-loop system has a time constant of approximately 0.25 min. The linear MPC design is based on a linear deviation model which is obtained from the nonlinear model via first-order Taylor series expansion about the nominal operating point in Table 1.1. The control actions are calculated by solving an open-loop optimal control problem at each sampling instant, as discussed in [10]. The MPC cost function includes weighting coefficients $Q = 4$ and $R = 2$ which penalize deviations of the output from the setpoint and deviations of the input from its target value, respectively. The controller is tuned with a sampling period $\Delta t = 0.05$ min and a control horizon $N = 16$.

In Figure 1.2 the linear and nonlinear controllers are compared for ±25 K changes in the temperature setpoint. The MPC controller produces a sluggish response for the negative setpoint change, while overshoot and small oscillations are obtained for the positive change. This type of response is commonly observed when a well tuned linear controller is applied to a highly nonlinear process. By contrast, the nonlinear controller provides rapid and smooth responses for both setpoint changes. Note that the responses are perfectly symmetrical, indicating that that the closed-loop behavior has been completely linearized. This is a fundamental property of linearizing con-

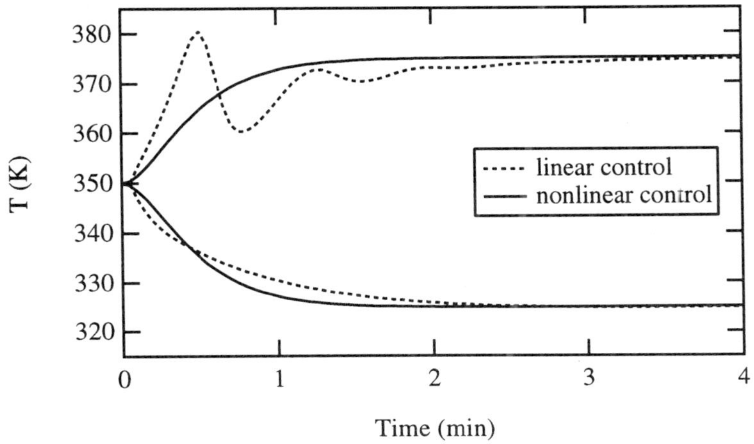

Figure 1.2: Closed-loop response for ±25 K setpoint changes.

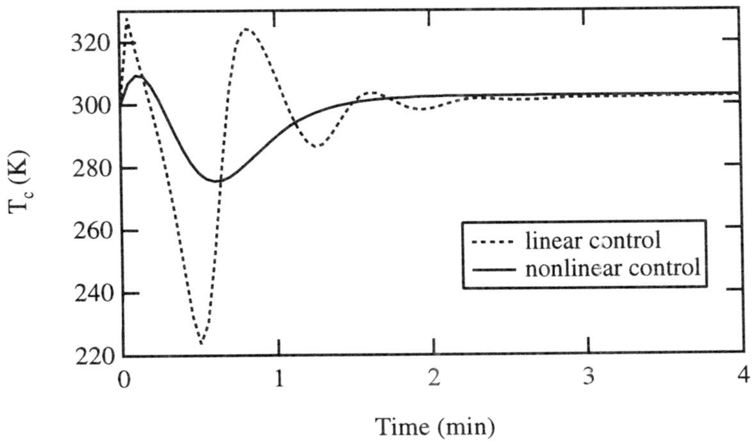

Figure 1.3: Control moves for +25 K setpoint change.

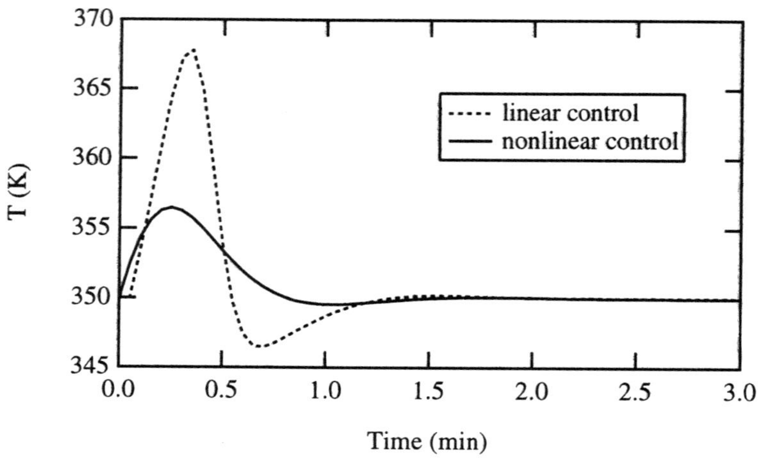

Figure 1.4: Closed-loop response for –150 K change in $\frac{E}{R}$.

trollers (see Chapter 4). Figure 1.3 shows the input moves produced by the controllers for the +25 K change. This result indicates that the superior performance of the nonlinear controller is not attributable to more aggressive control action, but rather to a more judicious use of the input.

Figure 1.4 compares the controllers for a –150 K step change in $\frac{E}{R}$. This type of unmeasured disturbance could be caused, for instance, by a sudden change in the reaction catalyst that results in a reduction of the activation energy. The linear controller yields large deviations from the setpoint, while the nonlinear controller provides very effective attentuation of the disturbance. This result demonstrates that nonlinear controllers can exhibit good robustness to modeling errors.

References

[1] K. J. Åström and B. Wittenmark. *Adaptive Control.* Addison-Wesley, Reading, MA, 1989.

[2] J. G. Balchen, B. Lie, and I. Solberg. Internal decoupling in non-linear process control. *Model. Ident. Control*, 9:137–148, 1988.

[3] R. D. Bartusiak, C. Georgakis, and M. J. Reilly. Nonlinear feedforward/feedback control structures designed by reference synthesis.

Chem. Eng. Sci., 44:1837–1851, 1989.

[4] B. W. Bequette. Nonlinear control of chemical processes: A review. *Ind. Eng. Chem. Res.*, 30:1391–1413, 1991.

[5] J. W. Eaton and J. B. Rawlings. Feedback control of chemical processes using on-line optimization techniques. *Comput. chem. Engng.*, 14:469–479, 1990.

[6] M. A. Henson and D. E. Seborg. An internal model control strategy for nonlinear systems. *AIChE J.*, 37:1065–1081, 1991.

[7] A. Isidori. *Nonlinear Control Systems.* Springer-Verlag, New York, NY, 1989.

[8] C. Kravaris and C.-B. Chung. Nonlinear state feedback synthesis by global input-output linearization. *AIChE J.*, 33:592–603, 1987.

[9] P. L. Lee and G. R. Sullivan. Generic model control (GMC). *Comput. chem Engng.*, 12:573, 1988.

[10] K. R. Muske and J. B. Rawlings. Model predictive control with linear models. *AIChE J.*, 39:262–287, 1993.

[11] S. J. Qin and T. A. Badgwell. An overview of industrial model predictive control technology. In *Proc. Chemical Process Control V*, Tahoe City, CA, in press.

[12] D. E. Seborg, T. F. Edgar, and D. A. Mellichamp. *Process Dynamics and Control.* Wiley, New York, NY, 1989.

[13] J.-J. E. Slotine and W. Li. *Applied Nonlinear Control.* Prentice Hall, Englewood Cliffs, NJ, 1991.

[14] D. A. White and D. A. Sofge. *Handbook of Intelligent Control – Neural, Fuzzy, and Adaptive Approaches.* Van Nostrand Reinhold, New York, NY, 1992.

[15] S. K. P. Wong and D. E. Seborg. Control strategy for single-input single-output non-linear systems with time delays. *Int. J. Control*, 48:2303–2327, 1988.

Chapter 2

Nonlinear Process Identification

RONALD K. PEARSON
BABATUNDE A. OGUNNAIKE

DuPont Central Science and Engineering
Experimental Station,
Wilmington, Delaware, 19880-0101

2.1 Introduction

Most, if not all, of the nonlinear process control strategies described in this book require an explicit mathematical model of process dynamics. This chapter is concerned with the development of models suitable for use in these control system design methodologies. Many of the key issues in the discussions that follow — motivation, methodology, open questions — may be related to the "process characterization cube" shown in Fig. 2.1.

The three axes of this cube correspond to the following three process characteristics:

1. degree of dynamic complexity;

2. degree of nonlinearity;

3. degree of interaction.

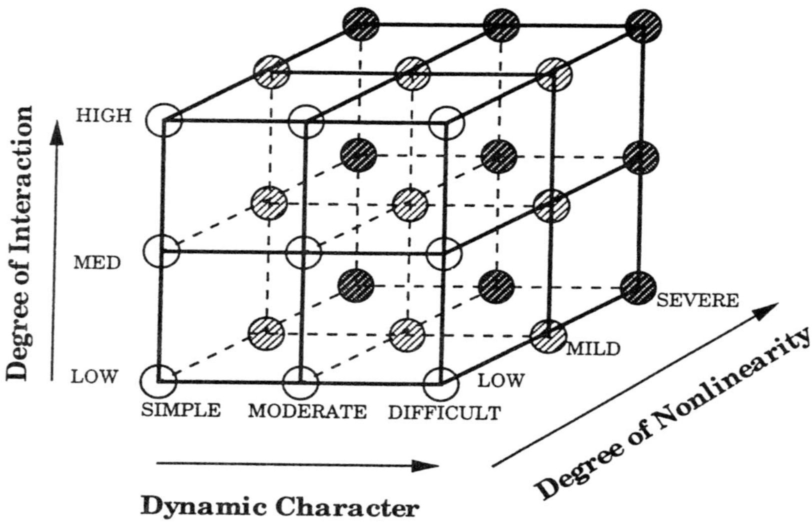

Figure 2.1: The Process Characterization Cube

This construct was originally proposed in reference [64], but a more detailed discussion, illustrated with specific process examples, is given in Chapter 30 of reference [66]. Our motivation for developing empirical process models becomes clear when we consider each of the following topics in terms of this cube:

1. real-world process dynamics;

2. fundamental models of process dynamics;

3. empirical models of process dynamics;

4. control-system design methodologies.

Specifically, the design of a process control system requires us to match the first item on this list with the last. In the development of model-based control system designs, this match is to be accomplished through a mathematical model that satisfies two important criteria. First, it must be compatible in structure and complexity with the requirements of the control system design methodology under consideration. As a specific example, the nonlinear Model Predictive Control (MPC) algorithms discussed by Rawlings [80] require discrete-time dynamic models and the computational complexity of the problem grows rapidly as the model complexity increases. The second criterion for a "good" process model is that it approximate the true process dynamics well enough that the resulting control system will perform adequately in practice.

In general, fundamental models — derived from known conservation laws, reaction kinetics, etc. — can be expected to describe process dynamics more completely than empirical models can. Conversely, since these models are fundamental descriptions of the process, we have limited control over their complexity, which can be substantial. For example, the detailed model given in [21] for the co-polymerization of methyl methacrylate and vinyl acetate entails approximately 50 differential and algebraic equations, mostly nonlinear, and about as many parameters. The advantage of empirical models is that we can specify model complexity explicitly, giving us at least partial control over the difficulty of the resulting control problem. On the other hand, the price we pay for this reduced model complexity is generally reduced model fidelity. Thus, to apply model-based control system design procedures effectively, it is necessary to develop empirical process models that are "close enough" in the process characterization cube to the dynamics of the real-world process, but that are also within the "domain" of the control system design methodology of interest. *The principal objective of this chapter is to describe the key issues that we must confront when we attempt to solve this problem in practice.*

Box and Jenkins proposed a systematic approach to linear time-series modeling, described in reference [10] and consisting of the following four steps:

1. selection of a general class $\mathcal{C}$ of empirical models for consideration;

2. identification of a specific subclass of models to be fit;

3. estimation of model parameters;

4. assessment of model adequacy.

The models considered by Box and Jenkins were *parametric* models, described by a finite number of adjustable parameters (see Section 5.1 for a brief discussion of parametric modeling and its alternatives). Consequently, step 2 of their procedure is primarily concerned with deciding how many model parameters to admit (i.e., how complex to allow the model to become). One of the key points of this chapter (Section 3.2) is that certain equivalences that hold between different linear models (e.g., discrete- and continuous-time models) ***do not hold*** for nonlinear models. Consequently, issues of model structure selection — corresponding to steps 1 and 2 of the Box-Jenkins procedure — are even more important in nonlinear modeling than in linear modeling. It is for this reason that we concentrate more on these issues in this chapter than on the "more traditional" topic of parameter estimation algorithms.

The remainder of this chapter is divided into sections, numbered 2 through 10. Sections 2 and 3 deal with continuous-time and discrete-time process models, respectively, providing useful background material for the discussions that follow. Sections 4 and 5 then discuss the empirical model building process and the problem of model structure selection, respectively, both closely related to the first step in the Box-Jenkins model development procedure. Because the subsequent three steps in this procedure have strong connections to statistics, Section 6 gives a brief overview of some useful statistical background material. Section 7 then discusses the problem of model parameter estimation, while Section 8 discusses the problems of cross-validation and input sequence design. Section 9 then illustrates some of the main ideas developed here with a case study (control of a high-purity distillation process) and Section 10 gives a brief summary of these ideas.

2.2 Continuous-Time Process Models

Fundamental process descriptions usually take the form of continuous-time models, so this section gives a brief overview of some of the most popular continuous-time model structures that appear in the control literature. All of these models are special cases of — or closely related to — the class of state-space models:

$$\begin{aligned} \dot{\mathbf{x}}(t) &= \mathbf{F}(\mathbf{x}(t), \mathbf{u}(t)) \\ \mathbf{y}(t) &= \mathbf{G}(\mathbf{x}(t)), \end{aligned} \tag{2.1}$$

Here, $\mathbf{x}$, $\mathbf{u}$, and $\mathbf{y}$ are state, input, and output vectors of dimensions n, m, and r, respectively. The vector function $\mathbf{F}$ maps the domain $R^n \times R^m$ into the range R^n, while the vector function $\mathbf{G}$ maps R^n into R^r.

The principal objective of this overview is to give some insight into the range of qualitative behavior these models can exhibit. This insight is important because in empirical modeling, we are attempting to develop a "moderate complexity" model that captures the *essential qualitative behavior* of the real-world process of interest.

2.2.1 Control-Affine Models

Kantor notes [47] that many first-principles models may be represented as *control-affine* special cases of Eq. (2.1). Here, the nonlinear vector map $\mathbf{F}(\mathbf{x}, \mathbf{u})$ has the special form:

$$\mathbf{F}(\mathbf{x}, \mathbf{u}) = \mathbf{f}(\mathbf{x}) + \mathbf{\Gamma}(\mathbf{x})\mathbf{u}. \tag{2.2}$$

This structural restriction on $\mathbf{F}(\mathbf{x}, \mathbf{u})$ necessarily limits the class of dynamic phenomena that may be represented, relative to the unconstrained state-space model (2.1). Despite these limitations, however, the discussion given here illustrates that this class of dynamic phenomena is still very broad. As a specific example, the second-order polynomial model considered in Section 4.2 is control-affine, but it is particularly interesting because it can exhibit a form of instability known as *finite escape time*. More generally, note that all linear models are included in the control-affine class, as are the bilinear models discussed in Section 2.2. In addition, Boyd and Chua [12] have shown that the *finite* Volterra models considered in Section 2.3 have bilinear realizations, implying that they, too, are equivalent to members of the class of control-affine models. Before proceeding to more detailed discussions of these important subsets of the control-affine model class, it is useful to briefly consider the following two examples. These examples are not members of the more restricted model classes discussed in Sections 2.2 and 2.3, and they illustrate the range of behavior that the control-affine class of dynamic models — but not all of its sub-classes — is capable of exhibiting.

Example 1 — Chaotic Dynamics

Chaos is a nonlinear dynamic phenomenon that has recieved much attention recently [26, 39, 96]. One of the simplest continuous-time models that is flexible enough to exhibit chaotic dynamics is the Lorenz equation [26]:

$$\begin{aligned} dx/dt &= -\sigma x + \sigma y \\ dy/dt &= -xz + rx - y \\ dz/dt &= xy - bz. \end{aligned} \tag{2.3}$$

Physically, these equations describe the convection-driven patterns that arise in fluid flow between rectangular plates. In this problem, the variables x, y, and z represent the intensity of convective motion, the temperature difference between ascending and descending currents in the fluid, and the deviation from linearity of the vertical temperature profile between the plates.

To see the connection between these equations and the control-affine model (2.2), define the state vector components $x_1 = x$, $x_2 = y$, and $x_3 = z$ and take r as the scalar control input u. In particular, note that r is a ratio of Rayleigh numbers and is proportional to the temperature difference between plates. Thus, if we consider the problem of controlling fluid motion by manipulating this temperature difference, the Lorenz equations may be re-written as a control-affine model by defining:

$$\mathbf{f}(\mathbf{x}) = \begin{bmatrix} -\sigma x_1 + \sigma x_2 \\ -x_1 x_3 - x_2 \\ x_1 x_2 - b x_3 \end{bmatrix}, \tag{2.4}$$

and

$$\mathbf{\Gamma}(\mathbf{x}) = \begin{bmatrix} 0 \\ x_2 \\ 0 \end{bmatrix}. \tag{2.5}$$

This example demonstrates that the class of control-affine models is "qualitatively rich enough" to exhibit chaotic dynamics.

Example 2 — Jump Phenomena

Jump phenomena are important in the theory of nonlinear oscillators [13, 62, 96]. While these phenomena will not be discussed in detail here, they have been widely studied and it is useful to note that the class of control-affine models can also exhibit this type of behavior. This point is significant since some of the interesting subsets of this model class lack this flexibility, as noted in subsequent discussions. Here, note that one of the models often used to illustrate jump phenomena is Duffing's equation:

$$\frac{d^2y}{dt^2} + c\frac{dy}{dt} + \alpha y + \beta y^3 = u(t). \tag{2.6}$$

This model may be cast in control-affine form by defining the state variables $x_1 = y$ and $x_2 = dy/dt$. The functions $\mathbf{f}(\mathbf{x})$ and $\mathbf{\Gamma}(\mathbf{x})$ appearing in Eq. (2.1) are then given by:

$$\mathbf{f}(\mathbf{x}) = \begin{bmatrix} x_2 \\ -\alpha x_1 - \beta x_1^3 - c x_2 \end{bmatrix}, \tag{2.7}$$

and

$$\mathbf{\Gamma}(\mathbf{x}) = \begin{bmatrix} 0 \\ 1 \end{bmatrix}. \tag{2.8}$$

2.2.2 Bilinear Models

For single-input problems (i.e., $\mathbf{u}(t) = u(t)$ is a scalar), if the control-affine model is further restricted to $\mathbf{f}(\mathbf{x}) = \mathbf{A}\mathbf{x}$ and $\mathbf{\Gamma}(\mathbf{x}) = \mathbf{N}\mathbf{x} + \mathbf{b}$, the result is a *bilinear* model. Here, $\mathbf{A}$ and $\mathbf{N}$ are $n \times n$ matrices and $\mathbf{b}$ is an n-vector. This class may be extended to MIMO problems by considering multiple terms of this general form, i.e.:

$$\mathbf{\Gamma}(\mathbf{x})\mathbf{u} = \sum_{i=1}^{m} u_i(t)\mathbf{N}_i\mathbf{x} + \mathbf{B}\mathbf{u}, \tag{2.9}$$

where $\mathbf{B}$ is an $n \times m$ matrix. Further, note that when $\mathbf{N}_i = \mathbf{0}$, for all i, the bilinear model reduces to the more familiar linear one. Because this structure is a special case of the more general control-affine structure, we can expect the class of behavior it is capable of representing to be correspondingly restricted. For example, Brockett has shown [13] that bilinear models cannot represent jump phenomena like that exhibited by the Duffing equation.

A point that will be emphasized repeatedly throughout this chapter is that the qualitative behavior of a nonlinear model can depend strongly on the specific inputs considered. The step response of a single-input, single-output bilinear model illustrates this point nicely. Specifically, consider the effect of the step input:

$$u(t) = \begin{cases} \alpha & t > 0 \\ 0 & t \leq 0 \end{cases} \tag{2.10}$$

on the bilinear system defined above. If $\mathbf{x}(t) = \mathbf{0}$ for $t \leq 0$, the bilinear model may be re-written as the "equivalent" linear model for $t > 0$:

$$\begin{aligned} \dot{\mathbf{x}}(t) &= (\mathbf{A} + \alpha\mathbf{N})\mathbf{x}(t) + \mathbf{b}u(t). \\ y(t) &= \mathbf{c}^T\mathbf{x}(t) \end{aligned} \tag{2.11}$$

The response of this model will be determined by the eigenvalues of the matrix $\mathbf{A} + \alpha\mathbf{N}$, which depend on the input amplitude α. For low-amplitude inputs (small α) and "weak" nonlinearities ("small" $\mathbf{N}$), these eigenvalues will be approximately the same as those of $\mathbf{A}$, so the effects of the nonlinearity will not be felt. As α increases, however, the qualitative behavior of the step response will come ultimately to be dominated by the eigenvalues of $\mathbf{N}$, which may be very different from those of $\mathbf{A}$.

2.2.3 Volterra Models and Fading Memory Systems

Another class of nonlinear models that has attracted significant interest is the class of Volterra models [12, 79, 88, 96]. For the single-input, single-output (SISO) case, this series is given by:

$$\begin{aligned} y(t) &= y_0(t) + \\ &\sum_{n=1}^{\infty} \int_{-\infty}^{\infty} \cdots \int_{-\infty}^{\infty} k_n(t; t_1, t_2, ..., t_n) u(t_1) \cdots u(t_n) dt_1 ... dt_n. \end{aligned} \tag{2.12}$$

Here, the functions $k_n(t; t_1, t_2, ..., t_n)$ are called the *Volterra kernels* and generalize the impulse response characterizing a linear system. Essentially, this series represents the response $y(t)$ in terms of the past history of the input $u(\tau)$ for all $\tau < t$ and the kernels must be constrained to enforce this *causality condition.* As a specific example, note that the standard linear convolution model may be represented as a Volterra model with all terms identically zero except the first-order kernel:

$$k_1(t; t_1) = \begin{cases} h(t - t_1) & t_1 < t \\ 0 & t_1 \geq t \end{cases} \tag{2.13}$$

To be useful in practice, Volterra models must be truncated to a finite number of terms. Boyd and Chua [12] consider the question of what class of system behavior may be approximated by such truncated Volterra models. They show that systems with *fading memory* on a particular set K of input sequences may be approximated arbitrarily well on that set by a truncated Volterra model of sufficiently high order. Roughly speaking, a fading memory system is one whose dependence on past inputs decreases "rapidly enough" with time. This concept is important because it provides a useful, broad partitioning of the enormous class of "nonlinear systems." Thus, we discuss it briefly here and will revisit it in subsequent sections, particularly in connection with the distinction between "autoregressive" and "moving average" discrete-time models and in connection with "mixing conditions" in statistics.

The precise definition of a fading memory system given by Boyd and Chua [12] is as follows. The set $\mathbf{C}(\mathbf{R})$ is defined as the space of bounded continuous functions on the real line $\mathbf{R}$, equipped with the supremum norm $||u|| = sup_{t\in\mathbf{R}} |u(t)|$. Let K be a subset of $\mathbf{C}(\mathbf{R})$ and let N be the nonlinear operator from $\mathbf{C}(\mathbf{R})$ to $\mathbf{C}(\mathbf{R})$ defined by the nonlinear system of interest. This operator has *fading memory* on the subset K if there exists a decreasing function $w(t) : \mathbf{R}_+ \rightarrow (0, 1]$ with $lim_{t\rightarrow\infty} w(t) = 0$ such that for each $u \in K$ and $\epsilon > 0$, there is some $\delta > 0$ such that for all $v \in K$,

$$
\begin{aligned}
sup_{t\geq 0} |u(-t) - v(-t)|w(t) &< \delta \\
&\Rightarrow |Nu(0) - Nv(0)| < \epsilon.
\end{aligned}
$$

Mathematically, this notion is a stronger version of continuity for the operator N, which would result if we took $w(t) = 1$ for all t [12]. Boyd and Chua prove that all finite-dimensional (exponentially stable) linear systems are fading memory systems.

The dependence on the set K in the definition of fading memory systems is important, for three reasons. First, it emphasizes the importance of explicitly considering the class of relevant inputs in modeling nonlinear systems, a point to which we shall return repeatedly. Second, and more immediately, Boyd and Chua's main approximation result is that if N has fading memory on the following specific set K:

$$
K = \{u \in \mathbf{C}(\mathbf{R}) \mid ||u|| \leq M_1, \; ||u(t-\tau) - u(t)|| \leq M_2\tau, \; \tau \geq 0\},
$$

it may be approximated with arbitrary accuracy by a finite Volterra series, for all $u \in K$. Physically, the set K consists of "slew-rate limited" signals that do not change "too rapidly" over short time intervals. In particular, piecewise-constant input sequences like steps are explicitly excluded from this set. Finally, the third point about the set K is that if N has fading memory on K and has a state-space representation, and if X is the set of states reachable from the input sequences in K, then N has a unique steady-state for all $u \in K$ and all initial conditions in X. Thus, fading memory systems are "well-behaved" in the sense that they cannot exhibit multiple steady-states or other related phenomena like chaotic responses. This point is important because chemical processes (in particular, reactors) can exhibit these phenomena. *Consequently, if fading memory models are to be developed, we can only expect them to be valid "locally," i.e., within the basin of attraction of a single steady state.*

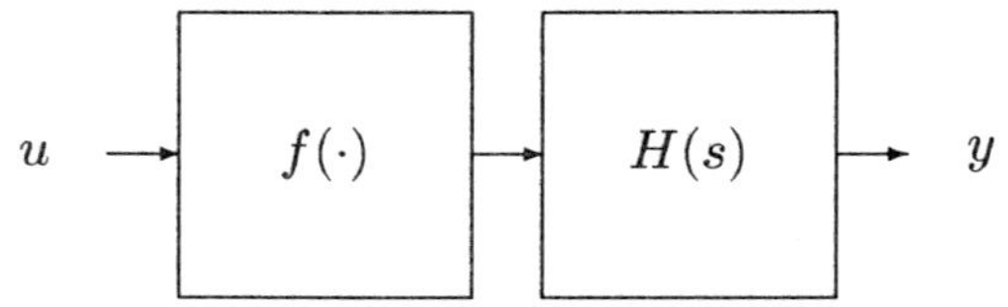

Static Nonlinearity   Linear Dynamics

Figure 2.2: Hammerstein Model Structure

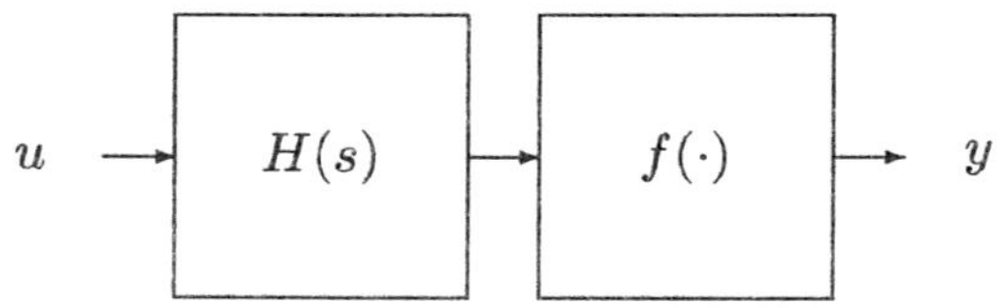

Linear Dynamics   Static Nonlinearity

Figure 2.3: Wiener Model Structure

2.2.4 Block-Oriented Nonlinear Models

Closely related to Volterra models are the *block-oriented* models, defined by cascade and/or parallel connection of static nonlinearities and linear dynamics. Probably the best-known member of this class is the *Hâmmerstein model* [3, 37] shown in Fig. 2.2. Because of its relatively simple structure, this model has become increasingly popular as a "next-step-beyond-linear-modeling" of chemical processes [31, 91]. In particular, note that this model combines linear dynamics with a nonlinear steady-state gain. If the order of these blocks is reversed — i.e., if the static nonlinearity *follows* the linear dynamics — the resulting system is called a Wiener model [3, 40, 36, 38], which has also been considered for process modeling applications [99, 100].

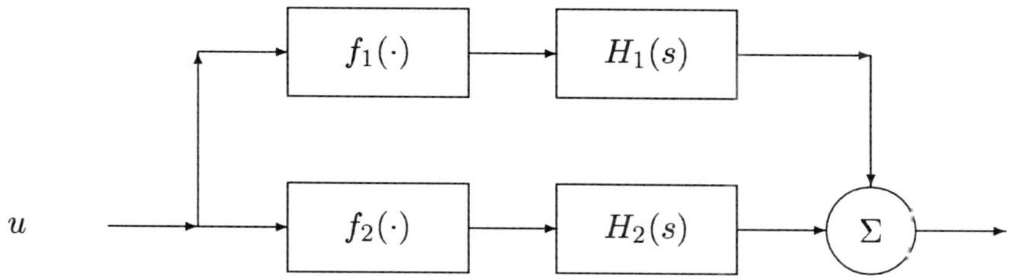

Figure 2.4: "Two channel" Uryson model

This model is shown in Fig. 2.3 and is *not* equivalent to the Hammerstein model [70], even though if $H(0) = 1$, the steady-state gain is specified by the function $f(\cdot)$ in both models and is therefore the same. Both of these structures are special cases of the more general "sandwich model" considered by Brillinger [15] and Greblicki and Pawlak [38], in which the static nonlinearity is "sandwiched" between two linear dynamic models. Still more general "block-oriented" nonlinear models have been investigated involving both series and parallel connections of static nonlinearities and linear dynamics [3, 38, 40]. As a specific example, the *Uryson model* consists of several Hammerstein models connected in parallel, driven by a common input, with their outputs summed [3], as shown in Fig. 2.4.

Chen [17] gives a reasonably detailed review of some of the recent electrical engineering and biological modeling literature on block-oriented nonlinear models. For SISO models, he introduces the following useful terminology — the letter "N" is used to indicate the placement of a static nonlinearity and the letter "L" is used to indicate the placement of linear dynamics. Thus, Hammerstein models are referred to as "NL" models, Wiener models are referred to as "LN" models, and sandwich models are referred to as "LNL" models. This nomenclature is extended to "parallel models," consisting of m of each of these models connected in parallel. Thus, the Uryson model composed of m Hammerstein models in parallel is designated "PNL_m." A number of useful results are presented in Chen's review and some of them will be discussed in subsequent sections here. It should be noted that his review does not cover any of the recent chemical engineering applications of block-oriented models, nor does it mention the growing literature on nonparametric identification approaches for these models, a topic introduced in Section 4 of this chapter and considered further in Section 7.

In any of these models, if the static nonlinearities are analytic, a Volterra representation may be developed from the Taylor series expansion of the nonlinearity and the convolution representation of the linear dynamics. To see this connection, consider the quadratic nonlinearity $f(x) = x^2$ in both the Wiener and the Hammerstein models. If $h(t)$ is the impulse response of the linear subsystem and $k_1(t;t_1)$ is defined as in Eq. (2.13), we have for the Wiener model:

$$\begin{aligned} y(t) &= \left[\int_{-\infty}^{\infty} k_1(t;t_1)u(t_1)dt_1 \right]^2 \\ &= \int_{-\infty}^{\infty}\int_{-\infty}^{\infty} k_2(t;t_1,t_2)u(t_1)u(t_2)dt_1dt_2, \end{aligned} \tag{2.14}$$

where $k_2(t;t_1,t_2)$ is the separable kernel:

$$k_2(t;t_1,t_2) = k_1(t;t_1)k_1(t;t_2). \tag{2.15}$$

Similarly, it is not difficult to show that the Hammerstein model is completely described by the second-order Volterra kernel:

$$k_2(t;t_1,t_2) = k_1(t;t_1)\delta(t_1 - t_2), \tag{2.16}$$

where $\delta(\cdot)$ is the Dirac delta function.

Analogous reasoning leads to finite Volterra representations for block-oriented models with polynomial nonlinearities of arbitrary order, and from there to infinite Volterra representations for analytic nonlinearities. Note that a general characteristic of analytic Hammerstein models is that they involve "diagonal kernels" of the form:

$$k_n(t;t_1,t_2,...,t_n) = \gamma(t;t_1)\delta(t_1 - t_2)\delta(t_3 - t_2)\cdots\delta(t_n - t_{n-1}). \tag{2.17}$$

Similarly, analytic Wiener models will involve separable kernels of the general form:

$$k_n(t;t_1,t_2,...,t_n) = \alpha_n\gamma(t;t_1)\gamma(t;t_2)\cdots\gamma(t;t_n). \tag{2.18}$$

One of the topics that Chen's review does treat fairly extensively is the relationship between the different Volterra kernels that must be satisfied by the more complex block-oriented models [17].

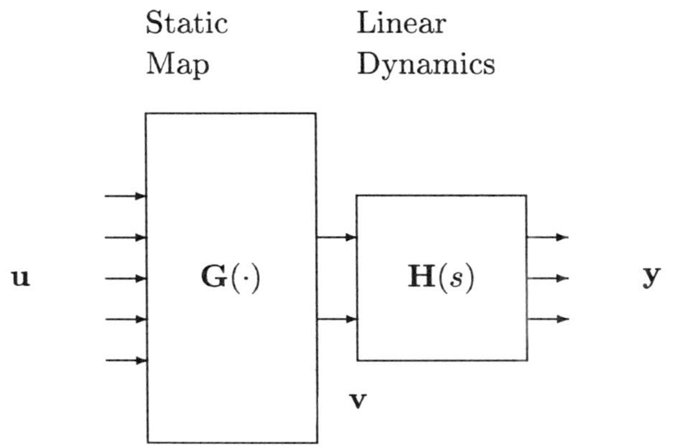

Figure 2.5: General "MIMO Hammerstein model" structure

2.2.5 MIMO Models and "Hidden Layers"

The preceeding discussions have focused on single-input/single-output models. These models have combined different forms of nonlinearity and dynamic complexity and thus occupy one face of the "process characterization cube" introduced in Section 1. To explore more of this cube, it is necessary to consider multivariable models, which are quite important in practice in spite of their added complexity. In particular, note that first-principles models are almost universally multivariable — specific control problems may focus on "single loop pairings," but this focus is an approximate one and sometimes an inappropriate one. In the linear case, multivariable state space models are simple extensions of SISO models, building on the same basic concepts of linear algebra, but these models only serve to map out the second face of the process characterization cube. To see the character of the interior of the cube, it is necessary to consider the simultaneous interplay between dynamics, nonlinearity, and multivariable character. In general, this region represents unexplored territory, but the following discussion of block-oriented MIMO models illustrates some important ideas. In addition, it will be seen that adopting a MIMO focus illuminates some important relationships between different classes of SISO models.

Block-Oriented MIMO Models

First, consider the "multivariable Hammerstein model," shown in Fig. 2.5. There, the scalar input variable u defining the standard Hammerstein model has been replaced with an m-vector $\mathbf{u}$ and the scalar output variable y has been replaced with an n-vector $\mathbf{y}$. In addition, an intermediate q-vector $\mathbf{v}$ is also shown, corresponding to the output from the first block of the diagram and the input to the second. This second block is a general q-input/n-output *linear* system, defined by the $n \times q$ transfer function matrix $\mathbf{H}(s)$. The first block is a $q \times m$ *nonlinear static map* $\mathbf{G}(\cdot)$, transforming the input vector $\mathbf{u}(t)$ at time t into an "intermediate" output vector $\mathbf{v}(t)$ at the same time instant. Each component of this "intermediate" output vector may depend arbitrarily on any or all of the components of the input vector, but there is no dependence on past inputs.

Multivariable Wiener models can be defined similarly, simply reversing the order of the two blocks appearing in Fig. 2.5; and "sandwich" or "LNL" models may be constructed by cascading two MIMO linear dynamic models with a multivariable static nonlinearity, just as in the SISO case. More complex block-oriented MIMO structures are discussed in some detail in Chen's review [17]. In all cases, the dimensions of the "internal" or "hidden" layers of the model (e.g., $\mathbf{v}$ in the MIMO Hammerstein model discussed above) represent useful "design variables," as the following subsection illustrates.

SISO Models with a "Hidden Layer"

Typical feedforward neural network structures involve an "input layer" of processing elements, one or more "hidden layers," and an "output layer" [85]. If we consider the special case of the MIMO Hammerstein model shown in Fig. 2.5 with $m = n = 1$ but $q > 1$, we see an interesting parallel: there is a q-dimensional "hidden layer" between the static nonlinear "input layer" and the linear dynamic "output layer." In particular, note that each component v_i of this "hidden layer" is defined by:

$$v_i(t) \;=\; G_i(u(t)), \tag{2.19}$$

where $G_i(\cdot) : R^1 \to R^1$ is a static nonlinearity. The "output layer" in this model is a linear MISO transfer function matrix $\mathbf{H}(s)$; that is, each component of the vector $\mathbf{v}$ passes through a linear SISO dynamic system and the results are summed to generate the response y. Combining these two layers results in the Uryson model discussed above, composed of q SISO

Hammerstein models connected in parallel, or in Chen's terminology, the PNL_q model. *An important point discussed further in Section 5.3 is that this model structure is closely related to the concept of "local linear modeling."*

If we reverse the order of the linear and nonlinear blocks in this Hammerstein model structure, we obtain a Wiener model that is also quite interesting. Specifically, for the "input layer," consider a linear state-space model of dimension $n \geq q \geq 1$ with a scalar input u and a q-dimensional output $\mathbf{v}$. This "hidden layer response" will be:

$$\begin{aligned} \dot{\mathbf{x}}(t) &= \mathbf{A}\mathbf{x}(t) + \mathbf{b}u(t) \\ \mathbf{v}(t) &= \mathbf{C}\mathbf{x}(t), \end{aligned} \tag{2.20}$$

where $\mathbf{A}$ is an $n \times n$ matrix, $\mathbf{b}$ is an n-vector, and $\mathbf{C}$ is a $q \times n$ matrix. The "output layer" then consists of the sum of q static nonlinearities, each applied to one component of $\mathbf{v}$, i.e.:

$$y(t) = \sum_{i=1}^{q} G_i(v_i(t)). \tag{2.21}$$

The two extremes of this representation are particularly interesting: taking $q = 1$, we recover the standard Wiener model, while Boyd and Chua have shown that any fading memory SISO system can be approximated arbitrarily closely by a model of this general structure with $q = n$ [12]. The question of when alternative structures might be advantageous — e.g., intermediate values of q in this "hidden layer Wiener model," the "hidden layer Hammerstein" structure considered above, "hidden layer sandwich models," or others — raise questions reminiscent of the debate over the optimal number of hidden layers and nodes in neural network approximations of static nonlinear maps [52].

2.3 Discrete-Time Process Models

While fundamental physical models are almost always developed in continuous-time, computer-based process control systems function in discrete-time: measurements are made and control actions are taken at discrete time instants, seconds, minutes, hours, or days apart. In addition, the input/output data we have available for model identification is generally only available at discrete time instants. Thus, while we can identify continuous-time models from this data, it is usually easier to identify discrete-time models and use these as a basis for designing discrete-time control systems for computer implementation. *This observation motivates our interest in discrete-time models, despite certain inherent differences between the behavior of discrete-time models and continuous-time models.*

2.3.1 ARMAX and NARMAX Models

The class of ARMAX models (AutoRegressive Moving Average models with eXogenous inputs) has been widely used as a basis for linear model identification [55]. These models relate an input sequence $\{u(k)\}$ to an output sequence $\{y(k)\}$ by the linear, constant-coefficient difference equation:

$$\begin{aligned} y(k) &= \sum_{j=1}^{p} a_j y(k-j) + \sum_{j=0}^{q} b_j u(k-j) \\ &+ \sum_{j=0}^{r} c_j e(k-j). \end{aligned} \tag{2.22}$$

This model originated in the time-series literature where the focus is on the statistical characterization of the output sequence $\{y(k)\}$ resulting when $\{e(k)\}$ is a "Gaussian white noise" sequence (see Section 6). The term "exogeneous input" for the sequence $\{u(k)\}$ refers to the fact that the ARMAX model is a generalization of the ARMA model obtained when $u(k) = 0$ identically. In the process control applications considered here, the emphasis is somewhat different. Specifically, it is the relationship between the sequences $\{u(k)\}$ and $\{y(k)\}$ that is of primary importance, while the sequence $\{e(k)\}$ represents a "modeling error" in this relationship, arising from the combined effects of measurement noise, unmeasured process disturbances, neglected nonlinearities, etc. The assumptions made regarding the nature of the two "model inputs" $\{u(k)\}$ and $\{e(k)\}$ will influence different aspects of the modeling problem in different ways and will be discussed from various perspectives in subsequent sections.

For nonlinear systems, a popular class of discrete-time models is the NARMAX (Nonlinear ARMAX) family, described by Billings and Voon [5]:

$$\begin{aligned} y(k) &= F(\, y(k-1), y(k-2), ..., y(k-p), \\ &\quad u(k), u(k-1), ..., u(k-q) \\ &\quad e(k-1), e(k-2), ..., e(k-r)) \; + e(k). \end{aligned} \tag{2.23}$$

Here, $F(\cdot)$ is a nonlinear function of the $p+q+r+1$ variables indicated and the sequences $\{y(k)\}$, $\{u(k)\}$, and $\{e(k)\}$ have the same definitions as in the ARMAX model described above. These models constitute an extremely broad class, including many other classes of nonlinear discrete-time models as special cases. To define a NARMAX model completely, it is necessary to specify the *order parameters* p, q, and r, and the form of the function $F(\cdot)$ appearing in Eq. (2.23).

In this section, we will focus on ideal input/output behavior for which $e(k) = 0$ identically, implying $r = 0$ by default. In contrast, the order parameters p and q are quite important. In particular, taking $p = 0$ and $q > 0$, the model output $y(k)$ depends only on the past history of the input sequence $\{u(k)\}$ and not on previous output values $y(k-j)$. Borrowing terminology from the time-series literature [76, 96], we will call this class the "Nonlinear Moving Average models with eXogeneous inputs," or NMAX models. Conversely, if we take $p > 0$ and $q = 0$, we obtain a model whose output $y(k)$ depends only on the current input $u(k)$ and the past history of the *output* sequence $\{y(k-j)\}$. Again borrowing terminology from the time-series literature, this class will be called the "Nonlinear AutoRegressive models with eXogeneous inputs" or NARX models. As a practical matter, $y(k)$ will generally depend on the input $u(k-d)$, delayed by a *single, fixed* number of samples d; this assumption does not change the nature of the model class — they may still be viewed as NARX models. What is *not* permitted in the NARX model class considered here is dependence on *several different* past inputs. If both $p > 0$ *and* $q > 0$, the output $y(k)$ at any time k will depend on both the input history and the output history, giving rise to the general NARMAX class, including both the NMAX and the NARX classes as special cases and exhibiting a correspondingly richer variety of dynamics. (It should be noted that Ljung [55] defines linear ARX models in a way that permits inclusion of "moving average" terms in the control input $u(k)$; due to the substantial qualitative behavior differences between nonlinear "autoregressive" and "moving average" models discussed below, we will restrict NARX models to exclude terms that are "moving average" with respect to the control input, as noted above.)

2.3.2 Non-Equivalent Representations

One unique feature of the linear problem is that if a continuous-time linear ordinary differential equation with constant coefficients is sampled at a constant sampling rate T_s, the samples may be described exactly by a discrete-time linear model [55, p. 20]. In addition, if T_s is sufficiently small to avoid "aliasing phenomena," this transformation is invertible. That is, given T_s and the discrete-time ARMAX model (2.22), we may derive an Nth-order constant-coefficient, linear ordinary differential equation relating $u(t)$ and $y(t)$ such that $y(t_k) = y(k)$ and $u(t_k) = u(k)$ for $t_k = (k-1)T_s$. Another characteristic feature of linear discrete-time ARMAX models is that they may be represented equally well by either "moving average" or "autoregressive" expressions. Specifically, note that the case $p = 0, q = \infty$ corresponds to a convolution model with $\{b_j\}$ representing the impulse response of the system. It is a standard result [55, p. 13] that any time-invariant, linear system may be completely characterized by its impulse response. Conversely, the case $p = \infty$, $q = 0$ represents an infinite-order autoregressive model, and it is not difficult to show that any time-invariant, linear system may be completely characterized by a model of this form as well [49, p. 112]. In other words, the response of any stable linear system may be represented either as a convolution model in terms of the past history of its *input* sequence or as an autoregressive model in terms of the past history of its *output* sequence.

These equivalences for linear models — continuous, discrete, autoregressive, and moving average — are summarized in Fig. 2.6. Unfortunately, *none* of these equivalences extend to nonlinear problems. For example, the logistic equation:

$$y(k) = ay(k-1)[1 - y(k-1)], \tag{2.24}$$

is a first-order nonlinear discrete-time model that has been extensively studied because it exhibits chaotic behavior [96]. In contrast, it is known that in continuous time, nonlinear ordinary differential equations must be at least third-order to exhibit chaotic behavior [39].

To see that NMAX and NARX models are not equivalent, consider the following NARX model [96]:

$$y(k) = \begin{cases} 2y(k-1) + u(k) & |y(k-1)| \leq 2 \\ u(k) & |y(k-1)| > 2 \end{cases} \tag{2.25}$$

If this model is driven by the input sequence $u(k) = (-1)^k$ — a periodic sequence with period 2, the response is a sequence with period 6. This lengthening of the period of the response relative to the period of the input is called *subharmonic generation* and is not possible in NMAX models [70]. That is, note that the response of an NMAX model to an input sequence $\{u(k)\}$ with period T is of the form:

$$y(k) \quad = \quad F(u(k), u(k-1), ..., u(k-m))$$

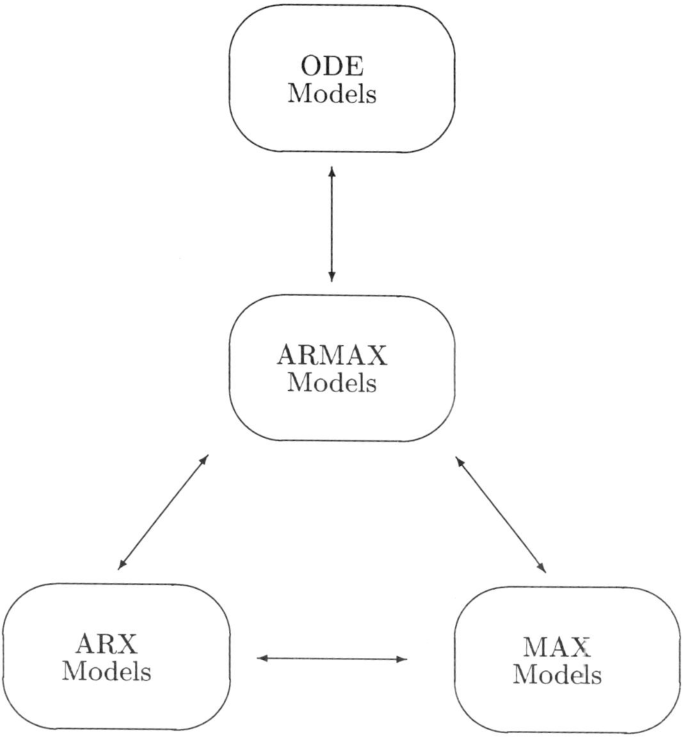

Figure 2.6: Equivalent Linear Models

$$= F(u(k+T), u(k+T-1), ..., u(k+T-m))$$
$$= y(k+T). \tag{2.26}$$

Thus, regardless of the form of the nonlinear function $F(\cdots)$, the response sequence $\{y(k)\}$ has the same period as the input sequence $\{u(k)\}$, excluding the possibility of subharmonic generation.

Since all of the equivalences noted in Fig. 2.6 hold for linear models, linear model structure selection is therefore largely a matter of convenience — it may be of substantial practical importance, but selecting the "wrong" model structure does not exclude any *qualitative* phenomena. Conversely, since *none* of these model structure equivalences hold for the nonlinear case, we must be more judicious in nonlinear model structure selection.

2.3.3 Special Classes of NARMAX Models

As noted above, the class of NARMAX models is an extremely large one, and in practice, it is usual to focus on various structurally defined subsets of this class. The first two steps in the four-step model-building procedure described in Section 1 then entail selection of a specific sub-class of NARMAX models and the specification of order parameters p, q, and r in Eq. (2.23). The selection of a sub-class of NARMAX models amounts to specifying a general form for the function $F(\cdots)$ in Eq. (2.23), a topic to be discussed further in Section 4. To facilitate that discussion, it is useful to consider first, briefly, the following special classes of NARMAX models.

NAARX Models

In theory, one approach to the model structure selection problem is the use of nonparametric statistical procedures, discussed briefly in Section 4. In practice, however, these procedures become increasingly difficult to apply as the number of variables involved increases. In Eq. (2.23), this number is $p + q + r + 1$, which is large enough to be a significant concern even for low-order models. Consequently, we must either impose a parametric structure on $F(\cdot)$ *a priori* or decompose the general structure into a collection of simpler sub-structures to which nonparametric techniques are applicable. The first approach is probably the most common in practice, typically taking $F(\cdot)$ as a polynomial of relatively low degree ℓ. The advantage of this choice is that it permits the use of statistical techniques like stepwise regression for model identification [3]. More recently, Billings and Zhu [7, 102] have applied similar procedures to the identification of *rational NARMAX* models, in which $F(\cdot)$ is the ratio of two polynomials. Alternatively, Tong [96] considers a variety of different models with the NARMAX structure in which the function $F(\cdot)$ is not a polynomial but instead involves discontinuous threshold functions.

The second alternative is illustrated by the class of NAARX models (Nonlinear Additive AutoRegressive models with eXogeneous inputs) defined by Chen and Tsay [16]:

$$y(k) \;=\; \sum_{i=1}^{p} f_i(y(k-i)) \;+\; \sum_{j=0}^{q} g_j(u(k-j)) \;+\; e(k). \tag{2.27}$$

Here, the functions $\{f_i(\cdot)\}$ and $\{g_j(\cdot)\}$ are scalar nonlinearities, making them well suited to the use of nonparametric procedures like those discussed in Section 4. This class of models is broad enough to exhibit a very wide range of behavior. In particular, note that this class includes the linear ARMAX models, along with many interesting nonlinear model classes considered later in this chapter. For example, note that both the logistic equation (2.24) and Tong's subharmonic generator model (2.25) are members of this class. Conversely, this model does not permit "cross-terms" involving products of inputs and outputs, or input or output values at different times. *An extremely interesting open question is what behavioral price we pay for imposing this analytically convenient structural restriction on a NARMAX model.*

Volterra and Block-Oriented Models

Discrete-time Volterra models may be defined by replacing the "multiple convolution integrals" appearing in Eq. (2.12) with "multiple convolution sums," i.e.:

$$\begin{aligned} y(k) = y_0 \;&+\; \sum_{j=0}^{\infty} a_j u(k-j) \\ &+ \sum_{i=0}^{\infty}\sum_{j=0}^{\infty} b_{i,j} u(k-i)u(k-j) \\ &+ \sum_{l=0}^{\infty}\sum_{i=0}^{\infty}\sum_{j=0}^{\infty} c_{l,i,j} u(k-l)u(k-i)u(k-j) \;+\; .... \end{aligned} \tag{2.28}$$

As in the case of continuous-time Volterra models, practical applications require that this series be truncated to a finite number of terms; these truncated series are known as *finite Volterra models.*

Extending their continuous-time results, Boyd and Chua [12] also define discrete-time fading memory systems and prove that these systems may be approximated arbitrarily well by finite Volterra models. Specifically, they consider nonlinear system operators N mapping ℓ^∞ into itself, where ℓ^∞ is the space of bounded sequences $\{u(k)\}$ equipped with the supremum norm $||u|| = sup_k \ |u(k)|$. The operator N has fading memory on a subset K of ℓ^∞ if the following conditions are met. First, there is a decreasing sequence $\{w(k)\}$ such that $0 < w(k) \leq 1$ for all k and $lim_{k\to\infty} \ w(k) = 0$. The operator N then has fading memory on K if, for each $u \in K$ and each $\epsilon > 0$, there is some $\delta > 0$ such that for all $v \in K$,

$$\begin{aligned} sup_{k\geq 0} \ |u(-k) - v(-k)|w(k) \ &< \ \delta \\ \Rightarrow \ |Nu(0) - Nv(0)| \ &< \ \epsilon. \end{aligned} \tag{2.29}$$

As in the continuous-time case, this condition implies that "the distant past has negligible influence on the present." Two important points are worth noting here, however. First, unlike the continuous-time case, the finite Volterra approximation result holds here for any set K of uniformly bounded sequences: i.e., $||u|| \leq M_1$. In particular, note that no "slew-rate limit" is imposed here, suggesting the discrete-time result may have wider validity than the continuous-time result. Conversely, the second point is that, as Boyd and Chua note [12], these finite Volterra models belong to the moving average class — i.e., they are polynomial NARMAX models with $p = 0$. Also, as in the case of continuous-time fading memory models, discrete-time fading memory models cannot exhibit output multiplicities [12]. It is known that the CSTR model can exhibit such multiple steady-state behavior [97], so it follows that any fading memory model (e.g., linear, Volterra, etc.) can only be a "reasonable" approximation over an operating range that is narrow enough to include just one steady-state. *This observation raises the interesting open question: what class of physical phenomena correspond to discrete-time models that do* **not** *have fading memory on the set $\mathcal{S}$ of input sequences of primary interest?*

Even when q is relatively small and only a few terms are retained in a finite Volterra model, the number of model parameters required to represent the series quickly becomes unreasonable. To address this problem in practice, various approaches to structural restriction have been proposed, including "pruning" (i.e., setting certain Volterra coefficients to zero) [74] and the use of orthogonal polynomial expansions [57]. Alternatively, this objection may be overcome by considering instead the discrete-time versions of the block-oriented models discussed in Section 2. As in the continuous-time case discussed there, models constructed from analytic nonlinearities may be viewed as structurally constrained Volterra models. Conversely, the block-oriented structure remains well-defined even if these nonlinearities are not analytic. This observation may be important in practice since the nonlinearity $f(\cdot)$ in either the Hammerstein or the Wiener model represents the steady-state gain of the model and analyticity is a fairly restrictive condition. For example, note that analytic functions cannot exhibit *hard saturation*: if $f(x) = c$ on some finite interval $[a, b]$ but this function is not identically constant, it is not analytic. This result follows from the fact that the behavior of an analytic function on any open interval completely determines its behavior [19, p. 87]. To distinguish between Hammerstein models with analytic nonlinearities from those with more general nonlinearities, it is useful to define the sub-class of *analytic-* or *A-Hammerstein* models. Similarly, Wiener models with analytic nonlinearities will be referred to as *analytic-* or *A-Wiener* models.

"Cross-Terms" and the PPOD Model

It is useful to note that the discrete-time Hammerstein model is a member of the NAARX class defined in Eq. (2.27). To see this connection, take $g_0(x) = f(x)$ as the static nonlinearity in the Hammerstein model and represent the linear dynamic part of the model by an ARMAX model with coefficients $\{a_i\}$ and $\{b_i\}$. The other functions appearing in the NAARX representation are then $f_i(x) = a_i x$ and $g_i(x) = b_i g_0(x)$. Conversely, note that Wiener models are *not* members of the NAARX class: "cross-terms" are present in the Wiener model. Again, an extremely interesting issue, both theoretically and practically, is what we gain or lose by including or excluding these terms from the general NARMAX structure.

Partially motivated by this question, Pawlak, Doyle and ourselves [69] have proposed a generalization of the Hammerstein model in which the nonlinearity itself is dynamic. This model structure will be abbreviated as the "PPOD model" and is of the form shown in Fig. 2.7. It is described by the

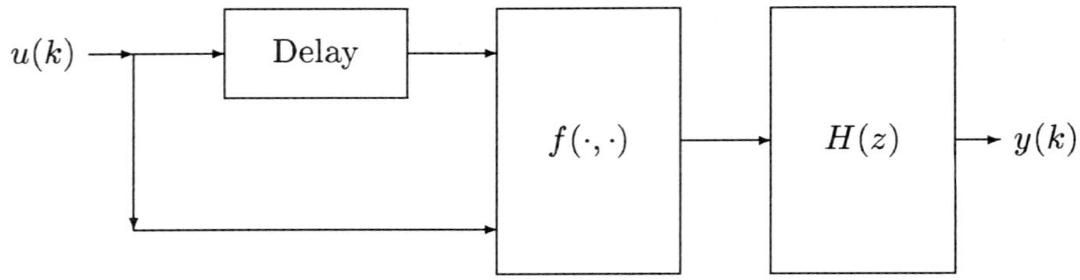

Figure 2.7: "PPOD" Model Structure Proposed by Pawlak, Pearson, Ogunnaike, and Doyle

following equation:

$$y(k) = \sum_{i=1}^{p} a_j y(k-j) + \sum_{j=0}^{q} b_j m(u(k-j), u(k-j-1)), \quad (2.30)$$

where $m(\cdot,\cdot) : R^2 \rightarrow R$ represents a dynamic nonlinearity, involving the lowest-order "cross-terms" missing from the NAARX models. In addition, by restricting the domain of this nonlinearity to R^2, we are still able to apply nonparametric regression procedures to gain some "hints" as to the nature of the cross-terms required to explain the input/output data [69].

It is instructive to consider some of the interesting special cases that are included in this general model class. First, note that if $m(x,y) = f(x)$ or $m(x,y) = g(y)$, this model reduces to the Hammerstein model. On the other hand, note that while the Hammerstein model is a member of the NAARX class, the PPOD model is not, unless $m(x,y) = f(x) + g(y)$. As a specific example, note that if $m(x,y) = g(ax+by)$, the result is a Wiener model consisting of a two-term moving average linear model followed by the static nonlinearity $g(\cdot)$. Finally, Zhu and Seborg [102] recently introduced a model of this form with $m(x,y) = \alpha x + p(y)$ where $p(y)$ is a polynomial. This model was motivated by practical considerations in implementing nonlinear model predictive control algorithms based on Hammerstein models and is discussed further in Section 4.4.

Relationships Between Different Model Classes

To conclude this section, Figs. 2.8 and 2.9 are Venn diagrams that attempt to illustrate the relationships between the different discrete-time nonlinear dynamic model classes considered here. These diagrams are based on similar ones appearing in [71], which discusses these relationships in somewhat greater detail. Note that the abbreviation "NAMAX" appearing in Fig. 2.9 refers to "Nonlinear Additive Moving Average models with eXogeneous inputs," defined by taking $p = 0$ in Eq. (2.27).

2.3.4 Added Complexity of MIMO Models

Mathematically, it is easy to extend the NARMAX model defined in Eq. (2.23) to the multiple-input/multiple-output (MIMO) setting. To handle multiple inputs, the arguments $u(k), u(k-1), ..., u(k-q)$ in the function $F(\cdot)$ are replaced with the larger set $u_1(k), u_1(k-1), ..., u_1(k-q_1), u_2(k), u_2(k-1), ..., u_2(k-q_2), ...$, etc. To handle multiple outputs, one equation of the form (2.23) is written for each output $y_i(k)$, with the arguments of each function $F_i(\cdot)$ further augmented to include the terms $y_j(k-1), y_j(k-2), ..., y_j(k-p_j)$ for the "other outputs" $j \neq i$. This generalization leads to combinatorial growth in model complexity since the number of *arguments* in each function $F_i(\cdot)$ for an n-input, m-output model will be $n_{arg} = p_1 + p_2 + \cdots + p_m + (q_1 + 1) + (q_2 + 1) + \cdots + (q_n + 1)$, assuming the past model error terms $e(k), e(k-1), ..., e(k-r)$ are omitted (as they often are in practice).

Even if each function $F_i(\cdot)$ is restricted to be a low-order polynomial, the number of possible terms in this model quickly becomes prohibitive. For example, consider a three-input, two-output model with $p_i = 2$ for all i and $q_j = 2$ for all j — each of the two functions $F_i(\cdot)$ then involves 13 arguments. Even if these functions are restricted to be cubic polynomials, the number of *possible* terms in each polynomial is 455, meaning that in the general case 910 model coefficients must be estimated to specify both functions fully. Increasing the polynomial order to 4 increases the total number of model coefficients to 14,560. In contrast, in the SISO case, for $p = 2$ and $q = 2$, $F(\cdot)$ is a function of 5 arguments and the total number of unknown model parameters is 35 for the cubic polynomial and 70 for the fourth-order polynomial case. Thus, as important as judicious model structure selection is in the SISO case, it is even more critical in the MIMO case. This particular point is illustrated in the case study in Section 9, in which a MIMO NARMAX model is developed for a high-purity distillation

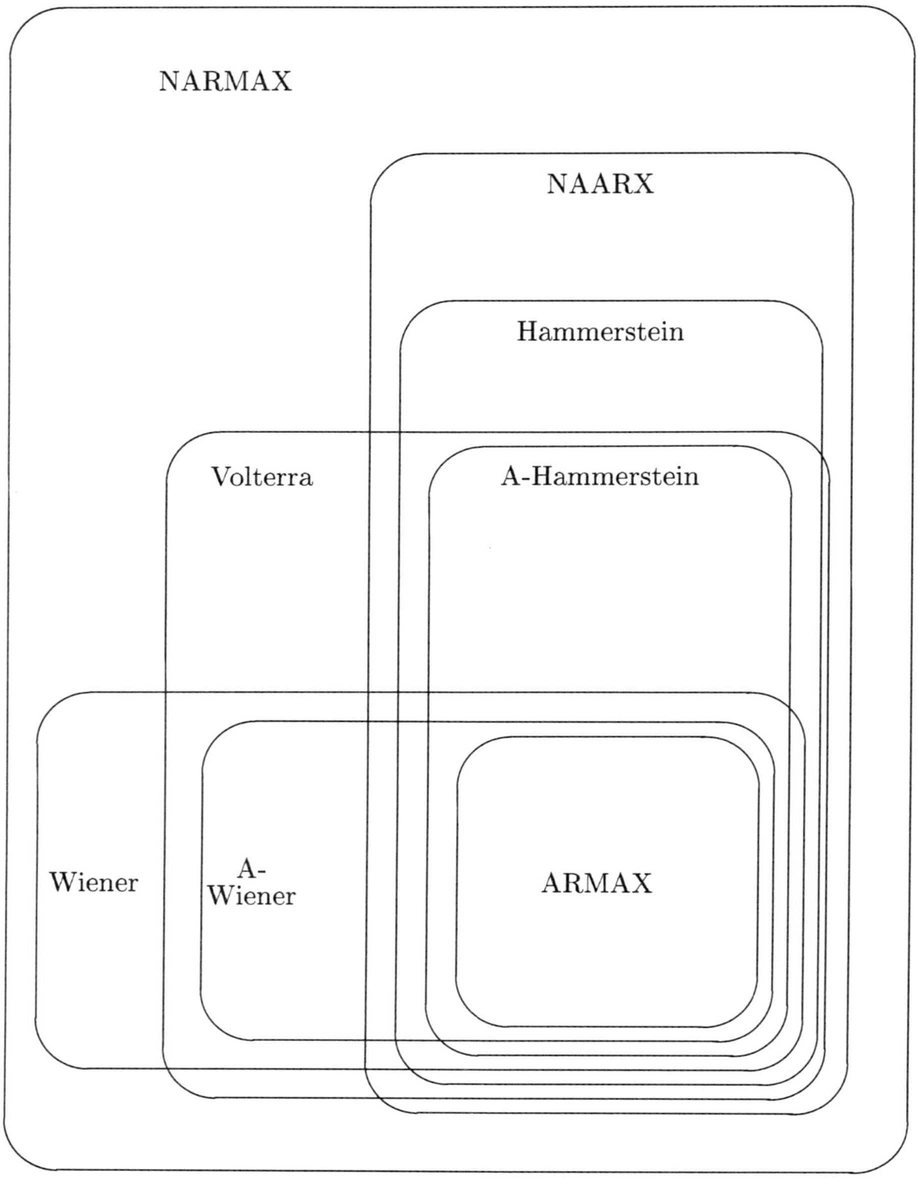

Figure 2.8: General taxonomy of NARMAX models

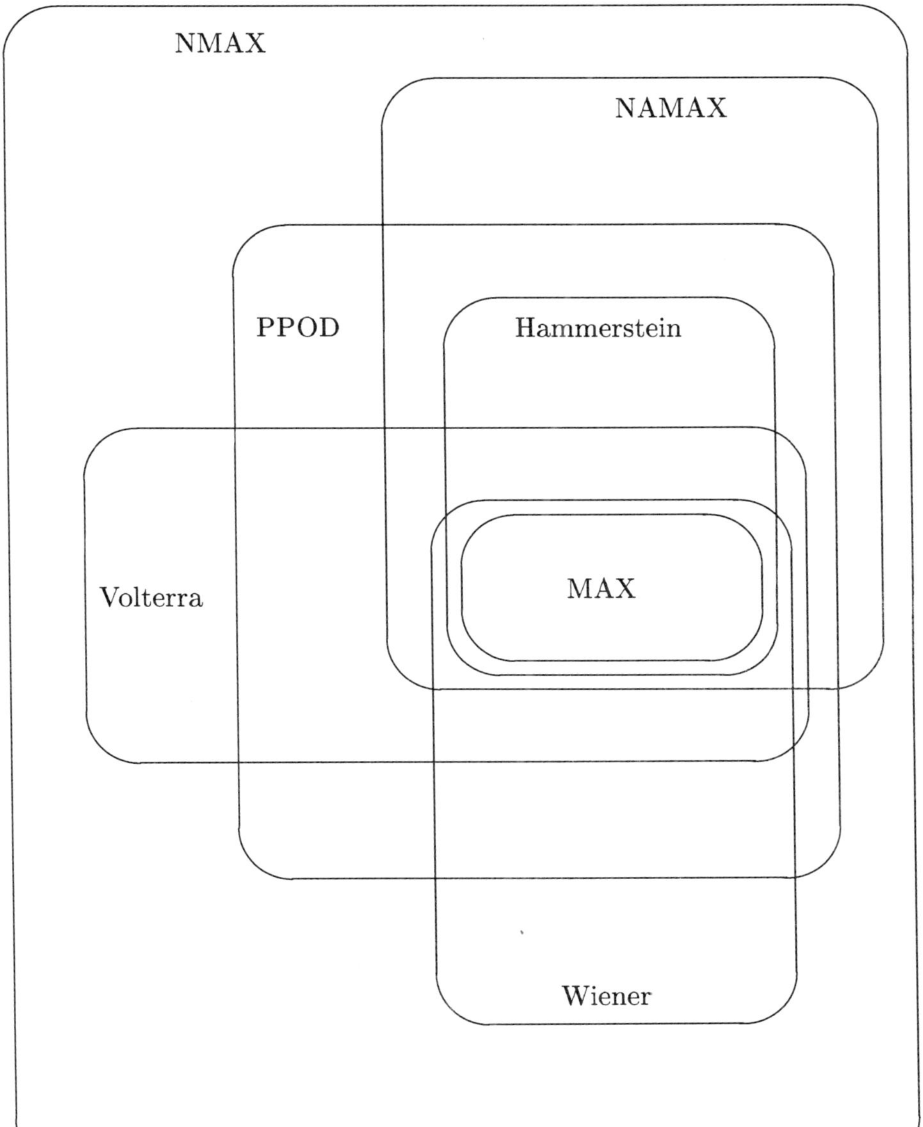

Figure 2.9: Taxonomy of NMAX models

column. One possible approach to the multivariable NARMAX modeling problem is to note its similarity to the multivariable regression problem [28]. There, polynomial models in which all possible terms are present are called ***saturated models*** [78] and approaches like stepwise regression [5, 44] have been developed to arrive at non-saturated models. "Regression-based" approaches like these will be considered briefly in Section 7 in connection with NARMAX model identification.

2.4 The Empirical Model-Building Process

The fundamental objective of empirical modeling is to construct a mathematical model $\mathcal{M}$ of a physical process $\mathcal{P}$. Given a set $\mathcal{S}$ of input sequences, a necessary — *but not sufficient* — condition for $\mathcal{M}$ to be a "good" model is that the predicted response $\mathcal{M}\mathbf{u}$ for any input $\mathbf{u} \in \mathcal{S}$ is approximately equal to the response $\mathcal{P}\mathbf{u}$ of the physical process to the same input. In symbols, we are seeking "solutions" $\mathcal{M}$ to the approximate equation:

$$\mathcal{M}\mathbf{u} \simeq \mathcal{P}\mathbf{u} \text{ for all } \mathbf{u} \in \mathcal{S}. \tag{2.31}$$

The "solution" we obtain for this "equation" will depend on the following factors:

- the process $\mathcal{P}$;
- the exact interpretation of the approximation symbol "$\simeq$";
- the set $\mathcal{S}$ of process inputs;
- the class $\mathcal{C}$ of empirical models $\mathcal{M}$ from which we are seeking the "best" representative.

Here, we assume that the process $\mathcal{P}$ is fixed, but that the other three items on this list are at least partially at the discretion of the model builder. Section 4.1 illustrates some of our options in interpreting the symbol "$\simeq$", while Section 4.2 considers specification of the set $\mathcal{S}$ of inputs over which we want our model to be valid. In practice, model-building is an iterative process in which we attempt either to "tighten" our notion of approximation, or to "enlarge" the model validity set $\mathcal{S}$. Section 4.3 briefly discusses this iteration process. The class $\mathcal{C}$ of candidate models to be considered must be specified at each stage of this iteration process; Section 5 is devoted to a detailed discussion of this model structure selection problem.

2.4.1 Interpretation of "$\simeq$"

The range of possible interpretations for the approximation symbol "$\simeq$" — and the impact of our choice on the final modeling result — is most easily illustrated with a simple example. Specifically, consider the problem of approximating an analytic function $f(\cdot)$ on a specified interval $[a, b]$. Three different "reasonable" approximation strategies will be considered and shown to lead to fundamentally different results: Taylor series expansion, Tchebyshev polynomial approximation, and Bernstein polynomial approximation.

Taylor Series Approximations

Since $f(\cdot)$ is analytic, it may be expanded as a Taylor series, i.e.:

$$f(x) = f(x_0) + (x - x_0)\left.\frac{df}{dx}\right|_{x=x_0} + (x - x_0)^2 \left.\frac{d^2 f}{dx^2}\right|_{x=x_0} + \ldots, \quad (2.32)$$

where x_0 is some reference point in the interval $[a, b]$. If we are to use this expansion in practice, it must be truncated to a finite number of terms n, resulting in an approximation error that depends on n. It is important to note that the underlying approximation philosophy here is a *local* one; in particular, the expansion is *exact* at $x = x_0$ and degrades as we move away from this point. To increase the approximation accuracy of the Taylor series on the interval $[a, b]$, we must increase the number of terms retained. Some Taylor series converge very slowly, however, such as the one for $f(x) = (1 + x)^{-1}$, valid for the range $-1 < x < 1$, but *very* slowly convergent as the limits of this range are approached.

Tchebyshev Polynomial Approximations

In contrast to the local nature of Taylor series expansions, Tchebyshev polynomials solve the problem of minimizing, for a fixed interval $[a, b]$ and a fixed polynomial order n, the maximum approximation error incurred on the interval [25, sec. 7.4]. That is, the Tchebyshev polynomial approximation $T_n(f; x)$ minimizes the maximum value of $|f(x) - T_n(f; x)|$ on the interval $[a, b]$. Thus, for given order n, the worst-case approximation error for the Tchebyshev polynomial approximation $T_n(f; x)$ is better than the worst-case Taylor series approximation error. Conversely, the *derivatives* of the function $f'(x)$, $f''(x)$, etc. are generally *not* well approximated by the corresponding derivatives of $T_n(f; x)$. This behavior is in contrast to the *local* behavior of the Taylor series, since the lowest-order derivatives of the truncated Taylor series are also exact at x_0, just like the function itself.

Bernstein Polynomial Approximations

By the Weierstrass approximation theorem [50], any continuous function $f(x)$ can be approximated uniformly to arbitrary accuracy on any finite, closed interval $[a, b]$ by a polynomial of finite degree. The polynomials often used to prove this theorem are the Bernstein polynomials $B_n(f; x)$, given by the explicit formula [25]:

$$B_n(f;x) = \sum_{k=0}^{n} f(k/n) \binom{n}{k} x^k(1-x)^{n-k}. \tag{2.33}$$

(This formula assumes the interval $[a, b]$ is $[0, 1]$; nothing significant is changed by re-scaling this expression to arbitrary finite intervals.)

It is important to note that the basis of the Weierstrass theorem is fundamentally different from the basis for Tchebyshev approximation. Specifically, Tchebyshev approximation constructs a *best approximation of fixed order* n, given the function and the approximation interval. On the other hand, in proving the Weierstrass approximation theorem, the maximum approximation error over the interval is specified and a Bernstein polynomial of *sufficiently high order* n is constructed to meet this approximation requirement. In fact, an unfortunate feature of the Bernstein polynomial approximations is that their convergence is rather slow [25], so the required polynomial order tends to be high. The primary advantage of these polynomials, however, is that the approximations preserve the *qualitative* character of the functions they approximate. In particular, low-order derivatives of Bernstein polynomial approximations are reasonable approximations of the derivatives of $f(x)$ and the approximation also preserves such general qualitative behavior as convexity and monotonicity [25].

Deciding Between Different Approximation Criteria

The key point of this discussion is that the "best nonlinear approximation" we obtain for the function $f(x)$ depends on both our "adequacy" criterion and the range over which we want this criterion to be met. In the above example, if our sole criteria is "goodness of fit," we will prefer the Tchebyshev approximation over either the Taylor series or the Bernstein polynomial approximation because, for a specified interval and polynomial order, it has the smallest possible worst-case approximation error. Conversely, the truncated Taylor series has a certain "fundamental feel" about it, since each term in the expansion is directly related to the derivatives of the function at a specified point. Note, however, that a truncated Taylor series expansion need not exhibit the general qualitative behavior of the function $f(x)$ on a given interval $[a, b]$. For example, consider the Taylor series expansion for the function $f(x) = (1+x)^{-1}$ on the interval $[-3/4, 3/4]$; expanded about $x = 0$ and truncated to second order, this series is $1 - x + x^2$. This approximation has a minimum in the interior of the approximation interval at $x = 1/2$, even though the original function $f(x)$ is monotonically decreasing on the whole interval. If we wish to achieve an approximation that retains these qualitative features over the entire interval of interest, we must accept a more complex model like the Bernstein polynomial approximation discussed above.

2.4.2 Specification of the Range of Model Validity

Overall, the "most reasonable interpretation" of the approximation symbol "$\simeq$" in Eq. (2.31) will depend on the following factors:

- the set $\mathcal{S}$ of process inputs;
- the class $\mathcal{C}$ of empirical models under consideration;
- our application (i.e., the process $\mathcal{P}$ and the control system design methodology under consideration).

The following paragraphs illustrate this dependence on the set $\mathcal{S}$ of process inputs, demonstrating the subtlety required in adequately defining this set. Dependence on the class $\mathcal{C}$ of models is related to the issue of model structure selection considered in Section 5. Finally, note that it is application-depencence that motivates recent work on "control-relevant" model identification [81]: "open loop" prediction error measures are replaced with frequency-weighted prediction error measures that penalize prediction errors with adverse control implications more than those with negligible control implications. It is not immediately obvious how to generalize these ideas to arbitrary nonlinear model identification problems because frequency-domain descriptions of nonlinear models are necessarily incomplete. *Still, the basic concept of making our approximation measure application-dependent seems like a useful one and this area appears to be a fruitful one for future research.*

Specification of the "range of validity" for a dynamic model $\mathcal{M}$ is more complex than specification of the range of a function $f(\cdot)$ mapping one finite-dimensional space into another. Specifically, note that the set $\mathcal{S}$ appearing in Eq. (2.31) is a set of *functions* $u(t)$ rather than a set of *values*. To illustrate this distinction, recall from the discussion of fading memory systems in Section 2.3 that Boyd and Chua's approximation result [12] was valid for a set K of slew-rate limited inputs. This restriction was necessary to make the Stone-Weierstrass approximation theorem apply to fading memory systems. The key point is that this restriction specifies *two* types of constraints on u:

1. the static amplitude limit $||u(t)|| \leq M_1$;
2. the dynamic slew-rate limit $||u(t-\tau) - u(t)|| \leq M_2\tau$.

This observation is important because specification of an input range condition like (1) alone is generally not sufficient to define the set $\mathcal{S}$. The following example provides a detailed description of another case where dynamic constraints are both important and non-obvious without further examination.

A Simple Nonlinear Continuous-Time Model

Consider the behavior of the first-order SISO model obtained by approximating the function $F(x(t), u(t))$ in Eq. (2.1) by the following *second-order* polynomial [65]:

$$F(x,u) \;\simeq\; (\phi_0 + \phi_1 x + \phi_2 x^2) \;+\; (\gamma_0 + \gamma_1 x + \gamma_2 x^2)u. \tag{2.34}$$

Note that this model is control affine but not bilinear unless $\gamma_2 = 0$ and $\phi_2 = 0$. This model structure was motivated by a desire to explore the "next-step-beyond-bilinear-models," retaining the first *two* terms of the Taylor series expansion of the nonlinearities appearing in the control-affine structure.

To obtain insights into the qualitative behavior of this model, it is useful to employ the same "trick" invoked earlier for the bilinear model. Specifically, note that the step response of this model is given by the solution of an equation of the form:

$$\frac{dx}{dt} \;=\; a + bx + cx^2, \tag{2.35}$$

where:

$$a \;=\; \phi_0 + \alpha\gamma_0$$

$$\begin{aligned} b &= \phi_1 + \alpha\gamma_1 \\ c &= \phi_2 + \alpha\gamma_2. \end{aligned} \tag{2.36}$$

An explicit solution to this equation may be obtained, but its form depends strongly on the constants a, b, and c. In particular, the character of the solution depends on the discriminant $\Delta = 4ac - b^2$: if this quantity is negative, nothing remarkable happens, but if it is positive, this model can exhibit *finite escape time*. That is, for $\Delta > 0$, the solution to Eq. (2.35) is:

$$x(t) = \frac{\sqrt{\Delta}}{2c}\left[\tan\left(\theta_0 + \frac{\sqrt{\Delta}}{2}t\right) - \tan\,\theta_0\right], \tag{2.37}$$

where $\theta_0 = \arctan(b/\sqrt{\Delta})$. Note that this solution diverges to $+\infty$ as the argument of the tangent function approaches $\pi/2$, which occurs at the escape time:

$$t^* = \frac{2}{\sqrt{\Delta}}\left[\frac{\pi}{2} - \theta_0\right]. \tag{2.38}$$

An extremely important point to note here is that the dominant qualitative behavior of this model — presence or absence of a finite escape time — depends on both the model structural parameters ϕ_i and γ_i and the input amplitude α. To see this point, expand Δ explicitly as:

$$\Delta = [4\phi_0\phi_2 - \phi_1^2] + [4(\gamma_0\phi_2 + \gamma_2\phi_0) - 2\gamma_1\phi_1]\alpha + [4\gamma_0\gamma_2 - \gamma_1^2]\alpha^2. \tag{2.39}$$

To force $\Delta < 0$ — and thus avoid the finite escape phenomenon — the following criteria must be met. First, to satisfy this condition for small input amplitudes ($\alpha \simeq 0$), the first term in brackets must be negative, representing an "inherent stability constraint." Similarly, to avoid finite escape at large input amplitudes ($\alpha \to \infty$), the last term in brackets must also be negative.

Note that both of these conditions are satisfied by the linear special case $\phi_0 = \phi_2 = 0$ and $\gamma_1 = \gamma_2 = 0$. In fact, for this case, $\Delta = -\phi_1^2 < 0$ for all α, correctly indicating that the linear case cannot exhibit finite escape time, regardless of the input amplitude. Similarly, note that both of these conditions are also satisfied for the bilinear case, for which the constraint $\gamma_1 = 0$ is relaxed. In fact, for the bilinear case, we have $\Delta = -(\phi_1 + \gamma_1\alpha)^2$, which is, again, non-positive for all possible input amplitudes and system parameters. Thus, the bilinear special case of this model cannot exhibit finite escape time, either. Conversely, note that if the first term in brackets is negative but the last term is positive, the model will exhibit a finite escape time for some sufficiently large input step amplitude α. As a specific example, consider $\phi_0 = 0$, $\phi_1 = 1$, $\phi_2 = 0$, $\gamma_0 = 1$, $\gamma_1 = 0$, and $\gamma_2 = 1$. Here, $\Delta = 4\alpha^2 - 1$, so the model cannot exhibit finite escape time for $\alpha < 1/2$, but it can for $\alpha > 1/2$. Finally, note that a perverse kind of "conditional stability" is also possible in this model: if the first and last terms in brackets are both negative but the middle term is positive, there may be an "intermediate amplitude" α that causes the middle term to dominate. In this case, $\Delta > 0$ and the model response will diverge in finite time. As a specific example, consider $\phi_0 = 0$, $\phi_1 = 1$, $\phi_2 = 1$, $\gamma_0 = 1$, $\gamma_1 = 1$, $\gamma_2 = 0$: it is easy to show that $\Delta > 0$ for α between $3 - \sqrt{8} \simeq 0.172$ and $3 + \sqrt{8} \simeq 5.828$.

Now, suppose we perform the following identification experiment: we wish to model a process $\mathcal{P}$ that is both well-behaved (i.e., does not exhibit finite escape times) and well approximated by a control-affine fundamental model. Suppose we excite this process with inputs from a set $\mathcal{S}$ and attempt to estimate the corresponding parameters for model (2.34) that best match the available data. Here, the set $\mathcal{S}$ consists of "high-frequency" piecewise-constant input sequences that assume arbitrary values in the interval $[-1, 1]$, but whose values switch rapidly enough that they never remain constant longer than 0.2 time units. For these inputs, the "conditionally stable" model just described is perfectly well-behaved — the worst-case (i.e., shortest) escape time for step inputs in the interval $[-1, 1]$ is $t^* = \pi/4 \simeq 0.7854$. Thus, it is possible that this model could be the "best fit" approximation of $\mathcal{P}$ from the available data, even though this model exhibits finite escape time but the process itself does not. *In particular, the point of this example is that it is the nature of the set $\mathcal{S}$ we choose that determines whether the model considered here is a "good" or a "pathologically poor" approximation of the process.* Specifically, the model considered here may be a good approximation on the set $\mathcal{S}$ of "high-frequency" piecewise-constant sequences used for identification, but it is completely inadequate when considered in terms of step responses since any positive step of amplitude between $3 - \sqrt{8}$ and $3 + \sqrt{8}$ will exhibit the finite escape instability. Note that the difference between the set $\mathcal{S}$ and the larger set of "all piecewise- constant input sequences" is that $\mathcal{S}$ excludes "low-frequency" signals (e.g., steps).

2.4.3 Sequential Model Building

In practice, model building is usually a sequential process, in which models of increasing complexity are developed from — and compared with — earlier models of lower complexity. In the development of nonlinear, dynamic models, two particularly useful and important points of comparison are:

- static (i.e., steady-state) models;
- linear dynamic models.

The nonlinear model structure selection problem will be considered in Section 5, but before considering this topic, it is worth considering the two "precursors" listed above, *specifically in their role as "guides" for the iterative nonlinear modeling process.*

Steady-State Behavior of Empirical Models

The idea of "steady-state agreement" between a candidate model and process behavior is often an important criterion for "model validity," but this issue is more subtle for nonlinear models than for linear ones. In particular, for linear models, the issue is whether the steady-state gain K is correct, but for nonlinear models, the notion of "steady-state agreement" between a model and a process is closely related to the issues of local *vs.* global modeling considered in Section 5. That is, for a linear SISO problem, the process input is typically represented as $U + u(t)$ where U is a steady-state reference value and $u(t)$ is the instantaneous deviation of the input from this reference value. Similarly, the process response is represented as $Y + y(t)$, where Y is steady-state response to the input U and $y(t)$ is the corresponding instantaneous deviation from this reference value. For stable, linear, time-invariant models, the following conditions hold:

$L1$ the steady-state response is $Y = KU$, independent of $u(t)$;

$L2$ the dynamic response is $y(t) = L\{u(t)\}$, independent of U,

where L is the linear model's dynamic response.

For nonlinear models, the situation is more complex, and we have the following possibilities:

$N1$ the steady-state value Y may fail to exist;

$N2$ when Y is well-defined, generally $Y = S\{U, u(t)\}$ is a function of both U and $u(t)$;

$N3$ the dynamic response $y(t) = D\{U, u(t)\}$ is a function of both U and $u(t)$.

The first of these possibilities (i.e., $N1$) is illustrated by the first-order threshold autoregressive model considered in [70]:

$$y(k) \;=\; a|y(k-1)| \;+\; bu(k-1).$$

In particular, if the amplitude of the input to this model is sufficiently large, the step response of this model is *chaotic* and never settles out to a steady-state value Y. As an example of nonlinear behavior $N2$, consider the response of a Hammerstein model with the quadratic nonlinearity $g(x) = x^2$ to the input sequence $U + u(t)$ where $u(t) = \alpha cos(2\pi f t)$. It follows from simple trigonometric identities that, if K is the steady-state gain of the linear part of the Hammerstein model, then $Y = K(U + \alpha^2/2)$. That is, the steady-state response depends on both the steady-state input U and the amplitude of the "fluctuations" $u(t)$ about steady-state. Note that this "rectification phenomenon" — i.e., conversion of "fluctuations" into a "steady-state offset" — necessarily arises in any model with an asymmetric nonlinearity. This phenomenon is commonly observed in chemical processes; for a specific example, refer to the high-purity distillation column discussed in Section 9.

Finally, the nonlinear behavior $N3$ listed above is illustrated by the following example. Consider a Hammerstein model with the saturation nonlinearity:

$$g(x) \;=\; \begin{cases} x & |x| \leq 1 \\ 1 & |x| > 1, \end{cases}$$

and suppose the input to this model is $U + \alpha cos(2\pi f t)$ where $\alpha = 1/2$. For $-1/2 < U < 1/2$, this model behaves linearly and its dynamic response will be of the form $H(f)\alpha cos(2\pi f t)$, where $H(f)$ is the amplitude response of the linear part of the Hammerstein model at the excitation frequency f. For $-3/2 < U < -1/2$ and $1/2 < U < 3/2$, the saturation nonlinearity $g(x)$ will cause harmonic generation at frequencies $3f$, $5f$, ..., and the intensities of these harmonics will depend on U. Finally, for $U < -3/2$ or $U > 3/2$, the input nonlinearity will be fully saturated, so the input to the linear part of the Hammerstein model will be constant at either $+1$ or -1. Consequently, the output of the Hammerstein model will saturate at $\pm K$. This dependence of the dynamic response $y(t)$ on both U and $u(t)$ is also commonly observed in chemical processes and is again illustrated in Section 9.

The key point here is that the "steady-state behavior" of a nonlinear model is usually inextricably linked with both dynamic responses and the class of inputs under consideration. In addition, even when well-defined steady-states exist for a given class of inputs, evaluating them may require significant effort. As a specific example, consider the task of characterizing the steady-state response of a NARMAX model of "moderate complexity" (e.g., 5 to 10 model parameters):

- does it exhibit chaotic regimes?
- are there pronounced differences in "steady-state gain" for different input sequence amplitudes?
- are there pronounced differences in "steady-state gain" for different input sequence "shapes"?
- etc.

Still, certain "steady-state characterizations" (e.g., "what does the step-response ultimately settle out to?") do yield useful insight into nonlinear model behavior and should be developed, if only as a prelude to a more complete dynamic characterization.

Development of a "Linear Skeleton"

Since the term "nonlinear models" automatically defines an alternative reference point — i.e., "linear models" — it is extremely useful to compare the performance of any nonlinear model with "nearby" linear ones. In addition, we may reverse this process, using linear models as a basis for guiding the development of nonlinear models. For example, in developing bilinear time-series models, Subba Rao and Gabr [92] advocate first constructing a linear model, taking advantage of such linear modeling tools as the Akaike Information Criterion (AIC) [14, 76, 96] to guide model order determination. Cinar [20] advocates a similar approach in developing NARMAX models: start with a "linear skeleton" that provides both guidance in selecting an initial model order and a basis for comparion in subsequent refinement steps. In addition, this approach may be taken a step further, "trimming" this "linear skeleton" through the use of techniques like stepwise regression to eliminate "marginal" model terms [92]; note that the "simplified" models developed by this procedure also serve as "secondary standards" for comparison in subsequent nonlinear model evaluations.

2.5 Nonlinear Model Structure Selection

As noted in Section 1, the first step in nonlinear process identification is the selection of an "appropriate" model structure. Unfortunately, this step is probably both the most difficult and the least amenable to systematic analysis. Consequently, this section offers brief summaries of the following four topics, to provide useful guidance:

1. parametric, nonparametric, and semi-parametric modeling;
2. variable selection and transformation;
3. local modeling;
4. control-motivated model structure selection.

2.5.1 Parametric, Nonparametric, and Semi-Parametric Models

Dynamic model identification is similar in some important respects to regression analysis. There, we are concerned with the identification of unknown functions (i.e., static maps) from available data. As a specific example, consider the scalar case: a real-valued function $f(\cdot)$ is to be identified such that the approximation $y \simeq f(x)$ is a "good" one for the available dataset $\mathcal{D} = \{(x_i, y_i)\}$. Three basic approaches have evolved for solving this problem — *parametric* regression, *nonparametric* regression, and *semi-parametric* regression. *Parametric* procedures postulate a general form for the function $f(\cdot)$, characterized by a p-vector of parameters θ. For example, an nth-order scalar polynomial is characterized by $n + 1$ coefficients. Given such a functional form, parametric procedures choose optimal values for these parameters by minimizing some measure of "lack of fit" (e.g., total squared error).

Nonparametric procedures do not assume an explicit form for the function, "allowing the data to choose." For example, kernel regression [42] constructs the following "smoothed" estimate of $f(x)$ from the available data:

$$\hat{f}(x) = \frac{\sum_{i=1}^{n} K(x - x_i) y_i}{\sum_{i=1}^{n} K(x - x_i)}. \tag{2.40}$$

Here, the function $K(\cdot)$ is a *smoothing kernel* that is typically of the form $K(x - x_i) = G(\frac{x - x_i}{h})$ where h is a constant parameter called the *bandwidth*. In practical terms, the end result of a nonparametric procedure is typically a plot of the smooth function $\hat{f}(x)$. The choice of both the form of the "basic kernel function" $G(\cdot)$ and the bandwidth h are important in practice, but practical advice on these topics is becomming increasingly available [42, 58].

Semi-parametric regression procedures decompose the unknown function $f(x)$ into a parametric part and a nonparametric part, applying the appropriate analytical procedures to each. That is, we write $f(x) = f_1(\theta; x) + f_2(x)$ and develop a procedure that both estimates the unknown parameter vector θ and "smooths" the data appropriately to obtain an estimate $\hat{f}_2(x)$ of the "nonparametric part" $f_2(x)$.

These ideas carry over to system identification. For example, in linear system identification, direct identification of ARMAX models is a parametric problem. That is, autoregressive and moving-average model orders p and q are chosen and the corresponding model parameters $\{a_i\}$ and $\{b_i\}$ that best fit the available data are estimated. Alternatively, frequency responses may be estimated nonparametrically by first computing classical estimates of the input autospectrum $S_{uu}(f)$ and the input-output cross-spectrum $S_{uy}(f)$ [55, 76]. The ratio $\hat{H}(f) = S_{uy}(f)/S_{uu}(f)$ then yields a nonparametric estimate of the frequency response of the linear model relating the input/output data. The use of semi-parametric techniques in Hammerstein model identification is discussed briefly in Section 7.2.

The advantage of nonparametric estimates is that they impose fewer "prior assumptions" on the resulting model, but there are two significant disadvantages. First, note that nonparametric models are not generally applicable to control system design directly because most model-based control schemes require parametric models. For example, to be useful in a linear pole-placement procedure, a nonparametric frequency response estimate would first have to be converted to a rational transfer function model. Typically, this conversion would be accomplished by fitting a ratio of low-order polynomials to the estimated frequency response; the coefficients in this parametric model would then be used for controller design. In spite of this ultimate need for a parametric model, nonparametric procedures can be quite useful in the exploratory initial stages of model development. This is particularly true in nonlinear process identification because the model structure selection problem is so difficult, a point discussed further in Section 7.2. The second disadvantage of nonparametric approaches is that estimation of *multivariable* nonlinear functions rapidly becomes difficult [42]. In particular, it is not practical at present to apply a general nonparametric regression procedure to the NARMAX model (2.23) for "reasonable" values of p, q, and r and expect meaningful results. Thus, while nonparametric procedures can be useful in the model structure selection process, they do not reduce the problem to a purely computational one.

2.5.2 Variable Selection and Transformation

Often, empirical process modeling leaves us a number of options regarding what variables to include in the model, and this choice can profoundly influence the nature of the model we ultimately develop. Specifically, if we have any discretion in the selection of variables to be included in a multivariable model, we may be able to influence the dynamic complexity, nonlinearity, and degree of interaction significantly, as the following simple example illustrates clearly. Consider a conical mixing tank with two feed streams — hot and cold water. The control objective is to maintain a fixed tank level and effluent temperature in the face of changes in effluent flow rate. Adopting a simple pairing scheme (e.g., controlling level by manipulating cold water flow rate, controlling temperature by manipulating hot water flow rate) leads to a nonlinear MIMO problem. If instead, we control tank level by manipulating total flow (i.e., the sum of hot and cold water flow rates) and control effluent temperature by manipulating an appropriately weighted difference between these flow rates, we obtain two weakly coupled SISO problems [66]. We can simplify this problem further by adopting the "extensive variable" control philosophy [34] and controlling not tank level, but the *volume* of material in the tank. That is, note that, given the height, minimum radius, and maximum radius of the tank, we can compute the volume of liquid in the tank from level measurements. The result of these changes in manipulated and controlled variables — based on physical insights — has taken us from a strongly nonlinear MIMO problem to two weakly coupled, weakly nonlinear SISO problems.

While this example is simple enough to be fairly obvious, it raises the point that the structure and complexity of an *empirical* model can depend dramatically on the exact choice of variables. Thus, if the physical variables available to us for empirical modeling are levels and individual flow rates, we may find that simpler models result if we consider such alternatives as the following:

- transform levels into approximate volumes;
- replace individual flow rates with total flow rates into and out of a vessel or "compartment";

- replace individual flow rates with ratios of individual flow rates;
- replace individual flow rates with ratios of individual flow rates to total flow rates.

Analytical procedures like nonparametric regression may also be useful in suggesting transformations to apply to individual variables to achieve simpler models. Finally, note that variable selection and transformations also influence robustness (i.e., sensitivity of model predictions to "outliers" or "bad data") [53].

2.5.3 Local Modeling

Another approach to the development of reduced complexity models is the development of multiple *local* models, each approximating process behavior "adequately" over some restricted operating range. The advantage of this approach is that the overall complexity of the collection of local models is often significantly less than that of the corresponding *global* model necessary to describe process dynamics over the entire operating range. For example, for many processes, if the range of control input variations is restricted sufficiently, it may be possible to approximate the process dynamics adequately by a linear model. Combining several linear models then leads to a model that is globally nonlinear but locally linear. This point is illustrated by Johansen and Foss [46] who develop a set of four local linear models to describe the dynamics of a batch fermentation reactor. The question of how *exactly* to specify the regime of validity of each model is an important one, as the following example illustrates.

Consider a single-input, single-output control problem and suppose the nonlinear process dynamics may be approximated "adequately" by a collection of m local linear models. Assume each model has a transfer function $H_i(s)$ and is associated with a set S_i of possible input values (e.g., intervals $[u_i^-, u_i^+]$). Define the m nonlinear functions:

$$G_i(u) = u\, \chi_i(u), \tag{2.41}$$

where $\chi_i(\cdot)$ is the *characteristic function* for the set S_i, i.e.:

$$\chi_i(u) = \begin{cases} 1 & u \in S_i \\ 0 & u \notin S_i. \end{cases} \tag{2.42}$$

Suppose we now construct an m-channel Uryson model (i.e., a PNL_m model in Chen's notation [17]) with the nonlinearity $G_i(\cdot)$ and the transfer function $H_i(s)$ in the ith channel for $i = 1, 2, ..., m$. If the model validity sets S_i are disjoint, then for any value of the input u, only one of the "intermediate variables" $G_i(u)$ can be non-zero, so the Uryson model output will be the response of the "appropriate" linear model.

This construction suggests that for Uryson models in general, the quantity $\nu_i(u) = G_i(u)/u$ may be viewed as a "model validity function" for the linear model $H_i(s)$. That is, if $\nu_i(u)$ — a measure of the nonlinearity of $G_i(\cdot)$ — approximates the characteristic function for some set S_i of possible input values, the "local linear modeling" interpretation holds. Alternatively, if the functions $\nu_i(u)$ exhibit broad peaks over some range of input values, it may be appropriate to interpret them as *fuzzy set* characteristic functions [29, 101], describing imprecise regions of model validity. At the other extreme, if $\nu_i(u) = 1$ for all i and all u, the Uryson model reduces to a linear model whose transfer function is $H(s) = \sum_{i=1}^{m} H_i(s)$. Intermediate between these limits, the Uryson model may be viewed as a collection of local linear models whose regions of validity — and influence — overlap strongly. While it is not equivalent to the nonlinear Uryson model, a useful frame of reference for this intermediate case is the local linear model:

$$\tilde{H}(s) \; = \; \sum_{i=1}^{m} \nu_i(u_0) \; H_i(s), \tag{2.43}$$

which should have approximately the same steady-state behavior for $u(t) \simeq u_0$.

The preceeding discussion illustrates the close connection between local linear modeling and Uryson models, but two significant points have not yet been addressed. The first is that the region of validity for a local linear model is often not directly expressible in terms of the input u alone. For example, in the fermentor models developed by Johansen and Foss, the fermentor is a batch process and the models describe the evolution of the batch from fixed initial conditions, so there is no "control input u." Instead, three different combinations of output variables and state variables for the fermentor are used to specify the region of model validity, based on a qualitative understanding of the evolution of the batch through different operating regimes. This observation suggests that, if the analogy between local linear models and block-oriented models is to hold for these more complex cases, the "validity functions" $\nu_i(u)$ must be replaced by multivariable functions $\nu_i(u, x_1, ..., x_r)$ of some set of r measurable "auxiliary variables." This change may destroy the block-oriented structure of the model, but investigation of the possibilities may yield useful insights into both block-oriented models and local modeling ideas.

The second point overlooked in the Uryson model discussion given above is the effect of transients that occur when switching from one region of model validity to another. That is, in switching from $u(t) \in S_i$ to $u(t') \in S_j$, both linear models i and j will have non-zero responses during the transient, which will be summed and appear in the Uryson model output. The question of how to combine local models is an important one and is considered in [2]. In addition, it is useful to note that the threshold autoregressive models discussed by Tong [96] are locally linear models. The fact that the subharmonic generating model discussed in Section 3.2 is of this form clearly illustrates that the behavior of "locally linear models" is quite distinct from that of "linear models."

Finally, it is important to note that, while the examples discussed here used local *linear* models, the local models need not be linear in general. In particular, recall that the fading memory systems discussed in Section 2 were defined on a specified set K of input sequences and exhibited a unique steady state for all inputs within that set. Further, for particular sets K (i.e., sets of bounded, "slew-rate limited" inputs), these systems could be approximated uniformly by finite Volterra models. This observation suggests that local Volterra modeling around each steady state may be a reasonable approach for systems (e.g., reactors) that are known to exhibit multiple steady states. Alternatively, local NARMAX models may be developed for different operating regimes of a process, resulting in a model with a wider overall range of validity and simpler structure than a global NARMAX model that attempts to describe all regimes simultaneously.

2.5.4 Control-Motivated Model Structure Selection

The primary motivation for empirical model identification given in Section 1 was the incompatibility of first-principles models with many control approaches due to the generally great complexity of such models. This observation provides a significant motivation for considering the "inverse problem": choosing a model structure that leads to a tractable control problem. Historically, this idea was one of the strongest motivations for the development of linear process models. More recently, this idea was one of the motivations for considering Volterra models — the model structure leads naturally to a special "nonlinear IMC" control approach [27].

Similarly, Zhu and Seborg [102] considered the "modified Hammerstein model" discussed earlier because it greatly simplifies the solution of the Model Predictive Control (MPC) problem relative to the standard Hammerstein model. Specifically, for the MPC problem, we ultimately need to solve for the control input $u(k)$ required to bring the predicted process output $\hat{y}(k)$ to some target value y^*. For a polynomial Hammerstein model, if $v(k)$ is the output of the static nonlinearity, it is given by:

$$v(k) = \sum_{i=1}^{L} \gamma_i [u(k)]^i, \tag{2.44}$$

while the modified Hammerstein model considered by Zhu and Seborg replaces this relation with the *dynamic* nonlinearity:

$$v(k) = \gamma_1 u(k) + \sum_{i=2}^{L} \gamma_i [u(k-1)]^i. \tag{2.45}$$

In the absence of either nonlinearity, the linear MPC problem gives the value of $v(k)$ such that $\hat{y}(k) = y^*$. To convert this result into a control input for the nonlinear problem, we must compute $u(k)$ from this value of $v(k)$. Zhu and Seborg proposed the modified Hammerstein model of Eq. (2.45) because inversion of Eq. (2.44) is a polynomial root-finding problem. In contrast, the inversion of Eq. (2.45) is accomplished by the simple analytical result:

$$u(k) = \frac{1}{\gamma_1} \left[v(k) - \sum_{i=2}^{L} \gamma_i [u(k-1)]^i \right], \tag{2.46}$$

which is directly computable since $u(k-1)$ is known at time k.

It was noted in Section 3 that this modified Hammerstein model is a member of the PPOD family of models. This observation raises a number of interesting questions. First and foremost is the question of how the class of dynamic nonlinearities described by this model differs from the range of behavior representable by the Hammerstein model. Zhu and Seborg's motivation for introducing it was the observation that if the sequence $\{u(k)\}$ is slowly varying, the control value $u(k)$ given by Eq. (2.46) should not differ too much from the solution to the unmodified Hammerstein problem. The more intriguing question is how the dynamics differ in cases where the difference between $u(k)$ and $u(k-1)$ is not small. In addition, referring back to the discussion of nonparametric approaches to Hammerstein modeling, note that Zhu and Seborg's model generalizes directly from Eq. (2.45) to:

$$v(k) = \gamma_1 u(k) + g(u(k-1)), \tag{2.47}$$

where $g(\cdot)$ is an arbitrary function. In particular, nonparametric estimation of this function might expand the applicability of this model considerably, as in the unmodified Hammerstein case (see Section 7.2 for a discussion of this point). Finally, note that the MPC problem would remain tractable for a still wider class of PPOD models. For example, consider the following "control affine" model:

$$v(k) = \gamma_1 u(k)\, g(u(k-1)), \tag{2.48}$$

where again, $g(\cdot)$ is an arbitrary function. As in the previous examples, this structure permits explicit computation of $u(k)$ from $v(k)$ as:

$$u(k) = \frac{v(k)}{\gamma_1\, g(u(k-1))}. \tag{2.49}$$

An important question raised by such control-motivated model structures is how well they can represent the process to be controlled. In the examples just described, the general answer is "at least as well as the linear sub-class they contain." That is, viewing this linear sub-class as a constrained form of the nonlinear model, the best unconstrained model within the class must represent the process dynamics at least as well as the best constrained (i.e., linear) model within the class. In general, we would expect the best unconstrained model to be a better representation of the process dynamics, but the practical question is "how much better?"

2.6 Statistical Considerations

Once we have selected a model structure, the remaining three steps in the empirical model development sequence discussed in Section 1 all have strong connections with statistics. For this reason, the following section summarizes some important statistical concepts that will be useful in subsequent discussions of these modeling steps. This section is necessarily brief and emphasizes concepts, restricting consideration to discrete-time stochastic processes. Probably the most important notion considered here is the distinction between the distribution of a sequence of random variables and its dependence structure, both of which are important in nonlinear model identification. For a more rigorous introduction to random variables and stochastic processes, see Billingsley [8]; for a more extensive "engineering-oriented" introduction, see Papoulis [68].

2.6.1 Working Definitions and General Concepts

The ideas presented here are mainly summarized from Rosenblatt [83], and rephrased slightly to emphasize the underlying concepts. The discrete-time stochastic processes considered here may be viewed as sequences $\{x_k\}$ of random variables where the index sequence k runs from $-\infty$ to $+\infty$. These sequences will be characterized in terms of the *joint densities* $p(x_j, x_{j+1}, ..., x_{j+m})$ for arbitrary subsequences $\{x_j, x_{j+1}, ..., x_{j+m}\}$. In particular, the following definitions are important.

The sequence $\{x_k\}$ is *statistically independent* if, for any j and m, the joint density $p(x_j, x_{j+1}, ..., x_{j+m})$ may be expressed as:

$$p(x_j, x_{j+1}, ..., x_{j+m}) = \prod_{i=0}^{m} p_{j+i}(x_{j+i}). \tag{2.50}$$

Here, the scalar function $p_{j+i}(\cdot)$ is the *marginal density* for the random variable x_{j+i}. A sequence that is *not* statistically independent will be called *dependent* and we will be very much concerned with its *dependence structure*, which is closely related to the ideas of dynamic modeling.

The sequence $\{x_k\}$ is *stationary* if, for any j and m, the joint density $p(x_j, x_{j+1}, ..., x_{j+m})$ is independent of j — i.e., if:

$$p(x_j, x_{j+1}, ..., x_{j+m}) = p(x_k, x_{k+1}, ..., x_{k+m}), \tag{2.51}$$

for arbitrary $k \neq j$. It follows that if $\{x_k\}$ is both statistically independent and stationary, the joint density is given by:

$$p(x_j, x_{j+1}, ..., x_{j+m}) = \prod_{i=0}^{m} p_0(x_{j+i}). \tag{2.52}$$

Here, all elements of the sequence $\{x_k\}$ are characterized by the single marginal density $p_0(\cdot)$, independent of the index k. Such sequences are called *independent, identically distributed* (i.i.d.) sequences, or "white noise" and are extremely important in the system identification literature.

Many standard results in statistics were first developed for i.i.d. sequences and then extended to sequences with "sufficiently weak" dependence structures. Commonly, any one of various "mixing conditions" are considered [83], which are somewhat analogous to the fading memory conditions discussed in Section 2. Specifically, these mixing conditions imply that the sequence $\{x_k\}$ is "asymptotically independent," so that, for example, $p(x_k, x_j) \simeq p(x_k)p(x_j)$ if $|j-k|$ is sufficiently large. A particularly strong mixing condition is the assumption of "m-dependence," which implies that $p(x_k, x_j) = p(x_k)p(x_j)$ *exactly* if $|j-k| > m$. The connection between both of these conditions and NARMAX models will be discussed below.

2.6.2 Gaussian Sequences and Linear Models

If the joint densities $p(x_j, x_{j+1}, ..., x_{j+m})$ are *Gaussian* for all j and m, then the sequence $\{x_k\}$ is called a *Gaussian stochastic process.* A characteristic feature of Gaussian stochastic processes is that they are completely characterized by their first two moments: the means $\mu_k = E\{x_k\}$ and the autocorrelations $R_{xx}(j,k) = E\{x_k x_j\}$. For a *stationary* Gaussian stochastic process, the means are constant — i.e., $\mu_k = \mu$ for all k — and the autocorrelations depend only on the distance $|j-k|$ between the elements of the sequence, i.e.:

$$R_{xx}(j,k) \;=\; R_{xx}(|j-k|). \tag{2.53}$$

Similarly, the *autocovariance function* [68] or *second cumulant* [63] for a stationary stochastic process is defined as:

$$c_2^x(\tau) \;=\; E\{[x(k)-\mu][x(k+\tau)-\mu]\}. \tag{2.54}$$

Equivalently, a stationary Gaussian stochastic process may be completely characterized by its mean μ and its *power spectral density* $S_{xx}(f)$, defined as the discrete Fourier transform of the autocorrelation function, i.e.:

$$S_{xx}(f) \;=\; \sum_{k=-\infty}^{+\infty} e^{-j2\pi k f T} R_{xx}(k). \tag{2.55}$$

Here, T is the time between successive samples of the sequence $\{x(k)\}$ and f is a frequency variable that ranges from $-1/2T$ to $+1/2T$.

Probably the best known example of a Gaussian stochastic process is "Gaussian white noise," an i.i.d. sequence $\{x(k)\}$ of zero-mean, Gaussian random variables. The autocorrelation function for this sequence is given by:

$$R_{xx}(\tau) \;=\; \begin{cases} \sigma^2 & \tau = 0 \\ 0 & \tau \neq 0, \end{cases} \tag{2.56}$$

and the power spectral density is:

$$S_{xx}(f) = \sigma^2 \tag{2.57}$$

for all frequencies $f \in [-1/2T, 1/2T]$. Viewed in the frequency domain, the fact that Gaussian white noise exhibits a constant power spectrum means that it has equal intensity at all frequencies. This characteristic is useful in linear system identification applications (discussed briefly in Section 7) and is closely related to the notion of a "persistently exciting" input sequence [55].

Gaussian stochastic processes are important here for two reasons. First, it follows from the central limit theorem [8, 68, 83] that many *estimators* approach Gaussian limiting distributions as the number of data points on which they are based becomes large. The second and more immediate reason they are interesting is that if the input sequence $\{u(k)\}$ to a linear, time-invariant, single-input, single-output system is a Gaussian stochastic process, the output sequence $\{y(k)\}$ will also be a Gaussian stochastic process. For linear models with zero-mean Gaussian input sequences, the output autocorrelation function $R_{yy}(\tau)$ depends only on the input autocorrelation function $R_{uu}(k)$ and the model's impulse response $\{h(k)\}$ [68]:

$$R_{yy}(\tau) = \sum_{j=0}^{\infty} \sum_{m=0}^{\infty} h(j)h(m)R_{uu}(\tau + j - m). \tag{2.58}$$

This result simplifies considerably if the input sequence is assumed to be i.i.d. In particular, substituting Eq. (2.56) into Eq. (2.58) yields the result:

$$R_{yy}(\tau) = \sigma^2 \sum_{j=0}^{\infty} h(j)h(j + \tau). \tag{2.59}$$

Similarly, if an i.i.d. input sequence (Gaussian or not) is applied to a static nonlinearity, the output sequence will have the same dependence structure (i.e., it will also be i.i.d.), but a different marginal distribution. *In general, if a stationary stochastic input sequence is applied to an arbitrary dynamical system, both the distribution and the dependence structure will change.* This observation is the key to stochastic system identification algorithms, as this section and the next will attempt to demonstrate clearly.

2.6.3 Higher-Order Statistics and Non-Gaussian Random Variables

Non-Gaussian random variables are typically characterized by their "deviation from Gaussian behavior." For example, the *skewness* γ and *kurtosis* κ are defined as [11]:

$$\gamma = E\{x^3\}/[E\{x^2\}]^{3/2}, \tag{2.60}$$

and

$$\kappa = E\{x^4\}/[E\{x^2\}]^2 - 3. \tag{2.61}$$

A key feature of both of these quantities is that they are identically zero for Gaussian random variables and may thus be viewed as "measures of non-Gaussianness." The third and fourth cumulants discussed below extend these ideas to sequences of random variables (i.e., stochastic processes) in the same way that the second cumulant defined in Eq. (2.54) generalizes the variance.

As in the Gaussian case, the simplest non-Gaussian stochastic processes are the i.i.d. sequences. Thus, for example, if the probability density is uniform, we have *uniform white noise*, while if this density is binary (i.e., restricted to two discrete values), we have *binary white noise*. A particular special case that will be of futher interest below is the i.i.d. *Laplace* or "double exponential" sequence, whose probability density function is $p(x) = \frac{1}{2}e^{-|x|}$. This distribution exhibits "heavier tails" than the Gaussian distribution, reflected by the positive kurtosis value $\kappa = 3$. The point is, the "whiteness" — i.e., the i.i.d. character of the sequence — is in no way related to the Gaussian or non-Gaussian density that describes each element of the sequence.

Similarly, we can define non-Gaussian *dependent* stochastic processes but here, the situation is a little more difficult for reasons of historical notation. That is, dependent processes are often called *correlated* because Gaussian stochastic processes may be completely described by their mean and autocorrelation function as noted earlier. If the sequence is i.i.d. — Gaussian or not — this autocorrelation function will be of the form (2.56), provided the variance σ^2 is finite (this need not be the case: useful *infinite variance* stochastic processes have been defined [8, 14, 33, 68]). The point is, if a sequence is i.i.d., it is necessarily uncorrelated in the sense defined by Eq. (2.56). The converse is not true — an uncorrelated sequence need not be independent, *unless it is Gaussian. Then*, the two notions are equivalent and the reverse implication also holds.

Dynamic Characterizations

To describe the dependence structure of non-Gaussian sequences, it is necessary to consider dynamic extensions of the higher-order moments discussed above. Typical higher-order auto-correlations include [63]:

$$T_{uuu}(\tau_1, \tau_2) = E\{u(k)u(k+\tau_1)u(k+\tau_2)\}, \tag{2.62}$$

$$T_{uuuu}(\tau_1, \tau_2, \tau_3) = E\{u(k)u(k+\tau_1)u(k+\tau_2)u(k+\tau_3)\}, \tag{2.63}$$

Note that, in the most general case, these expectations would depend on all of time indices $k, k+\tau_1, k+\tau_2, k+\tau_3$ appearing on the right-hand sides of Eqs. (2.62) and (2.63), but if the sequence $\{u(k)\}$ is stationary, these correlations will only depend on the time differences τ_1, τ_2, and τ_3 and not on the time origin k. Definitions of higher-order cumulants are somewhat more involved [63], but these quantities are generally easier to work with than the corresponding moments. In particular, the third- and fourth-cumulants are closely related to the skewness and kurtosis, respectively, when all time differences τ_i are set to zero.

Higher-order spectra are defined as the Fourier transforms of the corresponding cumulants. As a specific example, if $\{x(k)\}$ is a zero-mean, stationary sequence, the third cumulant $c_3^x(\tau_1, \tau_2)$ is equal to the third-order correlation $T_{xxx}(\tau_1, \tau_2)$ defined above. The *bispectrum* $B_x(f_1, f_2)$ of the sequence sequence $\{x(k)\}$ is defined as the Fourier transform of the third cumulant with respect to both time variables τ_1 and τ_2:

$$B_x(f_1, f_2) = \sum_{\tau_1=-\infty}^{\infty} \sum_{\tau_2=-\infty}^{\infty} c_3^x(\tau_1, \tau_2)\, exp\{-j2\pi(f_1\tau_1 + f_2\tau_2)T\}. \tag{2.64}$$

Since the Fourier transforms involved in this definition are invertible, there is no loss of information in converting from the third cumulant to the bispectrum or *vice versa*.

2.6.4 Four Simple Examples

Many of the ideas discussed in this section — and their relavence to nonlinear empirical modeling — may be seen by considering the following four simple examples.

The Gaussian $AR(1)$ Model

The Gaussian $AR(1)$ stochastic process $\{y(k)\}$ is defined by taking $p = 1$ and $q = 0$ in Eq. (2.23) with $e(k) = 0$ identically and assuming $\{u(k)\}$ to be a zero-mean, i.i.d. Gaussian sequence with variance σ^2. The impulse response of this linear model is:

$$h(k) = \begin{cases} b_0 \, a^k & k \geq 0 \\ 0 & k < 0 \end{cases} \tag{2.65}$$

where a is the first-order autoregressive coefficient a_1 in Eq. (2.24). It then follows from Eq. (2.59) that the autocorrelation function $R_{yy}(\tau)$ is:

$$R_{yy}(\tau) = \sigma_y^2 \, a^{|\tau|}, \tag{2.66}$$

where:

$$\sigma_y^2 = \sigma^2 \frac{b_0^2}{1 - a^2}. \tag{2.67}$$

Note that as $a \to 0$, the sequence $\{y(k)\}$ approaches the i.i.d. input sequence $\{u(k)\}$, while as $a \to 1$, this sequence becomes a *nonstationary* process called *Brownian motion* or "integrated white noise." In particular, it can be shown that, for this process, $E\{y(k)\} = 0$ for all k, but $R_{yy}(j, k) = \sigma^2 \, min(j, k)$, [22, p. 85] which is *not* a function only of the magnitude of the difference $|j - k|$. For $|a| < 1$, the sequence $\{y(k)\}$ is stationary and satisfies a variety of mixing conditions [83, p. 34], although it is *not* an m-dependent sequence for any finite m.

A Non-Gaussian $AR(1)$ Model

Suppose the i.i.d. Gaussian input sequence $\{u(k)\}$ used in defining the AR(1) process above is replaced by the i.i.d. Laplace sequence discussed in Section 6.4. Damsleth and El-Shaarawi [23] considered this model, deriving simple expressions for its moments and joint distributions for $(u(k), u(j))$. For $a = 0$, the output sequence $\{y(k)\}$ reduces to the i.i.d. Laplace input sequence $\{u(k)\}$, while as $a \to 1$, the output sequence becomes both "more dependent" and "more nearly Gaussian." In particular, the central limit theorem applies to the case $a = 1$, from which it follows that $\{y(k)\}$ approaches Brownian motion as $a \to 1$.

For $|a| < 1$, the sequence $\{y(k)\}$ is stationary and satisfies similar mixing conditions as in the Gaussian case, but unlike the Gaussian case, the shape of the marginal distribution changes with a. In particular, the kurtosis may be computed explicitly [23] as:

$$\kappa = 3\frac{1-a^2}{1+a^2}. \tag{2.68}$$

Note that this value decreases monotonically from the Laplace value $\kappa = 3$ for $a = 0$ to $\kappa = 0$ in the Gaussian limit $a = 1$. The point of this example is that for *non-Gaussian* inputs, even *linear* systems alter the probability density of the input sequence.

The Gaussian $MA(1)$ Model

The Gaussian $MA(1)$ stochastic process is defined by taking $p = 0$ and $q = 1$ in Eq. (2.31) and assuming the input sequence $\{u(k)\}$ is a zero-mean, i.i.d. Gaussian sequence. In this case, the impulse response sequence $\{h(k)\}$ is simply $\{b_0, b_1, 0, 0, 0, ...\}$, so the autocorrelation function reduces to:

$$R_{yy}(\tau) = \begin{cases} (b_0^2 + b_1^2)\sigma^2 & \tau = 0 \\ b_0 b_1 \sigma^2 & \tau = \pm 1 \\ 0 & \text{otherwise.} \end{cases} \tag{2.69}$$

The sequence $\{y(k)\}$ is a stationary Gaussian sequence with zero mean and is completely characterized by the two non-zero autocorrelation values $R_{yy}(0)$ and $R_{yy}(\pm 1)$. Unlike the two autoregressive models considered above, this sequence is m-dependent for $m = 1$.

A Second-Order Volterra Example

As an example of an *uncorrelated but dependent, non-Gaussian process*, consider the following second-order Volterra example [63]:

$$y(k) = u(k) + bu(k-1)u(k-2). \tag{2.70}$$

If the input sequence $\{u(k)\}$ is zero-mean Gaussian white noise with variance σ^2, then both the autocorrelation function $R_{uu}(\tau)$ of the input sequence $\{u(k)\}$ and the autocorrelation function $R_{yy}(\tau)$ of the output sequence $\{y(k)\}$ are of the form given in Eq. (2.56), characteristic of "white" or i.i.d. sequences. The output sequence $\{y(k)\}$ is *not* i.i.d., however, because there is an explicit statistical dependence between each element of this sequence and both of its nearest neighbors [63]. In particular, note that the sequence $\{y(k)\}$ is m-dependent for $m = 2$ because $y(k)$ and $y(k+j)$ are dependent for $|j| \leq 2$ but independent for $|j| > 2$.

It can be shown [63] that the "hidden" dependence structure that does not appear in the autocorrelation (or power spectral density) of $\{y(k)\}$ is completely described by the third cumulant (or bispectrum). In addition, it is interesting to note that the skewness γ for this sequence — i.e., its *static* third cumulant — is identically zero, even though the dynamic third moment provides a complete characterization. In fact, this is a very interesting example because it passes many tests for linearity for arbitrary values of b [70].

2.7 Model Identification Algorithms

The model identification literature is vast and we will not attempt to review it here. Instead, we will present a few "representative" results that convey the general flavor of the problem and illustrate some of the key issues that arise. For more extensive surveys, see references [3, 17, 32, 55, 96]. The results presented here will focus primarily on stochastic approaches to model identification because much more can be said about these results than in the deterministic case. We begin with a brief discussion of linear model identification because this is the source of much of our insight to date. The underlying idea behind either linear or nonlinear model identification is that a dynamical system changes the distributional character of its input sequence. Thus, identification of these models from input/output data amounts to characterizing the difference in distributional character (i.e., marginal densities and dependence structure) between the input sequence $\{u(k)\}$ and the output sequence $\{y(k)\}$.

2.7.1 Linear Model Identification

One approach to linear model identification with random inputs is based on the relation between the input-output cross correlation function $R_{uy}(\tau)$, the input autocorrelation function $R_{uu}(\tau)$, and the impulse response $h(k)$ of the linear system to be identified. This relationship is expressed by the convolution equation [68]:

$$R_{uy}(\tau) = \sum_{k=0}^{\infty} h(k)\, R_{uu}(\tau - k), \qquad (2.71)$$

where $R_{uy}(\tau)$ is defined by:

$$R_{uy}(\tau) = E\{u(k)y(k+\tau)\}, \tag{2.72}$$

and reduces to the autocorrelation function $R_{uu}(\tau)$ when $y = u$.

It is common practice (e.g., step response modeling for model predictive control) to truncate the infinite sequence $\{h(k)\}$ after some maximum lag M, beyond which $h(k) \simeq 0$. This truncation corresponds to the construction of the moving average model:

$$y(k) = \sum_{k=0}^{M} \alpha_j u(k-j), \tag{2.73}$$

where $\alpha_j = h(j)$ for $0 \leq j \leq M$. Under this approximation, the correlation expression in Eq. (2.71) becomes a system of $M+1$ simultaneous linear equations in $M+1$ unknowns — called the Yule-Walker equations [48] — which may be written as:

$$\mathbf{r}_{uy} = \mathbf{R}_{uu}\,\boldsymbol{\alpha}. \tag{2.74}$$

Here, α is an $M+1$-vector containing the impulse response coefficients $\alpha_0, \alpha_1, ..., \alpha_M$, $\mathbf{R}_{uu}$ is an $(M+1) \times (M+1)$ matrix whose j,k-element is the autocorrelation $R_{uu}(j-k)$, and $\mathbf{r}_{uy}$ is an $M+1$-vector containing the cross-correlations $R_{uy}(k)$. Eq. (2.74) may be solved explicitly for the parametric model coefficient estimate:

$$\hat{\boldsymbol{\alpha}} = \mathbf{R}_{uu}^{-1}\,\mathbf{r}_{uy}. \tag{2.75}$$

Note that the existence of this estimate depends on the existence of the inverse autocorrelation matrix $\mathbf{R}_{uu}^{-1}$ for the input sequence. For i.i.d. input sequences, $\mathbf{R}_{uu} = \sigma^2\mathbf{I}$, where $\mathbf{I}$ is the $(M+1) \times (M+1)$ identity matrix, so the inverse exists for any $\sigma > 0$. For highly correlated input sequences, it is possible for $\mathbf{R}_{uu}$ to be singular, so $\mathbf{R}_{uu}^{-1}$ will not exist. A more common practical problem is that $\mathbf{R}_{uu}$ may be non-singular but *poorly conditioned*, exhibiting a wide range of eigenvalues [90]. In this case, the estimated parameters tend to be extremely sensitive to errors in the data. This point is closely related to both the ideas of input sequence design discussed in Section 8.2 and the problem of multivariable model identification, discussed in Section 7.4.

Taking the Fourier transform of Eq. (2.71) converts the convolution sum into a simple multiplication:

$$S_{uy}(f) = H(f)\,S_{uu}(f). \tag{2.76}$$

Here, $H(f)$ is the frequency response of the linear model, defined as:

$$H(f) = \sum_{k=0}^{\infty} h(k)\, e^{-j2\pi kfT}, \tag{2.77}$$

and $S_{uy}(f)$ is the *cross-spectrum* between u and y, defined as:

$$S_{uy}(f) = \sum_{k=-\infty}^{+\infty} e^{-j2\pi kfT} R_{uy}(k). \tag{2.78}$$

Inverting Eq. (2.76) yields the nonparametric frequency response estimate:

$$\hat{H}(f) = S_{uy}(f)/S_{uu}(f). \tag{2.79}$$

This equation provides a practical way of estimating frequency responses, *provided the spectra $S_{uy}(f)$ and $S_{uu}(f)$ are estimated accurately.* A similar statement may be made in connection with the autocorrelation-based approach described earlier, but the problem is more severe with spectral estimates. This last point is important because spectral estimation has become a fairly sophisticated art [48, 76, 95] and the difference between "good" spectral estimates and "poor" ones can be dramatic [48, 95]. *In particular, the "unwindowed periodogram" obtained by taking the squared magnitude of the fast Fourier transform (FFT) of a random data sequence is a statistically inconsistent spectral estimator and should not be used.*

2.7.2 Nonlinear Model Identification

As noted in Section 4, it is not practical to consider the "general NARMAX model identification problem" without imposing *some* working constraints on the structure of the function $F(\cdot, \cdots, \cdot)$ and the order parameters p, q, and r appearing in Eq. (2.23). In fact, it has also been noted that this difficulty has motivated the definition of certain sub-classes of the NARMAX class (e.g., NAARX models). Consequently, this section will briefly discuss identification approaches for four special cases of the NARMAX class:

1. polynomial NARMAX models;
2. NAARX models;
3. Volterra models;
4. block-oriented models.

Polynomial NARMAX Model Identification

As noted in Section 3, one of the most popular classes of NARMAX models is obtained by taking $F(\cdot, \cdots, \cdot)$ to be a low-order polynomial in the past p output variables and the current and past q input variables. For a polynomial of order ℓ, the resulting saturated model would contain $\binom{p+q+1+\ell}{p+q+1}$ terms [78, p. 7], which grows very rapidly with increasing order parameters p, q, and ℓ. The same problem occurs in multivariable regression analysis, where the technique of stepwise regression has been developed to provide a systematic approach to building non-saturated models. Motivation for this technique comes from the observation that often, many of the terms in the saturated model do not contribute significantly to the overall goodness of fit. Stepwise regression starts with an initial (simple) model that excludes most of the terms in the saturated model and then tests each of these excluded terms in turn to see which one most improves the overall goodness of fit. As new terms are added, previously significant terms may be found to be no longer significant and dropped from the model. This procedure continues until no further "significant" improvements are possible. Billings and Voon [5] describe this procedure for NARMAX modeling and demonstrate it for four simple examples. The first is a simulated Hammerstein model, for which the "exact" NARMAX model is known. Starting with the correct model order parameters (i.e., $p = 2$, $q = 2$, $r = 2$, $\ell = 3$), they obtain a final model by stepwise regression that is very close to the true Hammerstein model. They also consider a Wiener model, but the exact NARMAX representation for this model is not obvious. Still, they obtain a final NARMAX model that appears to be of approximately the correct order, again with far fewer terms than the saturated model. Hernandez and Arkun used the stepwise regression approach in fitting NARMAX models built from polynomials, but including a "squashing function" type saturation nonlinearity to guarantee bounded model responses [44].

A variety of approaches to determining which possible terms should or should not be included in regression models has evolved since the basic idea was first proposed. For example, Mallows [56] describes an approach based on plots of an easily computed statistic called C_p *versus* the number of terms p included in the regression model. Examples are included to illustrate the effects of correlations among the possible explanatory variables, including cases where certain terms are only jointly significant, but not individually significant. Other useful regression model-buliding statistics are discussed and compared in [93, 94]. The point is that a variety of techniques have been proposed in the statistics literature, with different advantages and disadvantages in different applications (e.g., sensitivities to serial correlation in the data, outliers, etc.); all of them are potentially applicable to the problem of NARMAX modeling.

NAARX Model Identification

As noted in Section 3, Chen and Tsay introduced the class of NAARX models [16] to permit the use of nonparametric regression methods in time-series analysis. Their paper describes two algorithms in detail, both of which combine techniques from nonparametric regression [42] with the stepwise regression ideas described above. Specifically, these algorithms allow for the possibility that certain past input terms and certain past output terms may not be "significant enough" to include in the final model. The resulting algorithms are rather complex but extremely flexible and are demonstrated for two simulation examples and one real data example. The "correct functional shape" is clearly obtained for both of the simulation examples, and for the real data example, a model is obtained that appears to fit significantly better than two other models considered for the same datasets. This last example is also more closely related to typical process identification problems than the first two, since it includes two exogenoeous inputs. Specifically, a NAARX model predicting river flow from past river flow data, rainfall data, and temperature data is developed. The functional forms suggested for the nonlinearities $f_i(\cdot)$ and $g_i(\cdot)$ by this analysis were then used to construct a piecewise-linear model containing 13 terms. This model exhibited significantly better "out-of-sample" predictions for the river flow data than either a linear model constructed from transfer function analysis or a previously published threshold autoregressive model. While these techniques are not yet as readily accessible as procedures for polynomial NARMAX modeling, they appear much more flexible and certainly worth watching as they mature.

Volterra Model Identification

The parameter estimation algorithm described earlier for moving average models may be extended to the identification of second-order Volterra models [51, 73]. In fact, if all of the possible third moments of the input sequence vanish (i.e., if $E\{u(k)u(\ell)u(m)\} = 0$ for all k, ℓ, m), then the linear coefficients in the Volterra model may be estimated using Eq. (2.75) *independent of the values of the quadratic terms in the Volterra model.* This separation of the problem into a linear part and a quadratic part arises from the fact that the linear model terms preserve the symmetry (specifically, the "non-skewness") of the input sequence, while the quadratic model terms introduce a pronounced (dynamic) skewness. If the input sequence is Gaussian, the quadratic terms in the Volterra model may be estimated as [51]:

$$\mathbf{B} = \frac{1}{2}\mathbf{R}_{uu}^{-1}\,\mathbf{T}\,\mathbf{R}_{uu}^{-1}, \tag{2.80}$$

where $\mathbf{T}$ is an $(M+1) \times (M+1)$ matrix of third-order input/output correlations $T_{ij} = E\{y(k)u(k+i)u(k+j)\}$. Results for different classes of non-Gaussian input sequences are given in [73].

Block-Oriented Model Identification

It was noted in Section 4 that nonparametric frequency response estimates may be obtained for linear models from power spectral densities. This result can be extended to various classes of nonlinear systems using higher-order spectra. As a specific example, Brillinger [15] describes a nonparametric approach to the identification of the "sandwich model" discussed in Section 3. In particular, Brillinger presents nonparametric frequency response estimates for both of the linear subsystems appearing in the sandwich model. These estimates involve the input autospectrum $S_{uu}(f)$, the input/output cross-spectrum $S_{uy}(f)$ and an input/output cross-bispectrum $S_{uuy}(f_1, f_2)$, informationally equivalent to the cross-bicorrelation matrix $\mathbf{T}$ considered in the Volterra example above. Note that Brillinger's results did not assume any particular functional form for the static nonlinearity $f(\cdot)$ appearing in the sandwich model, but they did require that this nonlinearity exhibit neither even nor odd symmetry.

The simpler problem of Hammerstein modeling provides a nice illustration of the range of options offered by parametric, nonparametric, and semi-parametric modeling. Because this model consists of two parts — a static nonlinearity $f(\cdot)$ and linear dynamics — we may apply either parametric or nonparametric approaches to either subsystem. These choices lead to the following four overall modelling options:

1. parametric identification of both subsystems;

2. nonparametric identification of both subsystems;

3. parametric identification of the static nonlinearity and nonparametric identification of the linear dynamics;

4. nonparametric identification of the static nonlinearity and parametric identification of the linear dynamics.

More specifically, Option 1 would typically entail approximation of the static nonlinearity by a polynomial and fitting the linear dynamics to an ARMAX model. This "fully parametric" approach is the one taken by Eskinat, Johnson, and Luyben in their study of Hammerstein models for chemical process modeling [31]. Option 2 would involve the use of nonparametric regression techniques to estimate the static nonlinearity, as discussed by Greblicki and Pawlak [37]. Nonparametric estimation of the linear dynamics would then be based on the use of nonparametric spectral and bispectral estimates, as discussed by Brillinger [15] in his analysis of the "sandwich model" identification problem. Options 3 and 4 represent semi-parametric approaches to the Hammerstein model identification problem, combining parametric estimates of one subsystem with nonparametric estimates of the other. To see the advantages of nonparametric or semi-parametric procedures, note that Eskinat, Johnson, and Luyben conclude that low-order (e.g., $n \leq 4$) polynomial Hammerstein models cannot adequately describe high-purity distillation column dynamics [31]. Part of the reason for this difficulty is that polynomials cannot describe the saturation nonlinearity that ultimately limits product purity; the use of nonparametric estimates for this nonlinearity could begin to address the more subtle question of whether the Hammerstein model structure is inherently flexible enough to describe high-purity column dynamics adequately.

2.7.3 Identification of Multivariable Models

Even in the linear case, multivariable model identification poses some extremely challenging practical problems. To illustrate this point, consider the MISO generalization of the moving average model considered in Section 7.1. Eq. (2.73) is then replaced by the m-input, single-output model:

$$y(k) = \sum_{j=0}^{Q} \alpha_j v_j(k), \tag{2.81}$$

where $Q = Mm$ and $\mathbf{v}(\mathbf{k})$ is the vector of lagged inputs:

$$\mathbf{v}(k) = [\quad u_1(k)\, u_1(k-1) \, \cdots \, u_1(k-M)$$

$$u_2(k)\ u_2(k-1)\ \cdots\ u_2(k-M)$$

$$\cdot$$

$$\cdot$$

$$\cdot$$

$$u_m(k)\ u_m(k-1)\ \cdots\ u_m(k-M)\ ]^T. \tag{2.82}$$

The cross-correlation vector between $\mathbf{v}(k)$ and $y(k)$ is $\mathbf{r}_{vy}$, whose $Q+1$ elements are $E\{v_i(k)y(k)\}$ for $j = 0, 1, ..., Q$. Similarly, define $\mathbf{R}_{vv}$ as the $(Q+1) \times (Q+1)$ autocorrelation matrix for $\mathbf{v}(k)$, whose i, j element is $E\{v_i(k)v_j(k)\}$. As in the SISO case considered in Section 7.1, if $\mathbf{R}_{vv}^{-1}$ exists, we may estimate the vector ﬀ of $Q+1$ model coefficients in Eq. (2.81) by solving the multivariable Yule-Walker equations:

$$\mathbf{R}_{vv}\text{ﬀ} \quad = \mathbf{r}_{vy}. \tag{2.83}$$

As before, the existence of a unique solution $\hat{\alpha}$ to these equations depends on the non-singularity of the matrix $\mathbf{R}_{vv}$. The possibilities for singularity and ill-conditioning are greater for the multivariable problem than for the SISO problem. In particular, note that $\mathbf{R}_{vv}$ will be singular if any two of the m input sequences $\{u_j(k)\}$ and $\{u_\ell(k)\}$ are perfectly correlated (i.e., if $u_j(k) = \gamma u_\ell(k)$ for some constant γ). In practical terms, this observation means that the composite autocorrelation matrix $\mathbf{R}_{vv}$ will be ill-conditioned ("nearly singular") if *either* the individual input sequences are too highly correlated — as in the SISO case — *or* the cross-correlations between the different input sequences are too great.

This last difficulty is closely related to the frequency-domain notion of "low-gain" and "high-gain" directions in MIMO systems. In particular, strongly interacting process units like some distillation columns may exhibit radically different behavior in their "low-gain directions" from their "high-gain directions." To see this point, consider a two-input, two-output linear transfer function model $\mathbf{y}(s) = \mathbf{H}(s)\mathbf{u}(s)$. Note that this transfer function matrix will be described, at every frequency, by two singular values and their corresponding singular vectors. Geometrically, these singular vectors represent "preferred directions" in which the process input vector can move the process output vector, while the singular values represent their associated gains. If the gains associated with the two different directions are significantly different, it will be much easier to move the process in the "high-gain direction" than in the "low-gain direction." Because these directions will change with both frequency *and any imposed closed-loop control structure*, it is necessary to identify *both* of these modes with sufficient accuracy to correctly predict these dependences.

Morari and Skogestad [61] note that if the difference in gain between these two directions is large enough, it is not sufficient to identify each element of the transfer function matrix $\mathbf{H}(s)$ independently. That is, while the estimation errors in each individual matrix element may be "small," it is the relationship between the different elements that determines the low- and high-gain directions and their gains. Thus, "reasonable" estimates of the individual elements may not yield "reasonable" estimates of these directions. Physically, this problem arises because the "low-gain directions" are typically not excited enough by the identification input sequence to estimate their effects accurately. Jacobsen [45] has considered this problem and recommends two approaches. First, he advocates using full MIMO identification procedures to estimate the complete transfer function matrix $\mathbf{H}(s)$ instead of the MISO procedures more commonly (and more conveniently) employed. His second suggestion is the use of "gray-box" approaches to design identification input sequences. Specifically, he advocates the use of physical insights to determine "low-gain directions" approximately prior to modeling; multivariable input sequences may then be designed to excite these modes of the column preferentially.

While the ideas presented above were developed for linear MIMO model identification, they are all applicable to nonlinear MIMO model identification. Specifically, while nonlinear MIMO models are no longer amenable to linear algebraic analysis, the concepts of "low-gain directions" and "high-gain directions" are *geometric* ones that will occur in some more complicated form in nonlinear problems. For example, we can expect that as internal variables saturate in a MIMO nonlinear process, amplitude-dependent changes in the "low-gain" and "high-gain" directions can occur. It is interesting to speculate whether block-oriented models can describe these phenomena reasonably, or whether NARMAX models are required. Similarly, the use of "gray-box" modeling ideas to attempt to estimate these directions and their amplitude dependences should also be possible.

2.8 Cross-Validation and Input Sequence Design

Previous discussions have emphasized the importance of specifying the set $\mathcal{S}$ of input sequences over which we wish to model process behavior. In general, this set contains an uncountably infinite collection of individual input sequences, so it is not feasible either to estimate or to validate model parameters for "the whole set $\mathcal{S}$." Thus, to obtain models that are "reasonable" for this set of possible inputs, not only must we know what the set is, but we must also construct a finite subset $\mathcal{S}_1$ of input sequences that is "sufficiently rich" that if $\mathcal{M}\mathbf{u} \simeq \mathcal{P}\mathbf{u}$ for all $\mathbf{u} \in \mathcal{S}_1$, it follows that $\mathcal{M}\mathbf{u} \simeq \mathcal{P}\mathbf{u}$ for all $\mathbf{u} \in \mathcal{S}$. Naturally, the quality of the approximation on this larger set $\mathcal{S}$ will probably be poorer than on the smaller design set $\mathcal{S}_1$, so the challenge is to specify a design set $\mathcal{S}_1$ that is small enough to be practical for model identification, but "rich enough" so that the quality of approximation is "adequate" on the larger set $\mathcal{S}$.

The practical necessity of basing models on finite datasets gives rise to two dangers. In both cases, the model approximation condition $\mathcal{M}\mathbf{u} \simeq \mathcal{P}\mathbf{u}$ holds on the finite dataset $\mathcal{S}_1$, but not on the larger dataset of interest $\mathcal{S}$. The first problem is that of "over-fitting" in which the complexity of the model $\mathcal{M}$ is allowed to become "too great" relative to the size of the dataset $\mathcal{S}_1$. The basic cure for this problem is to impose a "complexity restriction" on the model $\mathcal{M}$ (e.g., limit model order parameters). In practice, a useful approach to this problem is the use of *cross-validation*, described below. Essentially, the idea is to use two different datasets: model parameters are estimated from a set $\mathcal{S}_1$ and model adequacy is assessed with respect to a second set $\mathcal{S}_2$. The second difficulty is that the input set $\mathcal{S}_1$ is not a "good approximation" to the larger set $\mathcal{S}$ of interest. While cross-validation addresses fundamental model limitations (e.g., excessive complexity), limitations of the set $\mathcal{S}_1$ are avoided through the techniques of *experimental design*, described in Section 8.2.

2.8.1 Cross-Validation

One of the easiest ways to understand cross-validation is to consider the problem of model order determination for linear autoregressive models. That is, given a dataset $\mathcal{D}$ of input/output pairs $(u(k), y(k))$, we wish to determine the order p of the ARX model that best approximates the available data. One approach to this problem is to consider such criteria as the Akaike Information Criterion (AIC) [55, 76, 96] that include both a prediction error term and a model order term. The basic idea is to penalize model complexity by minimizing this quantity, obtaining a model that is "reasonable" in terms of both prediction error variance and complexity, as measured by the number of model parameters. Several extensions of this idea have been proposed [96, sec. 5.4.4] and these ideas have been applied to a variety of nonlinear modeling applications [77, 92, 96].

One disadvantage of the AIC and other related approaches is that they require evaluation of likelihood functions, which may be impractical for complex models. Alternatively, cross-validation [30, 96] provides an approach to model order selection directly from data, and this approach is more readily extended to complex models. Here, the available dataset $\mathcal{D}$ is first subdivided into two mutually exclusive subsets $\mathcal{D}_1$ and $\mathcal{D}_2$. A sequence $\{\mathcal{M}_k\}$ of k-parameter models is then identified that each best fit the dataset $\mathcal{D}_1$ for all k in a specified range. Since k determines the number of degrees of freedom in the model, the corresponding sequence of prediction error variances decreases monotonically toward zero with increasing k. These models are next evaluated in terms of goodness of fit with respect to the *second* dataset $\mathcal{D}_2$; here, the prediction error will generally decrease with increasing model order k until some critical order k^*. Beyond this point, over-fitting of the original dataset is occurring and the performance on the second dataset begins to degrade, causing its associated prediction error variance to increase. Thus, the model order k^* that performs best with respect to this second dataset is accepted as the "best" model order p for the process generating the data. Tong discusses an application of this approach to model order selection for nonlinear autoregressive models [96, sec. 5.5.8]. The key point is that this approach explicitly addresses the inadequacy of goodness of fit criteria by themselves, augmenting it with an *implicit* penalty on model complexity: if the model is complex enough to be over-fitting the first ("training") dataset, its performance with respect to the second ("validation") dataset degrades correspondingly. These points are illustrated in the case study presented in Section 9.

2.8.2 Optimal Input Sequences for Linear Modeling

In practical empirical modeling, it may or may not be possible to specify the input sequence $\{u(k)\}$ on which the model identification process will be based. *Whenever it is possible to specify this sequence, the opportunity to do so should be taken, as the difference in models obtained with "good" input sequences and "poor" input sequences may be great.* This section will briefly review some of the notions of experimental design that have been found useful for linear model identification; for more thorough treatments of the underlying ideas, see references [1, 9, 78]. In particular, Atkinson [1, ch. 6] provides a very readable introduction to the basic objectives of experimental design and some of the general approaches that have emerged in this area.

The fundamental idea behind experimental design is to select the input sequence $\{u(k)\}$ to minimize some measure of the ultimate uncertainty in our estimated results (i.e., to make our "experiment" as "informative" as possible). These measures are generally based on the *Fisher information matrix*:

$$\mathcal{F}(\theta) = -E\left\{ \frac{d^2}{d\theta^2} \log f(\theta|\mathcal{D}) \right\}, \tag{2.84}$$

where θ is the vector of model parameters to be estimated, $\mathcal{D}$ is the dataset from which they are to be estimated, and $f(\theta|\mathcal{D})$ is the likelihood of the parameter vector θ given the dataset $\mathcal{D}$. The significance of this matrix is that if $\hat{\theta}(\mathcal{D})$ is an unbiased, asymptotically normal estimator of θ, its covariance matrix Σ satisfies the *Cramer-Rao* inequality [55] or *information inequality* [54]:

$$\Sigma \geq \mathcal{F}^{-1}, \tag{2.85}$$

where this matrix inequality means that the difference $\Sigma - \mathcal{F}^{-1}$ is non-negative definite.

In more practical terms, this inequality means that the inverse Fisher information matrix specifies both lower bounds for the variances of the individual parameter estimates and some measure of the independence of the parameter estimates. Specifically, the Cramer-Rao lower bound implies that $var\{\theta_i\} \geq [\mathcal{F}^{-1}]_{ii}$, while if the Fisher information matrix is diagonal, all of these parameter estimates are uncorrelated. The basis for experimental design is the observation that, since $\mathcal{F}$ depends on the available dataset $\mathcal{D}$, we can choose the "controllable" parts of $\mathcal{D}$ to make $\mathcal{F}$ "well-behaved." Specifically, typical design objectives would be to make $\mathcal{F}$ as close to diagonal as possible to minimize correlation between parameter estimates and to make the diagonal elements of $\mathcal{F}^{-1}$ as small as possible to minimize the achievable estimator variances for these parameters. These ideas lead to specific criteria like D-optimality [1, 55, 78] in which the determinant (i.e., the product of the eigenvalues) of $\mathcal{F}^{-1}$ is minimized.

These ideas have been applied to linear system identification and are described in detail in Ljung's book [55, ch. 14]. In particular, detailed expressions for the Fisher information matrix are given and it is noted that this matrix depends only on the *power spectrum* of the input sequence $\{u(k)\}$. This result is significant because the power spectrum alone does not completely characterize the input sequence unless it is also required to be a zero-mean Gaussian sequence. In practice, more popular input sequences are the Pseudo-Random Binary Sequences (PRBS) [55] and sums of sinusoids [81], which may also be tailored to have specified power spectra. For a more extensive treatment of the issue of input sequence design for linear system identification, see the book by Godfrey and his collaborators [35].

Sadegh *et al.* [86] extend this idea into the arena of gray-box modeling by adopting a Bayesian approach to the problem. Specifically, they note that if θ is given a Gaussian prior distribution with covariance matrix Σ_p and $\hat{\theta}$ is computed from a MAP (maximum *a posteriori*) estimation procedure, the inverse of the posterior covariance matrix Σ is given by:

$$\Sigma^{-1} = \Sigma_p^{-1} + \mathcal{F}. \tag{2.86}$$

Sadegh *et al.* then consider a physically-motivated model structure of relatively low complexity (i.e., SISO linear model, nonlinear in four unknown parameters of interest, with one unmeasured state variable, two unmeasured disturbances, and two "nuisance parameters") and use these ideas to develop optimal input sequences. Specifically, they estimate Σ_p on the basis of physical insights, compute $\mathcal{F}$ for the model structure under consideration, and determine an optimal sinusoidal input sequence by maximizing various information measures based on Σ^{-1}. Since all of these information measures essentially generalize the notion of D-optimality, the differences between them for different prior assumptions illustrate the influence of prior parameter uncertainties on the results of optimal input sequence design.

2.8.3 Input Sequence Design Issues for Nonlinear Modeling

Unfortunately, these input sequence design results are somewhat difficult to apply to nonlinear model identification for the same reason that AIC and related criteria are difficult to apply: they require explicit computation of the likelihood $f(\theta|\mathcal{D})$. While these computations are sometimes possible, they are frequently quite involved and they must be attempted for each class of models considered. Consequently, few results are available concerning experimental design for nonlinear dynamic models. For this reason, the following subsections will focus on these topics:

1. general guidelines for input sequence selection;

2. detailed results for identification of 2nd-order Volterra models;

3. specific examples of *poor* input sequences for nonlinear model identification.

General Guidelines

One of the primary distinguishing features between linear models and nonlinear ones is the amplitude-dependence of the qualitative behavior of nonlinear models. Specifically, if a model $\mathcal{M}$ is linear, its behavior is *homogeneous* [24]: if the response to the input sequence $\{u(k)\}$ is $\{y(k)\}$, the response to $\{\lambda u(k)\}$ is $\{\lambda y(k)\}$ for any real constant λ. Thus, if an input sequence $\{u(k)\}$ is "good" for linear model identification (e.g., persistently exciting [55]), any other sequence $\{\lambda u(k)\}$ is "equally good," for arbitrary λ, at least in theory. In practice, amplitude *does* matter because if the range of the input sequence is too small, the response sequence $\{y(k)\}$ will be difficult to distinguish from measurement uncertainties and other sources of modeling error. Thus, the objective of input sequence design for linear modeling is, roughly, to obtain a sequence that "puts input energy into those frequency bands where it will do the most good" and that is "sufficiently intense that the responses will be clearly distinguishable from noise." In practice, upper bounds are also often placed on input sequence amplitude, as noted by Ljung [55, p. 371].

In nonlinear model identification, it is necessary to consider explicitly model response over a *range* of input amplitudes, since the response $\{y_\lambda(k)\}$ to an input sequence $\{\lambda u(k)\}$ for $\lambda \neq 1$ may be qualitatively very different from that for $\lambda = 1$. In fact, this observation suggests the possibility of *multiple step* model identification algorithms, using different amplitude input sequences at each step. Pajunen [67] developed an algorithm for Wiener model identification based on this idea. Specifically, he proposed a two-pass algorithm in which the model is first excited with a low-amplitude input sequence, yielding a low-amplitude response by the linear subsystem in the Wiener model. Under the (frequently valid) assumption that the static nonlinearity satisfies $g(x) \simeq \alpha x$ for "small" x, this nonlinearity may be neglected in analyzing the "low amplitude response," yielding an estimate of the linear dynamics via standard linear model identification techniques. In the second pass of Pajunen's algorithm, the model is excited with a higher amplitude sequence that elicits a significant response from the static nonlinearity. The response $\{v(k)\}$ of the previously identified linear subsystem is then computed, giving an estimate of the "hidden layer signal" driving the nonlinearity. This nonlinearity is then estimated from *its* input/output data $\{(v(k), y(k))\}$. A somewhat more involved three-pass identification algorithm for 2nd-order Volterra models is described briefly in the next subsection.

Specific Results for 2nd-Order Volterra Identification

While these results are not "optimal input sequences" in the same sense as those discussed previously for the linear model identification problem, we have obtained *useful* results concerning input sequence design for the second-order Volterra model identification problem [60, 72, 74]. These results were obtained by observing that, for i.i.d. input sequences $\{u(k)\}$, the prediction error variance σ_p^2 associated with *perturbed* model parameters $a_i = a_i^o + \phi_i$ and $b_{i,j} = b_{i,j}^o + \Gamma_{i,j}$ is:

$$\sigma_p^2 \;=\; \sigma_o^2 + \sigma^2 \sum_{i=1}^{M} \phi_i^2 + (\kappa+2)\sigma^4 \sum_{i=1}^{M} \Gamma_{i,i}^2 + \sigma^4 \sum_{j=1}^{M} \sum_{j \neq i} \Gamma_{i,j}^2 . \quad (2.87)$$

Here, σ_o^2 is the prediction error variance associated with the optimal parameter estimates $\{a_i^o\}$ and $\{b_{i,j}^o\}$, σ^2 is the variance of the i.i.d. input sequence $\{u(k)\}$, and κ is its kurtosis. The coefficients in front of the three sums in this expression may be viewed as "weights" associated with estimation errors in the linear terms $\{a_i\}$, the diagonal quadratic terms $\{b_{i,i}\}$, and the off-diagonal quadratic terms $\{b_{i,j}\}$, respectively.

This interpretation leads to the following three observations:

1. for low-amplitude input sequences (i.e., $\sigma^2 << 1$), prediction errors are dominated by linear parameter perturbations;

2. for high-amplitude, low-kurtosis input sequences (i.e., $\sigma^2 >> 1, \kappa \simeq -2$), prediction errors are dominated by off-diagonal quadratic parameter perturbations;

3. for high-amplitude, high-kurtosis input sequences (i.e., $\sigma^2 >> 1, \kappa >> 1$), prediction errors are dominated by diagonal quadratic parameter perturbations.

These observations lead to the following three-pass model identification algorithm [72]:

1. excite the process with a low-amplitude, Gaussian i.i.d. input sequence $\{u_1(k)\}$; estimate linear model parameters $\{a_i\}$ from the response $\{y_1(k)\}$;

2. excite the process with a high-amplitude, high-kurtosis non-Gaussian i.i.d. input sequence $\{u_2(k)\}$; compute the first-order model residuals $r_1(k) = y_2(k) - \sum_{i=1}^{M} a_i u_2(k-i)$; estimate the diagonal model parameters $\{b_{i,i}\}$ from the data $\{(u_2(k), r_1(k))\}$;

3. excite the process with a high-amplitude, low-kurtosis non-Gaussian i.i.d. input sequence $\{u_3(k)\}$; compute the second-order model residuals $r_2(k) = y_3(k) - \sum_{i=1}^{M} a_i u_3(k-i) - \sum_{i=1}^{M} b_{i,i} u_3^2(k-i)$; estimate the off-diagonal model parameters $\{b_{i,j}\}$ from the data $\{u_3(k), r_2(k)\}$.

This algorithm, as described here, is not the one we ultimately desire in practice, since it is based on i.i.d. input sequences and does not explicitly account for input amplitude constraints and other practical limitations. It does, however, perform quite well in simulation, and it provides us significant insights into some of the underlying ideas behind input sequence design for nonlinear model identification. For example, one of the results to emerge from this investigation was the utility of symmetrically distributed input sequences: for second-order Volterra models, the response consists of a symmetric component, from which the linear coefficients may be estimated, and an asymmetric component, from which the quadratic coefficients may be estimated. Symmetric non-Gaussian input sequences thus permit separate identification of odd (i.e., linear) and even (i.e., quadratic) model parameters, as noted in Section 7.2. This general conclusion extends to higher-order Volterra models and suggests that separate characterization of symmetric and asymmetric responses of other nonlinear models to symmetric input sequences may yield useful insights. In addition, it is clear that for second-order Volterra model identification, both input amplitude and the kurtosis of the input sequence are important design variables. This last observation is related to the "shape" of the input sequence, which is not important in linear model identification — as noted in Section 8.2 — but which can be *critical* in nonlinear model identification, as the next example illustrates.

Poor Input Sequences for Nonlinear Model Identification

Rohatgi and Szekely [82] show that, for symmetric distributions (i.e., $\gamma = 0$), the kurtosis is bounded below by $\kappa \geq -2$, further noting that this lower bound is achieved by distributions supported on at most two points. In particular, the "binary white noise" sequence assuming values $\pm a$ for arbitrary $a > 0$ exhibits $\kappa = -2$. Substituting this result into Eq. (2.87), it follows immediately that if this input sequence is used for identification, the prediction error variance will be completely independent of errors in the diagonal quadratic model parameters $\{b_{i,i}\}$. In practical terms, this observation means that these model parameters cannot be identified from binary white noise input sequences. Also, note that this result is independent of the *amplitude* of the input sequence: the *shape* of the input sequence is inappropriate for identifying diagonal terms in the second-order Volterra model. Conversely, note that this sequence would be an excellent candidate for the final step of the three-pass identification algorithm described above.

In fact, it is not difficult to show that Pseudo Random Binary Sequences (PRBS) with *any* dependence structure (i.e., either the "binary white noise sequence" considered here or the more usual correlated binary sequences [55, 68]) are unsuitable for either second-order Volterra model identification or Hammerstein model identification. That is, if the input sequence $\{u(k)\}$ assumes only the values $\pm a$, it follows that $u^2(k-j) = a^2$ for all j. From this observation, it is not difficult to show that the Volterra model identification equations have no solution for the diagonal terms $\{b_{i,i}\}$ [74]. Similarly, for the Hammerstein model, note that if $\{u(k)\}$ is a binary sequence, then the sequence $\{v(k) = g(u(k))\}$ generated by the static nonlinearity $g(\cdot)$ is either also binary or constant. If this sequence is binary, the response $\{y(k)\}$ of the linear subsystem will be identical for any two nonlinearities $g_1(\cdot)$ and $g_2(\cdot)$ such that $g_1(a) = g_2(a)$ and $g_1(-a) = g_2(-a)$. As a specific example, note that if $a = 1$, it will not be possible to distinguish any members of the family $g_n(x) = x^{2n+1}$ of odd-order nonlinearities for $n = 0, 1, \ldots$. The situation is even worse for even-order nonlinearities, since the sequence $\{v(k)\}$ driving the linear dynamical system is now constant, so we cannot even identify the linear dynamics.

The real point of this section is not to argue against binary sequences, but rather to make the point that one of the unique features of nonlinear model identification is that the *shape* of the input sequence is critically important. This observation represents a marked contrast to the linear case, where optimal designs are determined completely by the power spectrum of the input sequence and are otherwise entirely independent of the shape of the input sequence.

2.9 Case Study: High Purity Distillation Process

Distillation is a very important unit operation in the chemical/petro-chemical industry; and high-purity distillation, in particular, poses difficult identification and control problems. The *specific* problem dealt with in this case study is as follows:

> A nonlinear discrete-time model, *to be used ultimately for model predictive control*, is to be developed for the high-purity distillation column studied by Chien and Ogunnaike [18].

A more detailed version of the discussion summarized here may be found in Sriniwas, *et al* [89].

The broader objective of the case study, however, is to use one specific example to illustrate the various aspects of the "art" involved in carrying out nonlinear model identification for chemical processes; and to show that there is in fact no single "right way" of carrying out nonlinear model identification in practice.

In what follows, a detailed first principle process model available in the literature is used in lieu of the actual process; the "plant datasets" used for empirical model identification are thus generated by simulation. In actual practice, because of the presence of process noise, measurement noise, outliers, and other such unavoidable and "naturally occuring gremlins," data gathered directly from the physical plant almost always requires preconditioning before they can be used for model identification.

The topic of "data cleaning" or "outlier detection and treatment" is vitally important in practice, but lies outside the scope of this chapter. Introductions to some of the key ideas and recent developments may be found in such books as the one by Hampel *et al.* [41] or Rousseeuw and Leroy [84]. For a discussion of some of these ideas in the context of system identification, see the paper by Poljak and Tsypkin [75]. More recently, Schick and Mitter [87] have described robust extensions of the Kalman filter, including a fairly detailed historical account of related developments. The simpler, but closely related problem of spectral estimation has been addressed by Martin and Thomson [59], who provide a fairly complex but reasonably effective "data cleaning" algorithm, designed to eliminate "non-Gaussian outliers" from a nominally Gaussian data sequence prior to spectral estimation. This algorithm is based on a robust Kalman filter formulation, also described in their paper. Here, the key point is that these issues inevitably arise in dealing with real data, but length limitations do not permit us to discuss them in enough detail to provide useful guidance. Consequently, the case study presented here uses simulation data, allowing us to focus entirely on the significant practical issues that must be faced given "reasonably clean" data.

2.9.1 The Process

The process under consideration is a 27-tray, Methanol-Ethanol distillation column, a rigorous model of which was developed by Weischedel and McAvoy [98]. The column is to be controlled with the configuration studied by Chien and Ogunnaike [18], in which the temperatures on Tray 21 and Tray 7 (used respectively as reasonable indicators of top and bottom product quality) are to be controlled by manipulating the reflux flow rate and the reboiler vapor rate.

As is often the case in many applications, it is convenient to represent the process variables in the following dimensionless form:

$$y_1 = \frac{T_{21} - T_{21ss}}{30.0} \quad ; \quad y_2 = \frac{T_7 - T_{7ss}}{30.0}$$
$$u_1 = \frac{L - L_{ss}}{6.0} \quad ; \quad u_2 = \frac{V - V_{ss}}{6.0} \tag{2.88}$$

where: T_{21}, T_7 = temperatures on tray 21 and tray 7 ($°R$), respectively
T_{21ss} = steady state temperature on tray 21 ($614.10°R$)
T_{7ss} = steady state temperature on tray 7 ($636.77°R$)
L = reflux flow rate (mol/min)
L_{ss} = steady state reflux flow rate ($3.384\, mol/min$)
V = reboiler flow rate (mol/min)
V_{ss} = steady state reboiler flow rate ($3.856\, mol/min$)

2.9.2 Model Structure Selection

As a first step in assessing the nature of the process behavior, Figures 2.10 - 2.13 show open-loop responses of the process for $\pm 1\%$ and $\pm 5\%$ step changes in the dimensionless reflux and vapor flows. These changes are well within the range of normal process operation. By mere inspection of these responses, it is clear that this multivariable process is interactive and nonlinear. Observe that a change in a single input variable affects *both* outputs; and mirror image input changes give rise to very distinctly asymetric responses.

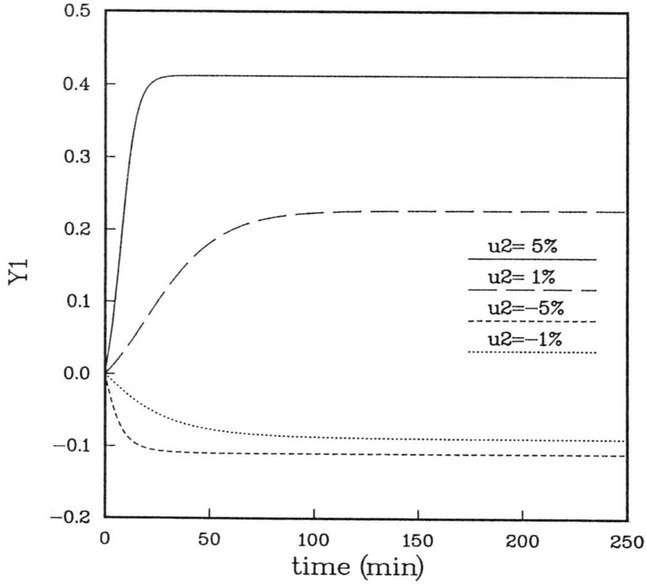

Figure 2.10: Response of plant output y_1 to step changes in u_1

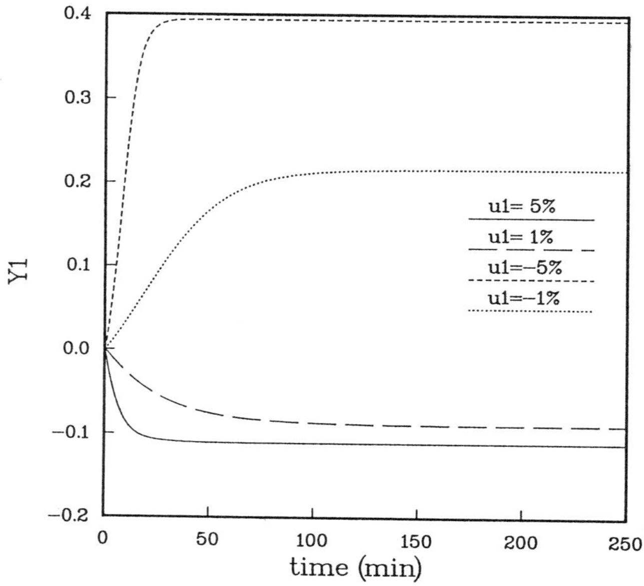

Figure 2.11: Response of plant output y_1 to step changes in u_2

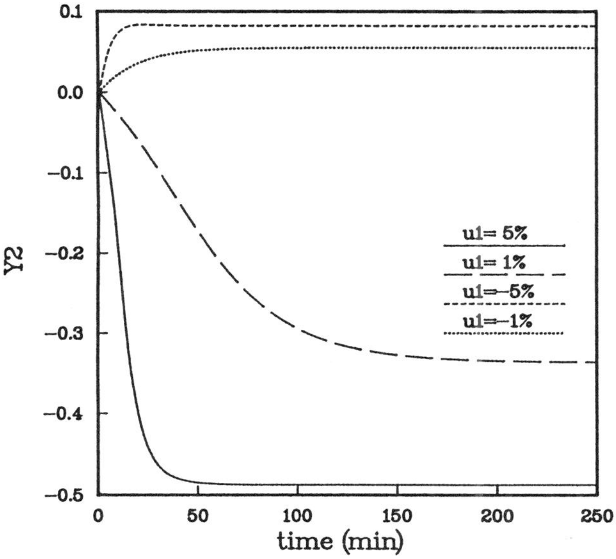

Figure 2.12: Response of plant output y_2 to step changes in u_1

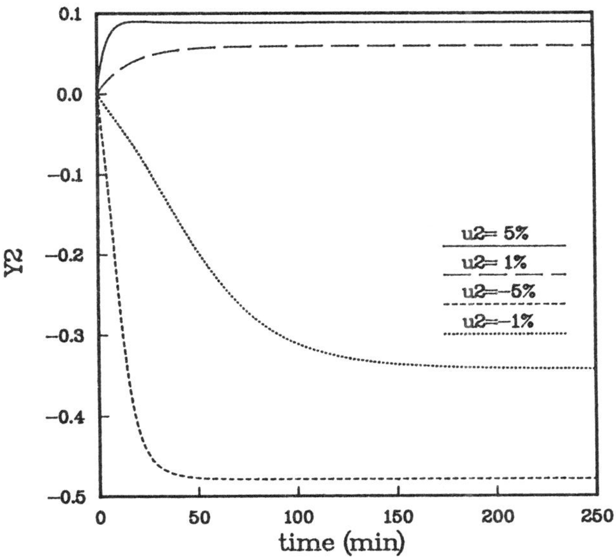

Figure 2.13: Response of plant output y_2 to step changes in u_2

A Nonlinearity test

While "subjective" process nonlinearity assessments such as the above "eyeball analyses" of step responses are often sufficient to motivate us to seek a nonlinear process model, it is also sometimes helpful to augment these with more rigorous nonlinearity tests. One such test, due to Billings and Voon [4], requires exciting the process with random input signals that have zero odd-order moments (for example, uniformly or normally distributed signals). It can be shown that in response to such input signals, if the system generating the output data $z(k)$ is linear, then

$$\begin{aligned} \phi_{zz^2}(\tau) &= E\left[(z(k-\tau) - \bar{z})(z(k) - \bar{z})^2\right] \\ &\approx \frac{1}{N}\sum_{k=\tau}^{N}(z(k-\tau) - \bar{z})(z(k) - \bar{z})^2 = 0 \quad \forall\, \tau \end{aligned} \tag{2.89}$$

where $\bar{z}(k)$ is the sample mean of the data sequence $z(k)$. According to this test, under the null hypothesis that the system generating the data is linear, $\phi_{zz^2}(\tau)$ has a Gaussian distribution. Consequently, at a confidence level of 95%, we have that for large sample sizes, the system generating the data is nonlinear if:

$$\left|\frac{\phi_{zz^2}(\tau)}{\sqrt{\phi_{zz}(0)}\sqrt{\phi_{z^2z^2}(0)}}\right| \geq \frac{1.96}{\sqrt{N - \tau}} \tag{2.90}$$

where ϕ_{zz} and $\phi_{z^2z^2}$ are computed from expressions similar to (2.89).

This test was applied to two sets of data obtained from the distillation column. In the first data set, the inputs are bounded between ± 0.01 (recall the smaller of the step inputs employed earlier). For comparison, in the second data set, the inputs are bounded between ± 0.001. In each case the inputs are drawn from a uniform distribution over the appropriate interval. The resulting outputs are shown in Figures 2.14 and 2.15, and Fig. 2.16 shows the results of the nonlinearity test. The indicated upper and lower acceptance limits correspond to $\pm 1.96(\sqrt{N - \tau})$. Observe that the values of $\phi_{zz^2}(\tau)$ for the first set of data (larger input excitation) clearly lie outside these limits for both outputs. Thus the null hypothesis, that the process generating the data is linear, is rejected, *for this input set.* (See Figs 2.10 - 2.13 for $\pm 1\%$ input step changes.)

It is interesting now to observe that for the second data set (much smaller amplitude input excitation), the value of the test statistic lies within the limits and the null hypothesis, that the data is generated by a linear system,

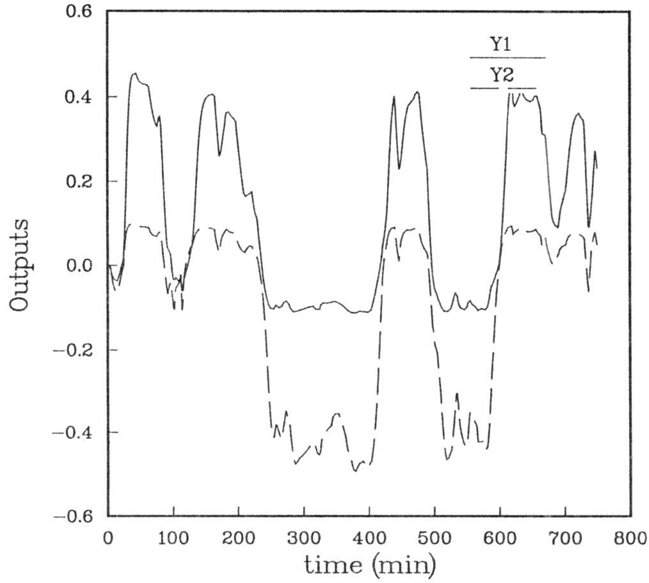

Figure 2.14: Nonlinearity test 1. Plant output data

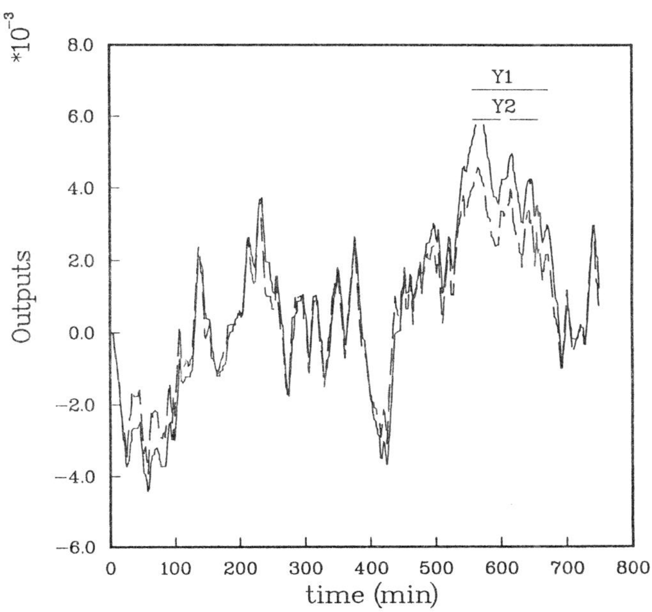

Figure 2.15: Nonlinearity test 2. Plant output data

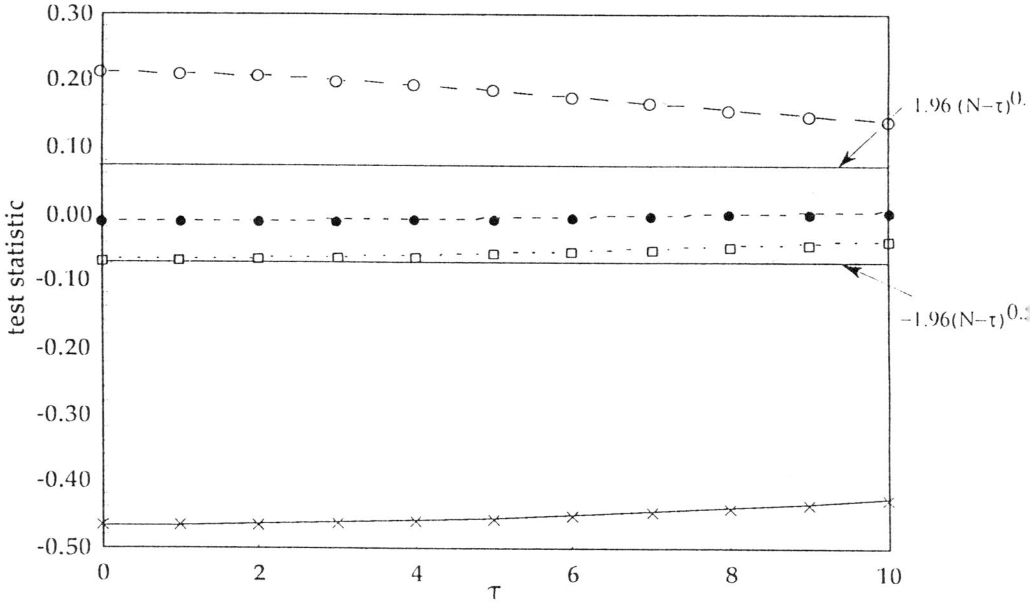

Figure 2.16: Nonlinearity test statistic. For input range ±0.01, y_1 : ∘ and y_2 : ×; For input range ±0.001, y_1 : • and y_2 : □.

is not rejected. Thus, in response to the lower amplitude input signal, the process exhibits essentially linear behavior.

This test thus confirms the fact that a linear model will be reasonable for the distillation column only for a limited range of operating conditions; for the operating conditions of interest, the process behavior is sufficiently nonlinear to indicate that linear models will be inadequate.

The Nonlinear Model

The model structure chosen for representing the dynamic behavior of this distillation column is the polynomial, nonlinear autoregressive moving average model with exogenous inputs (NARMAX):

$$
\begin{aligned}
y(k) &= \theta^0 + \sum_{i=1}^{n_y} \theta_i^1 y(k-i) + \sum_{i=\varrho}^{n_u} \theta_i^2 u(k-i) \\
&+ \sum_{i=1}^{n_y}\sum_{j=1}^{i} \theta_{(i,j)}^3 y(k-i)y(k-j) + \sum_{i=\varrho}^{n_u}\sum_{j=\varrho}^{i} \theta_{(i,j)}^4 u(k-i)u(k-j) \\
&+ \sum_{i=1}^{n_y}\sum_{j=\varrho}^{n_u} \theta_{(i,j)}^5 y(k-i)u(k-j) + \ldots \qquad (2.91)
\end{aligned}
$$

where k is the current sampling time, $y(k)$ is the current value of the output, $u(k)$ is the current value of the input, n_y and n_u are the number of output and input lags, respectively.

There are primarily two major reasons for this choice:

1. These models are more convenient for multivariable systems; and
2. The model parameters enter linearly, so that parameter estimation via a one-step-ahead-prediction error minimization can be formulated as a linear regression problem.

(See Hernández and Arkun [44] for additional rationale for utilizing these models for process control purposes.)

Determining à-priori, which terms to include in the polynomial model (i.e. model complexity) remains the main challenge with this class of models. This involves specifying values for n_u and n_y — the number of input and output lags to be included in the model — *and* specifying the maximum polynomial order to be considered. Even for modest lag values and polynomial orders, the total number of parameters contained in the model can be quite large. It is customary to employ standard statistical techniques to determine the significance of the estimated model parameters and then retain only those parameters that are "significant." (See Hernández [43] for additional discussions on this issue). For this specific case study, as we will discuss next in more detail, we specify *three* different sets of values for n_u and n_y; restrict the polynomial order to cubics; identify the resulting models from data; and employ various statistical techniques to validate the models and to discriminate among them.

2.9.3 Model Identification

Input Sequence

The *input sequences* used to generate the process data sets from which model identification was carried out were obtained as follows:

1. the scaled process inputs u_1 and u_2 are chosen to be two independent stochastic sequences restricted to lie between ± 0.05 — a range large enough to produce data in the operating range of interest;

2. the specific stochastic characteristics of the input sequences are obtained by allowing each input variable to change at each sampling time (every minute) to a new value with a predetermined *switching probability*, p; if the input is to change, the new value is drawn from a uniform distribution over the interval $(-0.05, 0.05)$.

Three specific input sequences (for $p = 0.1, 0.3, 0.8$) were used to generate three different sets of input-output data abbreviated as follows:

Data set	Switching Probability, (p)
I01	0.1
I02	0.3
I03	0.8

For purposes of illustration, Figures 2.17 and 2.18 show the input sequences for $p = 0.1$ and 0.8 respectively.

Candidate Models and Modeling Results

By specifying the following three sets of values for n_u and n_y

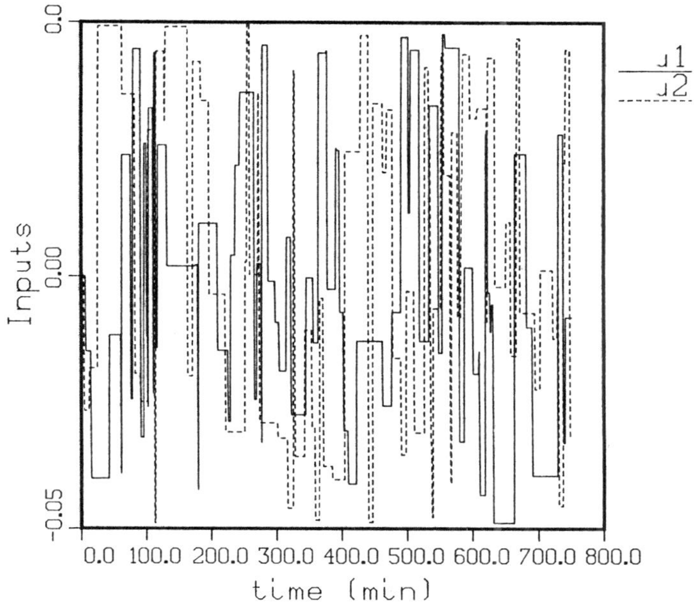

Figure 2.17: Input data with switching probability $p = 0.1$

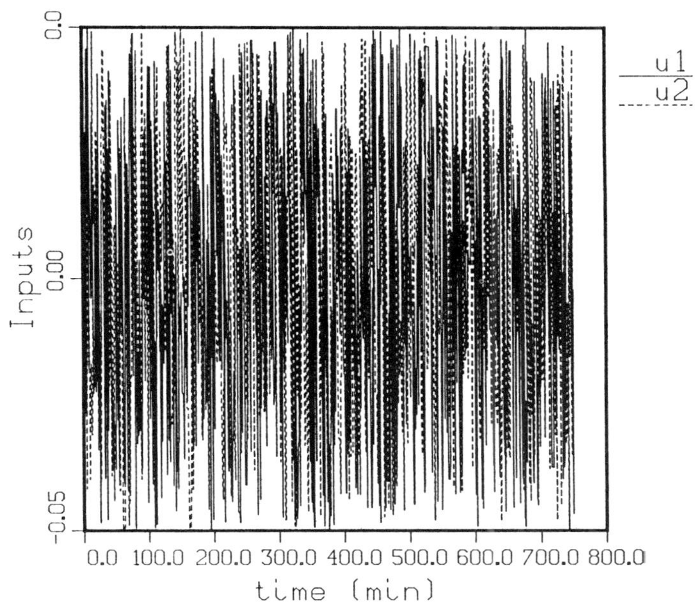

Figure 2.18: Input data with switching probability $p = 0.8$

1. $n_y = 1$ and $n_u = 1$

2. $n_y = 2$ and $n_u = 2$

3. $n_y = 5$ and $n_u = 1$

three different models were identified from *each* data set, making a grand total of *nine* models. Each data set consists of 800 pairs of input and output data, 500 of which were used for model identification with the remaining 300 reserved for cross-validation (see later). The model parameters were estimated by standard, linear stepwise regression techniques.

The modeling results are summarized in Table 2.1 where each model is identified by a code representative of the number of regressors it contains. For example, the code "14_9" implies that the model has 14 regressors for the first output (y_1) and 9 for the second output (y_2). The actual equations for one of the models (5_5) are presented in Table 2.1 as an example. For convenience, the models are identified by these codes from now on.

Note that many more candidate models could have been identified from the current data sets merely by specifying other values of n_y, n_u and by increasing the maximum polynomial order. Here we have limited the number of models to nine for further evaluation and model validation. One would typically re-identify new models only when the models in hand are deemed "inadequate" after model validation tests (and, in this case, after closed-loop control performance evaluation).

2.9.4 Model Comparison and Validation

We now seek to evaluate the quality and adequacy of the 9 models that have been identified from the three sets of process data. Among the various means by which the adequacy of a model can be tested, the following four methods will be used for this specific example [55]:

1. Statistical Information Criteria;

2. Cross validation: One-step ahead predictions;

3. Cross validation: "Pure" predictions;

4. Residual Analysis.

Delay	I01	I02	I03
$n_y = 1$ $n_u = 1$	8_2	8_8	6_5
$n_y = 2$ $n_u = 2$	10_2	14_9	8_5
$n_y = 5$ $n_u = 1$	7_2	9_8	5_5

Model 5_5
$y_1(k+1) = .0012 + 0.98y_1(k) - .18u_1(k) + 1.1y_1(k-2)u_2(k) - 1.8y_2(k)u_1(k)$
$y_2(k+1) = .0018 + 0.92y_2(k) - .22u_1(k) + 30.4y_2^2(k)u_2(k) - 1.7u_2^2(k)$

Table 2.1: Models identified from the data

Statistical Information Criteria

As mentioned in Section 8, the Akaike Information Criterion (AIC) and other similar criteria consist of two terms: a prediction error term and a model order term. The first is a measure of how well the model fits the data; the second a measure of the model complexity required to obtain the fit. These information criteria therefore try to strike a balance between low residual variance and excessive number of model parameters, with smaller values indicating comparatively more desirable models. The Akaike Information Criterion (AIC), Final Prediction Error (FPE), Bayesian Information Criterion (BIC), and the Law of Iterated Logarithms Criterion (LILC), are four of the more commonly employed criteria; each is defined below, and will be used to evaluate the 9 models obtained for the distillation column:

$$AIC = N \ln \left[\frac{1}{N} \sum_{k=1}^{N} \varepsilon(k)^2 \right] + 2\varphi$$

$$FPE = N \ln \left[\frac{1}{N} \sum_{k=1}^{N} \varepsilon(k)^2 \right] + N \ln \left[\frac{N+\varphi}{N-\varphi} \right]$$

$$BIC = N \ln \left[\frac{1}{N} \sum_{k=1}^{N} \varepsilon(k)^2 \right] + \varphi \ln(N)$$

$$LILC = N \ln \left[\frac{1}{N} \sum_{k=1}^{N} \varepsilon(k)^2 \right] + 2\varphi \ln(\ln(N)) \qquad (2.92)$$

Here, $\varepsilon(k) = y(k) - \hat{y}(k)$ is the residual at time k; with, $y(k)$ as the plant measurement and $\hat{y}(k)$ as the model prediction; φ is the total number of terms included in the model.

Note that for large values of N, the first two criteria, AIC and FPE, are equivalent, since:

$$\lim_{N \to \infty} N \ln \left[\frac{N + \varphi}{N - \varphi} \right] = 2\varphi$$

Table 2.2 presents the comparison between the nine models (with $N = 500$ in each case). For simplicity only information for models involving y_1 is shown as the corresponding values for y_2 are quite similar. Note that the values of AIC and FPE are virtually identical.

Data Set	Model	AIC	FPE	BIC	LILC
I01	8_2	-5094.2	-5094.2	-5079.4	-5056.3
	10_2	-5001.1	-5001.1	-4982.9	-4954.7
	7_2	-4734.9	-4734.9	-4721.7	-4701.2
I02	14_9	-4937.89	-4937.89	-4913.9	-4874.98
	8_8	-4881.77	-4881.76	-4866.94	-4844.02
	9_8	-4899.40	-4899.4	-4882.93	-4857.46
I03	6_5	-4446.3	-4446.3	-4434.8	-4416.8
	8_5	-4465.9	-4465.9	-4451.1	-4428.1
	5_5	-4442.7	-4442.7	-4432.8	-4417.4

Table 2.2: Information Criteria for the various models

At this point, these criteria appear to indicate that the models obtained from the data sets with smaller switching probabilities (the lower frequency input signal) are statistically superior. But this is hardly conclusive.

Model		Prediction Error (%) for identification data		Prediction Error (%) for cross-validation data	
		$y_1\, I_{id}$	$y_2\, I_{id}$	$y_1\, I_{val}$	$y_2\, I_{val}$
I01	8_2	.08	.18	.09	.28
	10_2	.07	.18	.10	.31
	7_2	.05	.17	.06	.28
I02	14_9	.464	.443	.617	.313
	8_8	.426	.442	.54	.310
	9_8	.399	.441	.534	.308
I03	5_5	1.43	6.65	5.1	13.5
	8_5	1.34	6.64	5.2	13.18
	6_5	1.40	6.6	5.05	13.61

Table 2.3: Single step prediction errors for all the identified models

Cross-Validation: One-Step Ahead Predictions

As mentioned earlier, for each model, 500 data points were used for identification and 300 additional data points were reserved for cross-validation. The model performance in predicting observed process behavior using both the "identification" data and the "validation" data may be assessed using the following evaluation criterion:

$$I = \frac{\sum_{k=1}^{N} [y(k) - \hat{y}(k)]^2}{\sum_{k=1}^{N} [y(k) - \bar{y}]^2} \times 100\% \tag{2.93}$$

where, $y(k)$ is the plant output at time k ; $\hat{y}$ is the model prediction at time k; $\bar{y}$ is the mean value of the plant output data, and N is the total number of data points considered. The results are presented in Table 2.3. As expected the models show better prediction capabilities for the identification data than the validation data. Again the trend is that the models that were identified from input data with smaller switching probabilities show better prediction capability.

"Pure" Prediction

In evaluating the "one-step ahead prediction" capabilities of the models, the models use actual past plant output data to predict the plant output behavior at the next sampling instant (i.e. one-step ahead). A more stringent test of the prediction capabilities of each model is to obtain model predictions without the use of available plant data, employing the model predictions themselves recursively in the model in lieu of the plant data.

When such "pure" predictions obtained from each of the nine models are compared with the entire data set (both the identification and cross-validation sets), using the same evaluation criteria I given in Eq. (2.93), the results are presented in Table 2.4. From the results in Tables 2.3 and 2.4, it is quite clear that the models that show the best "pure" prediction capabilities are those obtained from the data set I02.

Now, the information criteria given in Table 2.1 indicate that I02 models are better than I03 models. Even though for "one-step ahead predition" the IO1 models appear superior to the IO2 models, we must keep in mind that for model predictive control applications, the "pure" prediction capabilities of the model are far more important. Table 2.4 shows the superiority of the IO2 models in this regard. As a result we choose to retain only the three I02 models (8_8, 14_9 and 9_8) for further evaluation.

Of these three surviving models, by comparing the prediction errors for single-step and "pure" prediction tests, we further observe that the model 14_9 exhibits the largest error magnitude. This is shown in Figure 2.19 (for the "pure" prediction errors). We therefore also discard model 14_9 and further reduce the number of candidate models to two. The parameter values and the regressors for the final two models are given in Table 2.5. These models are now compared using residual analysis.

Residual Analysis

The classical tests consisting of the autocorrelation function of the residuals, and the cross-correlation between the residuals and the inputs, were originally developed for linear models. As such, they are not sufficient for nonlinear systems. To detect the presence of unmodeled linear *and* nonlinear terms, tests involving higher order correlations are needed [6]. These tests, and the expressions for obtaining the required higher order correlation *estimates*, are given as follows:

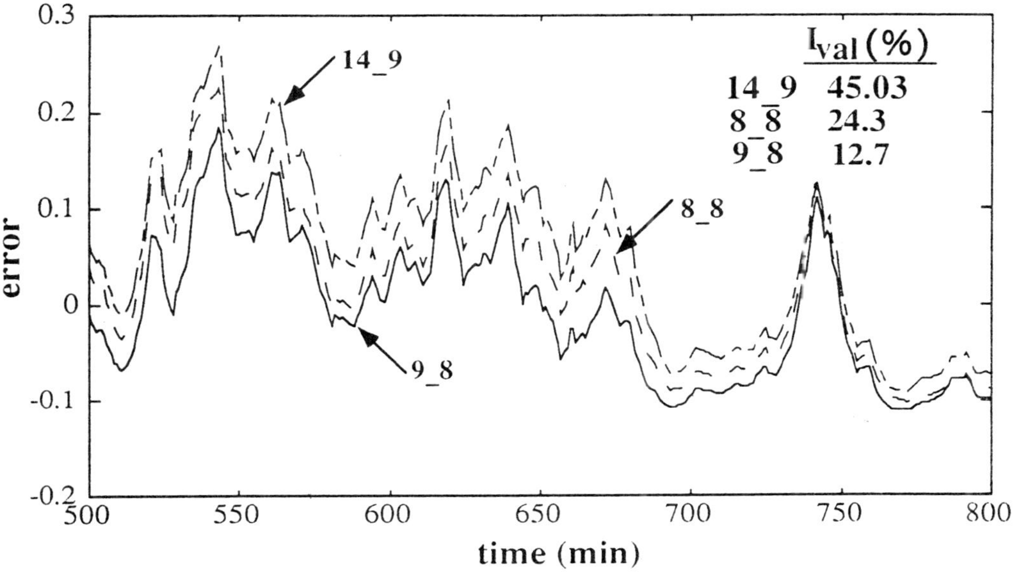

Figure 2.19: Model Validation Prediction errors: y_1 predictions from models 8_8, 14_9, and 9_8.

$$
\begin{aligned}
\phi_{(u^2)'\xi}(\tau) &= 0 \qquad \forall\tau \\
\phi_{(u^2)'\xi^2}(\tau) &= 0 \qquad \forall\tau \\
\phi_{\xi\xi u}(\tau) &= 0 \qquad \forall\tau \\
\phi_{u\xi}(\tau) &= 0 \qquad \forall\tau \\
\phi_{\xi\xi}(\tau) &= \delta(\tau) \quad \forall\tau
\end{aligned}
\tag{2.94}
$$

where:

$$
\phi_{xy}(\tau) = \frac{\sum_{k=\tau}^{N} (x(k) - \bar{x})(y(k-\tau) - \bar{y})}{\sqrt{\sum_{k=1}^{N} (x(k) - \bar{x})^2 \sum_{k=1}^{N} (y(k) - \bar{y})^2}}
$$

Model		Prediction Error (%) for identification data		Prediction Error (%) for cross-validation data	
		$y_1\, I_{id}$	$y_2\, I_{id}$	$y_1\, I_{val}$	$y_2\, I_{val}$
I01	8_2	15.8	230.9	33.0	342.7
	10_2	18.01	127.5	18.06	146.5
	7_2	28.9	94.01	18.58	105.6
I02	14_9	8.12	9.06	45.03	6.61
	8_8	6.85	8.7	12.7	6.5
	9_8	7.06	8.8	24.3	5.41
I03	5_5	36.47	69.88	63.2	124.46
	8_5	35.29	73.67	59.37	107.71
	6_5	37.0	70.19	57.37	93.37

Table 2.4: Pure prediction errors for all the identified models

and

$$\phi_{\xi\xi u}(\tau) = \frac{\sqrt{N}\sum_{k=\tau}^{N} \xi(k)\,\xi(k-1)\,(u(k-\tau)-\bar{u})}{\sum_{k=1}^{N}\xi^2(k)\sqrt{\sum_{k=1}^{N}(u(k)-\bar{u})^2}}$$

Billings and Voon [6] suggest that at a 95 % confidence level, the residuals have no unmodeled linear, nonlinear structure remaining, if the absolute value of each test statistic shown above is less than $1.96/\sqrt{N}$.

The results of these tests for the two models are presented in the Figures 2.20 – 2.23. These plots show that the test quantities for Model 9_8 lie within the 95 % confidence bands more often than the test quantities for Model 8_8. Thus, considering all the validation tests used so far, it is fair to conclude that Model 9_8 is the most promising model.

This model was subsequently used for closed-loop control of the distillation column. Additional details regarding the model development, evaluation, and control system performance are available in Sriniwas *et al* [89].

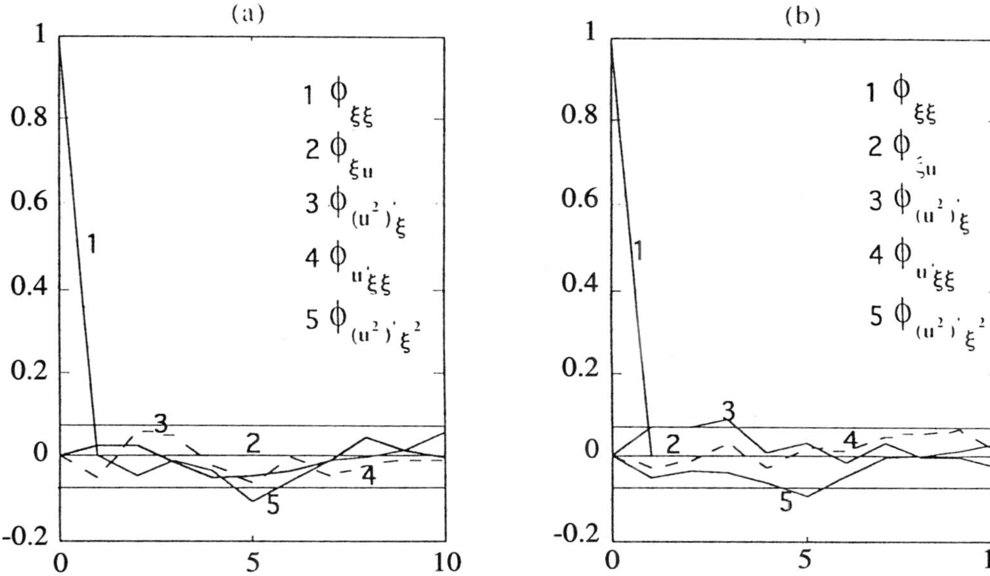

Figure 2.20: Correlation Tests for Model 8_8 based on residuals for (a) first output (ξ_1) and u_1; (b) first output (ξ_1) and u_2.

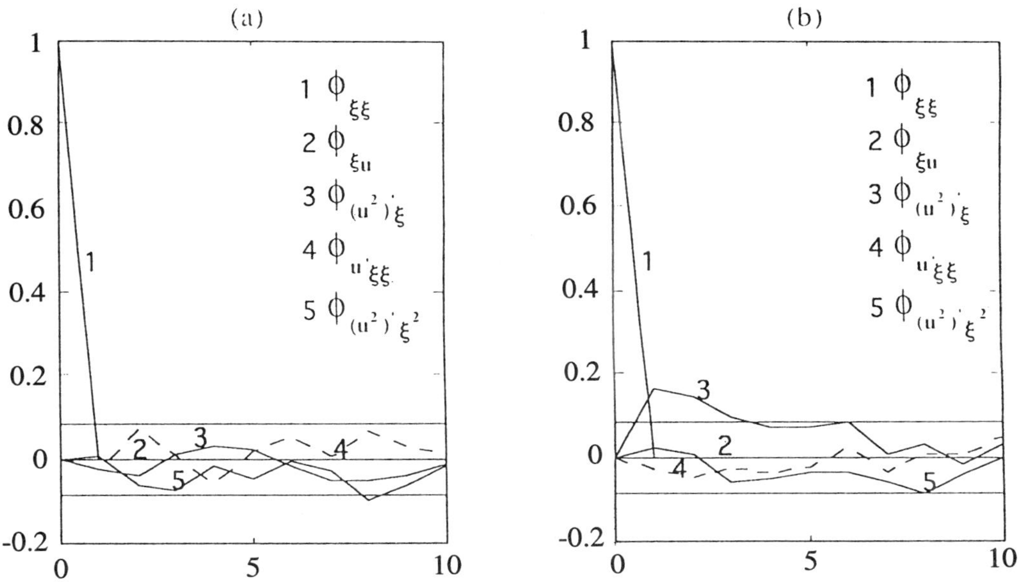

Figure 2.21: Correlation Tests for Model 8_8 based on residuals for (a) second output (ξ_2) and u_1; (b) second output (ξ_2) and u_2.

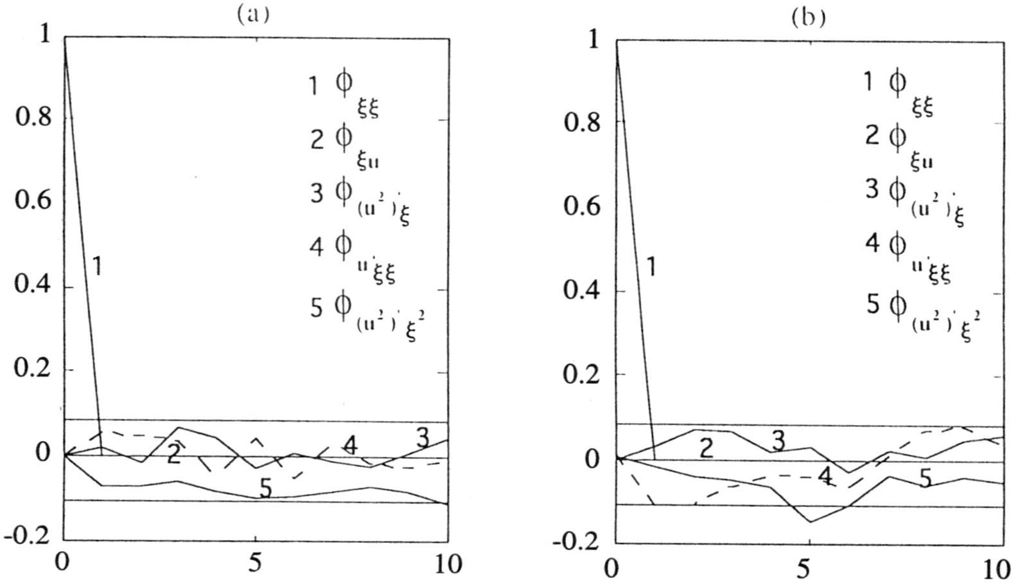

Figure 2.22: Correlation Tests for Model 9_8 based on residuals for (a) first output (ξ_1) and u_1; (b) first output (ξ_1) and u_2.

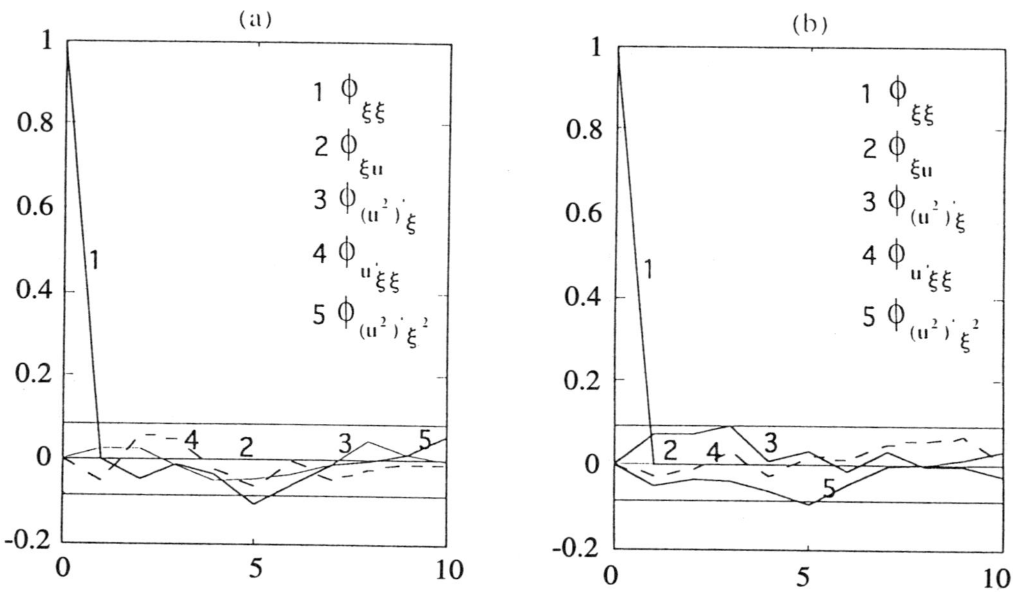

Figure 2.23: Correlation Tests for Model 9_8 based on residuals for (a) second output (ξ_2) and u_1; (b) second output (ξ_2) and u_2.

Output 1			Output 2		
Regressors	Θ		Regressors	Θ	
	9_8	8_8		9_8	8_8
1.0	1.4×10^{-4}	-5×10^{-5}	1.0	-1.7×10^{-3}	-1.2×10^{-3}
$y_1(k)$	.98	.99	$y_2(k)$	.99	.99
$u_1(k)$	-0.27	-0.27	$u_1(k)$	-0.32	-0.32
$u_2(k)$	0.21	0.18	$u_2(k)$	0.31	0.32
$y_1(k)u_1(k)$	-1.91	-1.98	$y_1(k)u_1(k)$	0.76	0.75
$y_1(k)u_2(k)$	1.44	1.05	$y_1(k)u_2(k)$	-1.02	-1.04
$y_1^2(k)u_2(k)$	0.0	5.24	$y_2^2(k)u_2(k)$	-1.22	-1.26
$y_1^3(k)$	0.0	-.316	$y_2(k)u_2^2(k)$	0.0	11.9
$y_1^2(k-1)u_1(k)$	5.96	0.0	$y_2(k-1)u_2^2(k)$	-12.1	0.0
$y_1^2(k-3)u_2(k)$	-3.71	0.0	-	-	-
$y_1^2(k-2)$	-0.11	0.0	-	-	-

Table 2.5: The regressors and parameters of models 9_8 and 8_8

2.10 Summary

Probably the key point of this review of nonlinear process identification is that the basic problem we are attempting to solve is *not* strictly a mathematical one. In particular, returning to the Box-Jenkins model-building procedure described in Section 1, only one of the four steps involved — estimation of model parameters from available data — is an inherently mathematical one like computing the square root of π. The other three steps — model structure selection, order parameter determination, and model validation — all involve considerable subjective judgement on the part of the model builder, as illustrated by the case study just presented. As a practical matter, it is *these* areas that seem to cause the greatest difficulty. Specifically, the following four problems are probably the greatest challenges in empirical process modeling, roughly in decreasing order of severity:

1. specification of a class $\mathcal{C}$ of "structurally adequate" models $\mathcal{M}$;

2. design of "good" input sequences for model identification;

Problem	"Truth Class"	Perturbation	Objective
1	$\mathcal{C}$	none	Algorithm design, basic verification
2	$\mathcal{C}$	"noise"	Algorithm design, influence of noise
3	$\mathcal{C}$	"outliers"	Algorithm design, robustness
4	$\mathcal{F}$	none	Model mismatch studies
5	$\mathcal{F}$	"disturbance"	Process disturbance studies
6	$\mathcal{L}$	minimal	Laboratory modeling studies
7	$\mathcal{S}$	moderate	Semi-works modeling studies
8	$\mathcal{P}$	arbitrary	Industrial process studies

Table 2.6: Eight distinct "modeling problems"

3. characterization of the domain of model validity;
4. determination of "adequate" model order parameters.

In order to appreciate the nature of these problems — and the differences between them — it is useful to consider the sequence of eight "modeling problems" listed in Table 2.6. The numbering indicates both an order of increasing difficulty and a recommended order of approach in attempting to solve them. The column labeled "Truth Class" indicates the class from which the *identification datasets* are obtained for the modeling study, while the column labeled "Perturbation" indicates the kind of deliberate or unavoidable "disturbances" that appear in this data. Finally, the column labeled "Objective" describes the principal objective motivating each of these modeling problems.

For problems 1, 2, and 3, the "Truth Class" is indicated as "$\mathcal{C}$," indicating that the datasets used in the studies are generated by a simulation model from the class $\mathcal{C}$ for which the modeling procedure was intended. Problem 1 then corresponds to algorithm validation: given a model $\mathcal{M}$ from class $\mathcal{C}$, how close is the identified model $\tilde{\mathcal{M}}$ to the one from which the datasets were generated? Problem 2 takes a first step toward greater realism, deliberately introducing "noise" as a perturbation and asking the same question as in Problem 1. Problem 3 takes the next step, replacing "noise" (typically assumed Gaussian) with a more severe perturbation that describes "outliers" in the data. *While there was not sufficient space to address this topic here, the issue of robustness — i.e., resistance to outliers — is a vital one in practice, as noted briefly at the beginning of Section 9; see the references cited there for an introduction to this subject.*

Note that, regardless of the details of the "noise" or "outlier" models used in Problems 1, 2, and 3, in all cases we have a *correct answer* with which to compare our estimated results. In problems 4 through 8, we lose this ability since the model generating the data is no longer a member of the class $\mathcal{C}$ of models we are fitting to the data. In particular, for Problems 4 and 5, the datasets under consideration are generated by a simulation model from the class $\mathcal{F}$ (possibly, though not necessarily, a fundamental model). Thus, while we have control over the perturbations we introduce into these datasets, the "best fit model $\mathcal{M}$ from the class $\mathcal{C}$" will no longer be the "correct" model that actually generated the data. The objectives of these problems are to understand the effects of this model mismatch (Problem 4) and additional mismatches that can arise due to "disturbances" or "process upsets" (Problem 5). In particular, note that the case study presented in Section 9 illustrates the issues addressed in Problem 4.

Finally, Problems 6, 7, and 8 involve data from one of three real-world processes: a laboratory process $\mathcal{L}$, a semi-works process $\mathcal{S}$, and an operating industrial process $\mathcal{P}$. Here, we are no longer *deliberately introducing* perturbations, but are going to varying lengths to *prevent* them. That is, datasets from the laboratory process $\mathcal{L}$ can be expected to exhibit fewer irregularities (i.e., outliers, missing data, etc.) than those from the semi-works process $\mathcal{S}$, which can in turn be expected to exhibit fewer irregularities than those from an operating industrial process $\mathcal{P}$.

While it is not necessary to solve every problem in this sequence to develop a useful empirical modeling procedure, it probably is necessary to solve several of them. The real point of this discussion is that successful empirical process modeling involves many different aspects — model structure selection, order determination, design and verification of estimation algorithms, verification of final results, etc. — and these different aspects each require different evaluation procedures. All of these aspects are critically important and they are probably too diverse in nature to be addressed by any one group of individuals. The challenges are substantial and can benefit from the best efforts of those in many different disciplines.

References

[1] A.C. Atkinson and A.N. Donev, *Optimum Experimental Designs*, Oxford University Press, New York, 1992.

[2] A. Bannerjee, Y. Arkun, R. Pearson, and B. Ogunnaike, *H_∞ control of Nonlinear Processes Using Multiple Linear Models*, Proc. 3rd European Control Conference, TP-8, Rome, 1995.

[3] S.A. Billings, "Identification of nonlinear systems — a survey," *IEE Proc.*, v. 127, pp. 272 - 285, 1980.

[4] Billings, S.A. and W.S.F. Voon, "Structure Detection and Model validity tests in the identification of nonlinear systems" *IEEE Proceedings*, Part D, pp. 130-193, 1983.

[5] S.A. Billings and W.S.F. Voon, "A prediction- error and stepwise regression algorithm for nonlinear systems," *Int. J. Control*, v. 44, pp. 803 - 822, 1986.

[6] S.A. Billings and W.S.F. Voon, "Correlation based model validity tests for nonlinear models," *Int. J. Control*, v. 44, pp. 235 - 244, 1986.

[7] S.A. Billings and Q.M. Zhu, "Rational model identification using an extended least squares algorithm," *Int. J. Control*, vol. 54, pp. 529 - 546, 1991.

[8] P. Billingsley, *Probability and Measure*, 2nd ed., John Wiley and Sons, New York, 1986.

[9] G.E.P. Box, W.H. Hunter, and J.S. Hunter, *Statistics for Experimenters*, Wiley, New York, 1978.

[10] G.E.P. Box, G.M. Jenkins, and G.C. Reinsel, *Time Series Analysis*, 3rd ed., Prentice-Hall, Englewood Cliffs, NJ, 1994.

[11] G.E.P. Box and G.C. Tiao, *Bayesian Inference in Statistical Analysis*, Addison-Wesley, Reading, MA, 1973.

[12] S. Boyd and L.O. Chua, "Fading Memory and the Problem of Approximating Nonlinear Operators with Volterra Series," *IEEE Trans. Circuits and Systems*, vol. 32, pp. 1150 - 1161, 1985.

[13] R.W. Brockett, "Convergence of Volterra series on infinite intervals and bilinear approximations," in *Nonlinear systems and applications*, V. Lakshmikathan, ed., Academic Press, New York, pp. 39 - 46, 1977.

[14] P.J. Brockwell and R.A. Davis, *Time Series: Theory and Methods*, 2nd ed., Springer Verlag, New York, 1991.

[15] D.R. Brillinger, "The identification of a particular nonlinear time series system," *Biometrika*, v. 64, pp. 509 - 515, 1977.

[16] R. Chen and R.S. Tsay, "Nonlinear Additive ARX Models," *J. Am. Stat. Assoc.*, v. 88, pp. 955 - 967, 1993.

[17] H.W. Chen, "Modeling and Identification of Parallel Nonlinear Systems: Structural Classification and Parameter Estimation Methods," *Proc. IEEE*, vol. 83, pp. 39 - 66, 1995.

[18] Chien, I-L., and B. A. Ogunnaike, "Modeling and Control of high-purity distillation columns" Paper 2a *Preprints of the 1992 Annual AIChE Meeting*, Miami-Beach, Florida, 1992.

[19] D.R.J. Chillingworth, *Differential topology with a view to applications*, Pitman, London, 1976.

[20] A. Cinar, personal communication.

[21] J.P. Congalidis, J.R. Richards, and W.H. Ray, "Feedforward and Feedback Control of a Solution Copolymerization Reactor," *AIChE Journal*, vol. 35, pp. 891 - 907, 1989.

[22] R.F. Curtain and A.J. Pritchard, *Functional Analysis in Modern Applied Mathematics*, Academic Press, New York, 1977.

[23] E. Damsleth and A.H. El-Shaarawi, "ARMA Models with Double-exponentially Distributed Noise," *J. Royal Statist. Soc.*, ser. B, vol. 51, pp. 61 - 69, 1989.

[24] H.T. Davis, *Introduction to Nonlinear Differential and Integral Equations*, Dover, New York, 1962.

[25] P.J. Davis, *Interpolation and Approximation*, Dover, New York, 1975.

[26] M.F. Doherty and J.M. Ottino, "Chaos in Deterministic Systems: Strange Attractors, Turbulence, and Applications in Chemical Engineering," *Chem. Eng. Sci.*, vol. 43, pp. 139 - 183, 1988.

[27] F.J. Doyle, III, B.A. Ogunnaike, and R.K. Pearson, "Nonlinear Model-based Control Using Second-order Volterra Models," *Automatica*, v. 31, pp. 697 - 714, 1995.

[28] N.R. Draper and H. Smith, *Applied Regression Analysis*, John Wiley and Sons, New York, 1981.

[29] D. Dubois and H. Prade, *Fuzzy Sets and Systems, Theory and Applications*, Academic Press, New York, 1980.

[30] B. Effron and R.J. Tibshirani, *An Introduction to the Bootstrap*, Chapman and Hall, New York, 1993.

[31] E. Eskinat, S.H. Johnson, and W.L. Luyben, "Use of Hammerstein Models in Identification of Nonlinear Systems," *AIChE Journal*, v. 37, pp. 255 - 268, 1991.

[32] P. Eykhoff, *System Identification*, John Wiley and Sons, New York, 1974.

[33] W.H. Feller, *An Introduction to Probability Theory and Its Applications*, vol. II, John Wiley and Sons, New York, 1966.

[34] C. Georgakis, "On the use of extensive variables in process dynamics and control," *Chemical Engineering Science*, vol. 41, pp. 1471 - 1484, 1986.

[35] K. Godfrey, *Perturbation Signals for System Identification*, Prentice-Hall, New York, 1993.

[36] W. Greblicki, "Nonparametric Identification of Wiener Systems," *IEEE Trans. Inform. Theory*, v. 38, pp. 1487 - 1493, 1992.

[37] W. Greblicki and M. Pawlak, "Nonparametric Identification of Hammerstein Systems," *IEEE Trans. Inform. Theory*, v. 35, pp. 409 - 418, 1989.

[38] W. Greblicki and M. Pawlak, "Nonparametric Identification of Nonlinear Block-Oriented Systems," *Preprints, 9th IFAC/IFORS Symposium*, Budapest, Hungary, pp. 720 - 724, 1991.

[39] J. Guckenheimer and P. Holmes, *Nonlinear Oscillations, Dynamical Systems, and Bifurcations of Vector Fields*, Springer Verlag, New York, 1983.

[40] R. Haber and H. Unbehauen, "Structure Identification of Nonlinear Dynamic Systems — A Survey on Input/Output Approaches," *Automatica*, v. 26, pp. 651 - 677, 1990.

[41] F.R. Hampel, E.M. Ronchetti, P.J. Rousseeuw, and W.A. Stahel, *Robust Statistics: The Approach Based on Influence Functions*, Wiley, New York, 1986.

[42] W. Hardle, *Applied Nonparametric Regression*, Cambridge University Press, New York, 1990.

[43] E. Hernandez, *Control of Nonlinear Systems using Input-Output Information*, PhD Thesis, Georgia Institute of Technology, August, 1992.

[44] E. Hernandez and Y. Arkun, "Control of Nonlinear Systems Using Polynomial ARMA Models," *AIChE Journal*, v. 39, pp. 446 - 460, 1993.

[45] E.W. Jacobsen, "Identification for Control of Strongly Interactive Plants," presented AIChE Annual Meeting, San Fransisco, November, 1994.

[46] T.A. Johansen and B.A. Foss, "State-space Modeling Using Operating Regime Decomposition and Local Models," *Proc. 12th IFAC World Congress*, Sydney, Australia, July, 1993.

[47] J.C. Kantor, "An Overview of Nonlinear Geometrical Methods for Process Control," in *Shell Process Control Workshop*, D.M. Prett and M. Morari, eds., Butterworths, New York, 1987.

[48] S.M. Kay and S.L. Marple, "Spectrum Analysis — A Modern Perspective," *Proc. IEEE*, vol. 69, pp. 1380 - 1419, 1981.

[49] S.M. Kay, *Modern Spectral Estimation: Theory and Application*, Prentice-Hall, Englewood Cliffs, NJ, 1988.

[50] G. Klambauer, *Mathematical Analysis*, Marcel Dekker, New York, 1975.

[51] T. Koh and E.J. Powers, "Second-Order Volterra Filtering and Its Applications to Nonlinear System Identification," *IEEE Trans. Acoustics, Speech, and Signal Processing*, v. 33, pp. 1445 - 1455, 1985.

[52] M.A. Kramer, "Nonlinear Principal Component Analysis Using Autoassociative Neural Networks," *AIChE Journal*, vol. 37, pp. 233 - 243, 1991.

[53] C. Leger and N. Altman, "Assessing Influence in Variable Selection Problem," *J. Am. Statist. Assoc.*, v. 88, pp. 547 - 556, 1993.

[54] E.L. Lehman, *Theory of Point Estimation*, John Wiley, New York, 1983.

[55] L. Ljung, *System Identification: Theory for the User*, Prentice-Hall, New York, 1987.

[56] C.L. Mallows, "Robust Methods — Some Examples of Their Use," *The American Statistician*, vol. 33, 1979, pp. 179 - 184, 1979.

[57] V.Z. Marmarelis, "Identification of Nonlinear Biological Systems Using Laguerre Expansion of Kernels," *Annals of Biomedical Engineering*, vol. 21, pp. 573 - 589, 1993.

[58] J.S. Marron, "Automatic smoothing parameter selection: a survey," *Empirical Econ.*, vol. 13, pp. 187 - 208, 1989.

[59] R.D. Martin and D.J. Thomson, "Robust-resistant spectrum estimation," *Proc. IEEE*, vol. 70, pp. 1097 - 1114, 1982.

[60] G.R. McCullough, "Identification of Second-Order Volterra Series Approximations to Nonlinear Process Systems," BSChE thesis, Purdue University, Department of Chemical Engineering, 1993.

[61] M. Morari and E. Zafiriou, *Robust Process Control*, Prentice-Hall, Englewood Cliffs, NJ, 1989.

[62] A.H. Nayfeh and D.T. Mook, *Nonlinear Oscillations*, John Wiley and Sons, New York, 1979.

[63] C.L. Nikias and A. Petropulu, *Higher-Order Spectra Analysis*, Prentice-Hall, New York, 1993.

[64] B.A. Ogunnaike, R.K. Pearson, and F.J. Doyle, "Chemical Process Characterization: with Application in the Rational Selection of Control Strategies," Proc. 2nd European Control Conference, pp. 1067 - 1071, Groningen, Netherlands, 1993.

[65] B.A. Ogunnaike, R.K. Pearson, N. Samardzija, and J.D. Bomberger, "Low-order Empirical Modeling for Nonlinear Systems," *Proc. ADCHEM'94*, Kyoto, May, pp. 41 - 46, 1994.

[66] B.A. Ogunnaike and W.H. Ray, *Process Dynamics, Modeling, and Control*, Oxford University Press, New York, 1994.

[67] G.A. Pajunen, "Application of a model reference adaptive technique to the identification and control of Wiener type nonlinear processes," PhD thesis, Acta Polytechnica Scandinavica, no. 52, Helsinki University of Technology, Department of Electrical Engineering, Finland, 1984.

[68] A. Papoulis, *Probability, Random Variables, and Stochastic Processes*, McGraw-Hill, New York, 1965.

[69] M. Pawlak, R.K. Pearson, B.A. Ogunnaike, and F.J. Doyle, III, "Nonparametric Identification of a Class of Nonlinear Time Series Systems," in preparation.

[70] R.K. Pearson, "Nonlinear Input/Output Modeling," *J. Procecss Control*, 5, (**4**), 197-211, 1995.

[71] R.K. Pearson, "An Incomplete Taxonomy of NARMAX Models," submitted for publication.

[72] R.K. Pearson, B.A. Ogunnaike, and F.J. Doyle, III, "Identification of Discrete Convolution Models for Nonlinear Processes," *Preprints of the 1992 Annual AIChE Meeting*, Miami-Beach, Florida 1992.

[73] R.K. Pearson, B.A. Ogunnaike, and F.J. Doyle (1994), "Identification of 2nd-order Volterra models for chemical processes," in preparation.

[74] R.K. Pearson, B.A. Ogunnaike, and F.J. Doyle, III, "Identification of Structurally Constrained Second-Order Volterra Models," submitted for publication.

[75] B.T. Poljak and JA. Z. Tsypkin, "Robust Identification," *Automatica*, v. 16, pp. 53 - 63, 1980.

[76] M.B. Priestley, *Spectral Analysis and Time Series*, Academic Press, New York, 1981.

[77] M.B. Priestley, *Nonlinear and Nonstationary Time Series Analysis*, Academic Press, New York, 1988.

[78] F. Puckelsheim, *Optimal Design of Experiments*, John Wiley and Sons, New York, 1993.

[79] N.S. Rajbman and V.M. Chadeev, *Identification of Industrial Processes*, North-Holland, New York, 1980.

[80] J.B. Rawlings, E.S. Meadows, and K.R. Muske, "Nonlinear Model Predictive Control: A Tutorial and Survey," *Proc. ADCHEM'94*, Kyoto, pp. 203 - 214, 1994.

[81] D.E. Rivera, S.V. Gaikwad, and X. Chen, "CONTROL-ID: A Demonstration Prototype for Control-Relevant Identification," *Proc. American Control Conf.*, Baltimore, MD, pp. 2055 - 2059, 1994.

[82] V.K. Rohatgi and G.J. Szekely, "Sharp Inequalities between Skewness and Kurtosis," *Statistics and Probability Letters*, vol. 8, pp. 297 - 299, 1989.

[83] M. Rosenblatt, "Dependence and Asymptotic Independence for Random Processes," in *Studies in Probability Theory*, M. Rosenblatt, ed., Mathematical Association of America, 1978.

[84] P.J. Rousseeuw and A.M. Leroy, *Robust Regression and Outlier Detection*, Wiley, New York, 1987.

[85] D.E. Rumelhart, *et al.*, *Parallel Distributed Processing*, vol. 1, MIT Press, Cambridge, MA, 1986.

[86] P. Sadegh, H. Melgaard, H. Madsen, and J. Holst, "Optimal Experiment Design for Identification of Grey-Box Models," *Proc. American Control Conf.*, Baltimore, MD, June, pp. 132 - 137, 1994.

[87] I.C. Schick and S.J. Mitter, "Robust recursive estimation in the presence of heavy-tailed observation noise," *Annals of Statistics*, v. 22, pp. 1045 - 1080, 1994.

[88] J.H. Seinfeld and L. Lapidus, *Mathematical Methods In Chemical Engineering, vol. 3: Process Modeling, Estimation, and Identification*, Prentice-Hall, Englewood Cliffs, NJ, 1974.

[89] Sriniwas, G.R., Y. Arkun, I.-L. Chien, and B. A. Ogunnaike, "Identification and Control of a High-purity Distillation Column: A Case Study", *J. Process Control*, 5 (**3**), 149-162, 1995.

[90] G. Strang, *Introduction to Applied Mathematics*, Wellesley-Cambridge Press, Wellesley, MA, 1986.

[91] H.T. Su and T.J. McAvoy, "Integration of Multilayer Perceptron Networks and Linear Dynamic Models: A Hammerstein Modeling Approach," *Ind. Eng. Chem. Res.*, v. 32, pp. 1927 - 1936, 1993.

[92] T. Subba Rao and M.M. Gabr, *An Introduction to Bispectral Analysis and Bilinear Time Series Models*, Springer-Verlag, New York, 1984.

[93] M.L. Thompson, "Selection of Variables in Multiple Regression: Part I. A Review and Evaluation," *International Statistical Review*, vol. 46, pp. 1 - 19, 1978.

[94] M.L. Thompson, "Selection of Variables in Multiple Regression: Part II. Chosen Procedures Computations, and Examples," *International Statistical Review*, vol. 46, pp. 129 - 146, 1978.

[95] D.J. Thomson, "Spectrum Estimation and Harmonic Analysis," *Proc. IEEE*, vol. 70, pp. 1055 - 1096, 1982.

[96] H. Tong, *Non-linear Time Series*, Oxford University Press, New York, 1990.

[97] A. Uppal, W.H. Ray, and A.B. Poore, "On the Dynamic Behavior of Continuous Stirred Tank Reactors," *Chem. Eng. Sci.*, vol. 29, pp. 967 - 985, 1974.

[98] Weischedel, K., and T. J. McAvoy, "Feasibility of decoupling in conventionally controlled distillation columns," *Ind. Eng. Chem. Fund.*, 19, 379, 1980.

[99] T. Wigren, "Recursive Identification Based on the Nonlinear Wiener Model," doctoral thesis, Uppsala University, 1990.

[100] T. Wigren, "Recursive Prediction Error Identification Using the Nonlinear Wiener Model," *Automatica*, vol. 29, pp. 1011 - 1025, 1993.

[101] L.A. Zadeh, "Fuzzy Sets," *Information and Control*, vol. 8, pp. 338 - 353, 1965.

[102] X. Zhu and D.E. Seborg, "Nonlinear Predictive Control Based on Hammerstein Models," presented at PSE '94 Conference, Kyongju, Korea, June, 1994.

[103] Q.M. Zhu and S.A. Billings, "Identification of Polynomial and Rational NARMAX Models," *Proc. SYSID'94*, Copenhagen, pp. 295 - 300, 1994.

Chapter 3

Nonlinear Systems Theory

FRANCIS J. DOYLE III
School of Chemical Engineering
Purdue University
West Lafayette, IN 47907-1283

MICHAEL A. HENSON
Department of Chemical Engineering
Louisiana State University
Baton Rouge, LA 70803-7303

In this chapter, we introduce the machinery of differential geometry and related concepts as analysis tools for nonlinear process control systems. This chapter will also serve as background for the nonlinear controller synthesis material that is discussed in Chapter 5. There are several texts and review articles which pursue the details of the differential geometric approach in greater depth than is presented here. Notable are the texts by Isidori [18] and Nijmeijer and Van der Schaft [25]. In the process control area, there is a thorough treatment of the material in the tutorial article by Kravaris and Kantor [21]; in addition, there is a good overview of general nonlinear approaches to process control system design by Bequette [1].

3.1 Introduction

Throughout this chapter, the discussion is restricted to the class of nonlinear systems which are linear with respect to the manipulated input (control-affine systems); more general nonlinear systems are discussed, for example, in [25]. In the single-input single-output, time invariant case, the nonlinear state-space model can be written as follows:

$$\begin{aligned}
\dot{x}_1 &= f_1(x_1,\ldots,x_n) + g_1(x_1,\ldots,x_n)u \\
&\vdots \\
\dot{x}_n &= f_n(x_1,\ldots,x_n) + g_n(x_1,\ldots,x_n)u \\
y &= h(x_1,\ldots,x_n)
\end{aligned}$$

where $x_1,\ldots,x_n$ are state variables, u is the manipulated input, and y is the controlled output. For simplicity, the nonlinear functions $f_1,\ldots,f_n$, $g_1,\ldots,g_n$, and h are assumed to C^∞ functions; *i.e.*, their partial derivatives of any order exist and are continuous. This equation can be written in the following, more compact, vector form:

$$\begin{aligned}
\dot{x} &= f(x) + g(x)u \\
y &= h(x)
\end{aligned} \tag{3.1}$$

We will make frequent comparisons to the linear approximation of this dynamical system, for which we will employ the standard state-space notation [4]:

$$\begin{aligned}
\dot{x} &= Ax + bu \\
y &= cx
\end{aligned} \tag{3.2}$$

3.2 Concepts from Differential Geometry

The material presented in this chapter draws largely from the field of differential geometry. There has been an explosive growth of research in this area over the past decade, and in this chapter we will attempt to lay the groundwork to enable an understanding of those results. In effect, these tools can provide both a means for nonlinear system *analysis* as well as nonlinear controller *synthesis* (the latter is the subject of Chapter 5). In the analysis context, these tools can be used to address questions such as: (i) does the system have a stable inverse?; (ii) are particular states reachable from the initial conditions?; and (iii) are outputs which result from different inputs distinguishable? In the synthesis context, the primary issue will be the design of a nonlinear change of coordinates and nonlinear feedback that make the system behave in a linear manner.

3.2.1 Manifolds

A manifold is a topological space, usually denoted M, which has special properties [27] that are useful for the results that follow. Most notably, a manifold is locally Euclidean. Consider the mapping of a point p in some neighborhood U of M to a point $\phi(p)$ in some open subset of $\mathcal{R}^n$. The mapping ϕ and its inverse (ϕ^{-1}) are assumed to be C^∞ functions. We can define a *coordinate chart* as the pair (U, ϕ). It is often useful to represent ϕ as a set $(\phi_1, \ldots, \phi_n)$, where $\phi_i : U \to \mathcal{R}$ is called the ith *coordinate function*. The set of real numbers $(\phi_1(p), \ldots, \phi_n(p))$ is called the set of *local coordinates* of p in the coordinate chart (U, ϕ).

3.2.2 Vector Fields

A *vector map* associates a point $x = (x_1, \ldots, x_n)$ on an open subset of $\mathcal{R}^n$ with the vector:

$$f(x_1, \ldots, x_n) = \begin{pmatrix} f_1(x_1, \ldots, x_n) \\ f_2(x_1, \ldots, x_n) \\ \vdots \\ f_m(x_1, \ldots, x_n) \end{pmatrix} \tag{3.3}$$

in $\mathcal{R}^m$. A scalar map or *function* merely maps some open subset of $\mathcal{R}^n$ to $\mathcal{R}$. A *vector field*, $f(x)$, on $\mathcal{R}^n$ is the mapping which assigns to every point $p \in M$ a tangent vector $f(p)$ in the tangent space to M. A vector mapping, f, is a *global diffeomorphism* if [18]: (i) f is invertible for all $x \in \mathcal{R}^n$, and (ii) f and its inverse (f^{-1}) are C^∞ functions. If these properties are only locally valid in the neighborhood of an equilibrium point, then f is called a *local diffeomorphism*.

3.2.3 Inverse and Implicit Function Theorems

The following two theorems will prove useful in the derivations that follow. In particular, they will find application in guaranteeing the existence of solutions to nonlinear algebraic equations.

Theorem 1 (Implicit Function Theorem) *[18] Let $A \subset \mathcal{R}^m$ and $B \subset \mathcal{R}^n$ be open sets. Let $F : A \times B \to \mathcal{R}^n$ be a C^∞ mapping. Let $(x,y) = (x_1, \ldots, x_m, y_1, \ldots, y_n)$ denote a point of $A \times B$. Suppose that for some $(x_0, y_0) \in A \times B$*

$$F(x_0, y_0) = 0$$

and the matrix

$$\frac{\partial F}{\partial y} = \begin{pmatrix} \frac{\partial f_1}{\partial y_1} & \cdots & \frac{\partial f_1}{\partial y_n} \\ \vdots & \ddots & \vdots \\ \frac{\partial f_n}{\partial y_1} & \cdots & \frac{\partial f_n}{\partial y_n} \end{pmatrix}$$

is nonsingular at (x_0, y_0). Then there exists open neighborhoods $A_0 \subset A$ of x_0 and $B_0 \subset B$ of y_0 and a unique C^∞ mapping $G : A_0 \to B_0$ such that

$$F(x, G(x)) = 0$$

for all $x \in A_0$.

Thus, the implicit function theorem provides sufficient conditions that guarantee the existence of a local solution $y = G(x)$ to the nonlinear algebraic equation $F(x,y) = 0$.

Theorem 2 (Inverse Function Theorem) *[18] Let A be an open set of $\mathcal{R}^n$ and $F : A \to \mathcal{R}^n$ a C^∞ mapping. If $\left[\frac{\partial F}{\partial x}\right]_{x_0}$ is nonsingular at some $x_0 \in A$, then there exists an open neighborhood $U \subset A$ of x_0 such that $V = F(U)$ is open in $\mathcal{R}^n$ and the restriction of F to U is a diffeomorphism onto V.*

The inverse function theorem provides sufficient conditions which guarantee the existence of a locally-defined inverse function F^{-1} for the function F.

3.2.4 The Lie Algebra

The derivative of a smooth (*i.e.*, C^∞) function $\lambda(x)$ along a smooth vector field f is defined to be the *Lie Derivative.* The notation we use is $(L_f\lambda)(p) = (f(p))(\lambda)$; *i.e.*, the Lie Derivative is equal to the value of the tangent vector

$f(p)$ at the point p. In local coordinates, one can represent this operator in the following manner:

$$L_f\lambda(x_1,\ldots,x_n) = \begin{pmatrix} \frac{\partial\lambda}{\partial x_1} & \cdots & \frac{\partial\lambda}{\partial x_n} \end{pmatrix} \begin{pmatrix} f_1(x_1,\ldots,x_n) \\ \vdots \\ f_n(x_1,\ldots,x_n) \end{pmatrix}$$

The following notation is used for repeated Lie Derivatives: $L_f(L_f\lambda(x)) = L_f^2\lambda(x)$.

Consider the vector space $V(M)$ of all smooth vector fields on a manifold M. In addition, consider the following binary operation (product) on V : $[V,V] \equiv V \times V \to V$. Assume that the binary operator satisfies the following properties:

1. The operator is skew commutative: $[v_1, v_2] = -[v_2, v_1]$.
2. The operator is bilinear over $\mathcal{R}$: $[\alpha_1 v_1 + \alpha_2 v_2, v_3] = \alpha_1[v_1, v_3] + \alpha_2[v_2, v_3]$, where $\alpha_1, \alpha_2 \in \mathcal{R}$.
3. The operator satisfies the Jacobi Identity: $([v_1,[v_2,v_3]] + [v_2,[v_3,v_1]] + [v_3,[v_1,v_2]]) = 0$.

In this case, the vector space $V(M)$ is a *Lie Algebra* with some very interesting structural properties. This binary operator on $V(M)$ is defined as the *Lie Bracket.* For two vector fields $f(x), g(x) \in V(M)$:

$$[f,g](x) = \frac{\partial g}{\partial x} f(x) - \frac{\partial f}{\partial x} g(x) \tag{3.4}$$

In local coordinates, this is given by:

$$\begin{pmatrix} \frac{\partial g_1}{\partial x_1} & \cdots & \frac{\partial g_1}{\partial x_n} \\ \vdots & \ddots & \vdots \\ \frac{\partial g_n}{\partial x_1} & \cdots & \frac{\partial g_n}{\partial x_n} \end{pmatrix} \begin{pmatrix} f_1 \\ \vdots \\ f_n \end{pmatrix} - \begin{pmatrix} \frac{\partial f_1}{\partial x_1} & \cdots & \frac{\partial f_1}{\partial x_n} \\ \vdots & \ddots & \vdots \\ \frac{\partial f_n}{\partial x_1} & \cdots & \frac{\partial f_n}{\partial x_n} \end{pmatrix} \begin{pmatrix} g_1 \\ \vdots \\ g_n \end{pmatrix} \tag{3.5}$$

The notation that is typically employed for this operator is as follows:

$$\begin{aligned} ad_f^0 g(x) &\equiv g(x) \\ ad_f^1 g(x) &\equiv [f,g](x) \\ ad_f^2 g(x) &\equiv [f,[f,g]](x) = [f, ad_f^1 g](x) \\ &\vdots \end{aligned}$$

It is interesting to note the interpretation of the Lie Derivative and the Lie Bracket for a linear dynamical system. In this case, $f(x) = Ax$, $g(x) = b$, and $h(x) = cx$. The repeated Lie Derivative of the output function $h(x)$ along the $f(x)$ vector is given by $L_f^k h(x) = cA^k x$. Thus, one obtains the rows of the observability matrix of the linear system. If one applies the Lie Derivative to the output function along $g(x)$, preceded by repeated derivatives along $f(x)$, one obtains $L_g L_f^k h(x) = cA^{k-1}b$; in other words, the Markov parameters for the linear system. Finally, consider the Lie Bracket of the $f(x)$ and $g(x)$ vector fields: $ad_f^k g(x) = (-1)^k A^k b$. In this case, we obtain the columns of the controllability matrix of the linear system multiplied by $(-1)^k$.

3.2.5 Coordinate Transformations

A general objective in the synthesis of feedback linearizing controllers (see Chapter 5) is the derivation of coordinate transformations which convert the original nonlinear system into a system that is "simpler" in the sense that controller synthesis is more straightforward. For example, a linear system is "simpler" than a nonlinear system. Before presenting the main result for this problem (the Frobenius Theorem), a few background comments on nonlinear coordinate transformations will be given.

Using the inverse function theorem in Section 3.2.3, one can show that a nonlinear coordinate transformation, $z = \Phi(x)$ from $\mathcal{R}^n$ to $\mathcal{R}^n$, is invertible if and only if the Jacobian matrix:

$$\begin{pmatrix} \frac{\partial \Phi_1}{\partial x_1}(x) & \cdots & \frac{\partial \Phi_1}{\partial x_n}(x) \\ \vdots & \ddots & \vdots \\ \frac{\partial \Phi_n}{\partial x_1}(x) & \cdots & \frac{\partial \Phi_n}{\partial x_n}(x) \end{pmatrix} \tag{3.6}$$

is invertible. Under this invertible coordinate transformation, a scalar function $h(x)$ is mapped to a new function:

$$\tilde{h}(z) = [h(x)]_{x=\Phi^{-1}(z)} \tag{3.7}$$

and a vector field $g(x)$ is mapped to the new vector field:

$$\begin{aligned} \tilde{g}_1(z) &= [< d\Phi_1(x), g(x) >]_{x=\Phi^{-1}(z)} \\ &\vdots \\ \tilde{g}_n(z) &= [< d\Phi_n(x), g(x) >]_{x=\Phi^{-1}(z)} \end{aligned} \tag{3.8}$$

where $< \cdot, \cdot >$ denotes the inner product.

For nonlinear systems, the following interpretation of the Lie Bracket is useful. Recall that the integral curve σ of f is the solution to the equation: $\frac{d\sigma}{dt} = f(\sigma(t))$. The Lie Bracket "measures" the extent to which the integral curves of f and g can be used to form the coordinates of a system. For example [27], consider the integral curve of f starting from a point p for some time δt. From the new location, follow the integral curve of g for time δt. From the new location, reverse the integral curve of f, and follow this for time δt. Finally, follow the reverse of the integral curve of g for time δt. If we denote the integral curve of f by ϕ and the integral curve of g by ψ, then the final location of the curve described above is denoted by $c(t) = \psi_{-\delta t} \circ \phi_{-\delta t} \circ \psi_{\delta t} \circ \phi_{\delta t}(p)$. The Lie Bracket provides insight into the ability to close this curve under the operations on the integral curves of f and g, thus validating their usefulness as coordinate lines for the system. One can show [27] that $\ddot{c}(0)$ is equal to $2[f, g]$ evaluated at p. Furthermore, if $[f, g] = 0$ in a neighborhood of p, then $c(t) = p \; \forall \; t$, *i.e.*, the curve is closed [27].

3.2.6 Distributions and Frobenius Theorem

Before introducing the main result of this section, a few definitions are required. First, the concept of a distribution is formalized:

Definition 1 (Distribution) *A distribution D is the vector space which consists of the span of some vectors f on some open set U of $\mathcal{R}^n$:*

$$D(x) \quad = \quad \text{span } \{f_1(x), \ldots, f_d(x)\}$$

A distribution D is non-singular if there exists an integer d such that $dim(D(X)) = d$ for all $x \in \mathcal{R}^n$

An involutive distibution is defined as follows:

Definition 2 (Involutivity) *A distribution D is involutive if for all vector fields $\tau_1, \tau_2 \in D$ the following holds: $[\tau_1, \tau_2] \in D$.*

A related property used later in this chapter is the notion of invariance for a distribution:

Definition 3 (Invariant Distribution) *A distribution D is invariant under a vector field f if for all vector fields $\tau \in D$ the following holds: $[\tau, f] \in D$.*

The Frobenius Theorem is concerned with solvability of a particular set of partial differential equations that arise in the derivation of coordinate transformations that "simplify" or "straighten out" a given nonlinear system. To see this, consider the transformation of the vector fields given in the previous section in (3.8). In order for the transformed vector field $\tilde{g}$ to align with e_n, the nth coordinate of the natural basis for linear coordinate systems, one requires that the first $n-1$ elements of $\tilde{g}$ vanish. In other words:

$$\frac{\partial \Phi_i(x)}{\partial x} g(x) = 0 \quad \text{for } i = 1, \ldots, n-1 \tag{3.9}$$

Thus one seeks $n-1$ independent solutions to this set of partial differential equation. When this is possible, one says that the distribution spanned by the single vector field, $\{g(x)\}$, is *completely integrable.* In a more general setting (the state feedback linearization problem), one is interested in the guaranteeing the *integrabilty* of a general distribution, $\{g_1(x), g_2(x), \ldots, g_d(x)\}$. For that problem, one seeks $n-d$ independent solutions of the following set of partial differential equations:

$$\frac{\partial \Phi_i(x)}{\partial x} [g_1(x), g_2(x), \ldots, g_d(x)] = 0 \quad \text{for } i = 1, \ldots, n-d \tag{3.10}$$

The key result which links this problem to the coordinate transformation construction results in the last section is given by the Frobenius Theorem [18]:

Theorem 3 (Frobenius Theorem) *A nonsingular distribution is completely integrable if and only if it is involutive.*

The proof of this theorem is rather involved, and the interested reader is referred to [18, 27] for the details.

3.2.7 Example – Continuous Fermentor

Many continuous fermentators can be described by the following unstructured model [14]:

$$\dot{X} = -DX + \mu X \tag{3.11}$$

$$\dot{S} = D(S_F - S) - \frac{1}{Y_{X/S}} \mu X \tag{3.12}$$

$$\dot{P} = -DP + (\alpha\mu + \beta)X \tag{3.13}$$

where X is the biomass concentration, S is the substrate concentration, P is the product concentration, D is the dilution rate, S_F is the feed substrate concentration, μ is the specific growth rate, $Y_{X/S}$ is the cell-mass yield, and α and β are kinetic parameters. The specific growth rate is modeled as:

$$\mu = \frac{\mu_m(1 - \frac{P}{P_m})S}{K_m + S + \frac{S^2}{K_i}} \tag{3.14}$$

where μ_m is the maximum specific growth rate, P_m is the product saturation constant, K_m is the substrate saturation constant, and K_i is the substrate inhibition constant.

Consider the problem of controlling the productivity ($y = DP$) using the feed substrate concentration as the manipulated variable ($u = S_f$). In order to check the involutivity condition for this system, we will define the state vector $x^T = [X \quad S \quad P]$. Thus, the vector fields of interest are:

$$f(x) = \begin{pmatrix} -Dx_1 + \mu(x)x_1 \\ -Dx_2 - \frac{1}{Y_{X/S}}\mu(x)x_1 \\ -Dx_3 + [\alpha\mu(x) + \beta]x_1 \end{pmatrix} \tag{3.15}$$

$$g(x) = \begin{pmatrix} 0 \\ D \\ 0 \end{pmatrix} \tag{3.16}$$

where:

$$\mu = \frac{\mu_m(1 - \frac{x_3}{P_m})x_2}{K_m + x_2 + \frac{x_2^2}{K_i}} \tag{3.17}$$

Furthermore, we can check that:

$$ad_f^0 g(x) = g(x) = \begin{pmatrix} 0 \\ D \\ 0 \end{pmatrix} \tag{3.18}$$

$$ad_f^1 g(x) = \frac{\partial g}{\partial x} f(x) - \frac{\partial f}{\partial x} g(x) = \begin{pmatrix} \frac{\partial \mu(x)}{\partial x_2} x_1 \\ -D - \frac{1}{Y_{X/S}} \frac{\partial u(x)}{\partial x_2} x_1 \\ \alpha \frac{\partial \mu(x)}{\partial x_2} \end{pmatrix} \tag{3.19}$$

To demonstrate involutivity, we need to show that $[ad_f^0 g(x), ad_f^1 g(x)]$ lies in the distribution spanned by $\left\{ad_f^0 g(x), ad_f^1 g(x)\right\}$:

$$[ad_f^0 g(x), ad_f^1 g(x)] \quad = \quad \begin{pmatrix} \frac{\partial^2 \mu(x)}{\partial x_2^2} x_1 \\ -\frac{1}{Y_{X/S}} \frac{\partial^2 \mu(x)}{\partial x_2^2} x_1 \\ \alpha \frac{\partial^2 \mu(x)}{\partial x_2^2} \end{pmatrix} \tag{3.20}$$

Through straightforward, though tedious, algebra one can show that the involutivity condition holds for $\forall\ x \in M$ where:

$$M \equiv \left\{x : x_2 \neq \sqrt{K_m K_i}\right\} \tag{3.21}$$

It is interesting to note that the point $x_2 = \sqrt{K_m K_i}$ corresponds to the optimum substrate concentration (*i.e.*, the substrate concentration corresponding to the maximum productivity).

3.3 Nonlinear Inversion

3.3.1 Linear Systems

Initially, we turn to linear systems analysis to develop concepts that will be useful later. Consider the state-space representation of the single-input single-output, linear system in (3.2). An equivalent input-output representation for this system is the following transfer function:

$$\frac{y(s)}{u(s)} \quad = \quad c(sI - A)^{-1} b = \frac{c\text{Adj}(sI - A)^{-1} b}{\det(sI - A)} \tag{3.22}$$

The *pole-zero excess* for this operator is defined to be the order of the denominator polynomial minus the order of the numerator polynomial. This quantity is also called the *relative degree* and can range from 1 to n. The relative degree also can be calculated from the Markov parameters, which represent the coefficients of the terms in a series expansion of the transfer function:

$$c(sI - A)^{-1} b \quad = \quad \frac{cb}{s} + \frac{cAb}{s^2} + \frac{cA^2 b}{s^3} + \cdots$$

Definition 4 (Relative Degree) *The relative degree of the system (3.2) is the smallest integer r for which $cA^{r-1} b \neq 0$.*

The relative degree also can be interpreted in the context of successive derivatives of the linear system output (y):

$$\begin{aligned}\dot{y} &= cAx \\ \ddot{y} &= cA^2x \\ &\vdots \\ \frac{d^{r-1}y}{dt^{r-1}} &= cA^{r-1}x \\ \frac{d^r y}{dt^r} &= cA^r x + cA^{r-1}bu\end{aligned}$$

From this analysis, it is clear that the relative degree corresponds to the lowest order derivative of the output that depends explicitly on the input. This equivalent definition will be utilized in defining the relative degree for a nonlinear system of the form in (3.1).

Next, we define a *normal form* representation for the linear system given in (3.2). A linear state transformation is sought which represents the dynamics in an insightful and compact form. Consider the following transfer function:

$$G(s) = k\frac{\beta_0 + \beta_1 s + \cdots + \beta_{n-r-1}s^{n-r-1} + s^{n-r}}{\alpha_0 + \alpha_1 s + \cdots + \alpha_{n-1}s^{n-1} + s^n} \tag{3.23}$$

A minimal state-space realization of this system is represented by the following matrices:

$$A = \begin{pmatrix} 0 & 1 & 0 & \cdots & 0 \\ 0 & 0 & 1 & \cdots & 0 \\ \vdots & \vdots & \vdots & \ddots & \vdots \\ 0 & 0 & 0 & \cdots & 1 \\ -\alpha_0 & -\alpha_1 & -\alpha_2 & \cdots & -\alpha_{n-1} \end{pmatrix} \quad b = \begin{pmatrix} 0 \\ 0 \\ \vdots \\ 0 \\ k \end{pmatrix}$$

$$c = (\,\beta_0 \quad \beta_1 \quad \cdots \quad \beta_{n-r-1} \quad 1 \quad 0 \quad \cdots 0\,) \tag{3.24}$$

To obtain a suitable normal form, the following transformation is proposed:

$$\begin{pmatrix} \xi_1 \\ \xi_2 \\ \xi_3 \\ \vdots \\ \xi_r \\ \eta_1 \\ \vdots \\ \eta_{n-r} \end{pmatrix} = Tx = \begin{pmatrix} cx \\ cAx \\ cA^2x \\ \vdots \\ cA^{r-1}x \\ x_1 \\ \vdots \\ x_{n-r} \end{pmatrix} \tag{3.25}$$

It is straightforward to check that this coordinate transformation is nonsingular, and leads to a transformed system of the following form [18]:

$$\begin{aligned} \dot{\xi}_1 &= \xi_2 \\ \dot{\xi}_2 &= \xi_3 \\ &\vdots \\ \dot{\xi}_{r-1} &= \xi_r \\ \dot{\xi}_r &= R\xi + S\eta + ku \\ \dot{\eta} &= P\xi + Q\eta \\ y &= \xi_1 \end{aligned} \tag{3.26}$$

where R and S are row vectors and P and Q are matrices. In this structure, the rth order subsystem representing the direct effect of the manipulated variable, u, on the output, y, is hierarchically decomposed. This structure will prove useful for nonlinear systems, and in that context is referred to as the Byrnes-Isidori Normal Form. The matrix Q has special significance: its eigenvalues are exactly equal to the zeros of the transfer function given in Equation (3.23).

It is relatively straightforward to construct an inverse for the linear system (3.2), such that the cascade of the system and the inverse system form the identity operator. Using the original state-space representation in (3.2), an inverse can be realized as:

$$\begin{aligned} \dot{Z} &= \left(A - \frac{bcA^r}{cA^{r-1}b}\right) Z + \frac{b}{cA^{r-1}b}\frac{d^r y}{dt^r} \\ u &= -\frac{cA^r}{cA^{r-1}b} Z + \frac{1}{cA^{r-1}b}\frac{d^r y}{dt^r} \end{aligned} \tag{3.27}$$

where Z is the state vector of the inverse system. Note that the inverse system takes the rth derivative of y as an input, and generates the value of u that originally was introduced to the plant. From the transfer function operator for the system, it is clear that the inverse system can have a minimal order of r. However, the previously derived operator has a dynamic order of n, and therefore it is not a minimal realization.

Now consider the derivation of an inverse system using the normal form representation in (3.26). In this case, it is easy to show that an inverse can be realized as:

$$
\begin{aligned}
\dot{z}_1 &= z_2 \\
&\vdots \\
\dot{z}_{r-1} &= z_r \\
\dot{z}_r &= \frac{d^r y}{dt^r} \\
\begin{pmatrix} \dot{z}_{r+1} \\ \vdots \\ \dot{z}_n \end{pmatrix} &= P \begin{pmatrix} z_1 \\ \vdots \\ z_r \end{pmatrix} + Q \begin{pmatrix} z_{r+1} \\ \vdots \\ z_n \end{pmatrix} \\
u &= \frac{1}{k} \left(\frac{d^r y}{dt^r} - R \begin{pmatrix} z_1 \\ \vdots \\ z_r \end{pmatrix} - S \begin{pmatrix} z_{r+1} \\ \vdots \\ z_n \end{pmatrix} \right)
\end{aligned} \tag{3.28}
$$

However, this system can be represented in a more compact form as:

$$
\begin{aligned}
\begin{pmatrix} \dot{\zeta}_1 \\ \vdots \\ \dot{\zeta}_{n-r} \end{pmatrix} &= P \begin{pmatrix} y \\ \vdots \\ \frac{d^{r-1} y}{dt^{r-1}} \end{pmatrix} + Q \begin{pmatrix} \zeta_1 \\ \vdots \\ \zeta_{n-r} \end{pmatrix} \\
u &= \frac{1}{k} \left(\frac{d^r y}{dt^r} - R \begin{pmatrix} y \\ \vdots \\ \frac{d^{r-1} y}{dt^{r-1}} \end{pmatrix} - S \begin{pmatrix} \zeta_1 \\ \vdots \\ \zeta_{n-r} \end{pmatrix} \right)
\end{aligned} \tag{3.29}
$$

where we have introduced the following change of variables: $[\zeta_1, \ldots, \zeta_{n-r}] = [z_{r+1}, \ldots, z_n]$. Note that the derived inverse system has order r, and thus it represents a *minimal* realization.

3.3.2 Nonlinear Normal Form

The notion of relative degree, introduced earlier for linear systems, can be extended in a natural way to include nonlinear systems [16]:

Definition 5 (Relative Degree) *The relative degree of the nonlinear system (3.1) is the smallest integer r for which $L_g L_f^{r-1} h(x) \neq 0$ and $L_g L_f^{r-2} h(x) = 0 \ \forall \ x$ in some neighborhood of the defined operating point x_0.*

A useful interpretation of the relative degree for nonlinear systems can be obtained by calculating derivatives of the output:

$$
\begin{aligned}
\dot{y} &= \frac{dh}{dt} = \frac{\partial h}{\partial x}\frac{dx}{dt} = L_f h(x) \\
\ddot{y} &= L_f^2 h(x) \\
&\vdots \\
\frac{d^{r-1} y}{dt^{r-1}} &= L_f^{r-1} h(x) \\
\frac{d^r y}{dt^r} &= L_f^r h(x) + L_g L_f^{r-1} h(x) u
\end{aligned}
$$

As with linear systems, the relative degree characterizes the lowest order derivative of the output, y, that is explicitly dependent on the input, u. It is important to note that there may exist points, x_0, at which the relative degree is not well defined. For example, consider a one-dimensional system for which $L_g L_f^{r-1} h(x) = x$. In this case, the relative degree is not well defined as the point $x_0 = 0$. The presence of such *singular points* complicates the subsequent analysis considerably [17, 18], and therefore is not considered here.

Using the definition of the relative degree, one can derive a nonlinear analog of the normal form in (3.26). This system, called the *Byrnes-Isidori* normal form [3], is derived using the following nonlinear coordinate transformation:

$$\begin{pmatrix} \xi \\ \eta \end{pmatrix} = \Phi(x) = \begin{pmatrix} h(x) \\ L_f h(x) \\ \vdots \\ L_f^{r-1} h(x) \\ t_1(x) \\ \vdots \\ t_{n-r}(x) \end{pmatrix} \tag{3.30}$$

The $t_i(x)$ functions are obtained as the solution to the partial differential equation:

$$L_g t(x) = 0$$

As long as the relative degree is well defined, the existence of $t_i(x)$ which satisfy this equation is guaranteed [18]. This coordinate transformation produces the Byrnes-Isidori normal form:

$$\begin{aligned} \dot{\xi}_1 &= \xi_2 \\ \dot{\xi}_2 &= \xi_3 \\ &\vdots \\ \dot{\xi}_{r-1} &= \xi_r \\ \dot{\xi}_r &= \alpha(\xi, \eta) + \beta(\xi, \eta) u \\ \dot{\eta}_1 &= q_1(\xi, \eta) \\ &\vdots \\ \dot{\eta}_{n-r} &= q_{n-r}(\xi, \eta) \\ \\ y &= \xi_1 \end{aligned} \tag{3.31}$$

where $\alpha = \left[L_f^r h(x)\right]_{x=\Phi^{-1}(\xi,\eta)}$, $\beta = \left[L_g L_f^{r-1} h(x)\right]_{x=\Phi^{-1}(\xi,\eta)}$, and $q_i = \left[L_f \Phi_i(x)\right]_{x=\Phi^{-1}(\xi,\eta)}$.

3.3.3 Hirschorn Inverse

For nonlinear systems, the order of the cascaded system consisting of the plant operator and an appropriately defined inverse is important, and it is necessary to make a distinction between the two possible cases [16]:

- The *left inverse* reconstructs the input from the plant output, its derivatives, and the state variables of the inverse (Figure 3.1).
- The *right inverse* produces the input history required to obtain a particular output (*i.e.*, an ideal feedforward controller) using the plant output and the state variables of the inverse (Figure 3.2).

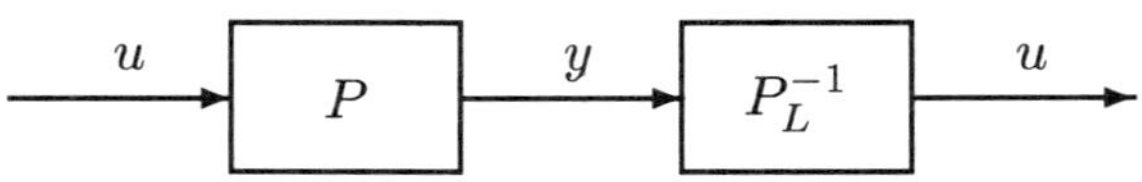

Figure 3.1: Left inverse of nonlinear operator

Following the derivation of the linear inverse, a nonminimal left inverse for the nonlinear system (3.1) can be written as:

$$\begin{aligned} \dot{Z} &= f(Z) + g(Z)\frac{\frac{d^r y}{dt^r} - L_f^r h(Z)}{L_g L_f^{r-1} h(x)} \\ u &= \frac{\frac{d^r y}{dt^r} - L_f^r h(Z)}{L_g L_f^{r-1} h(x)} \end{aligned} \tag{3.32}$$

In analogy to the linear case, a minimal realization can be achieved by exploiting the Byrnes-Isidori normal form:

$$\dot{\zeta}_1 = q_1\left(y, \dot{y}, \ldots, \frac{d^{r-1}y}{dt^{r-1}}, \zeta_1, \ldots, \zeta_{n-r}\right)$$

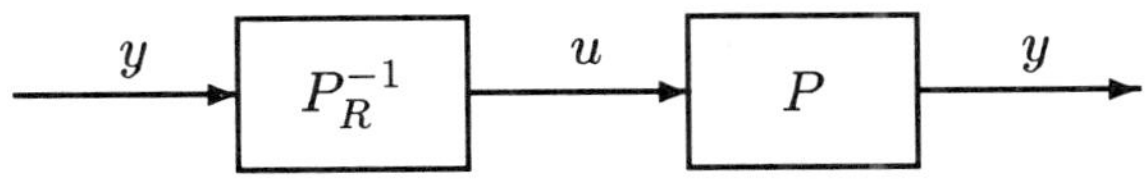

Figure 3.2: Right inverse of nonlinear operator

$$
\begin{aligned}
&\vdots \\
\dot{\zeta}_{n-r} &= q_{n-r}\left(y, \dot{y}, \ldots, \frac{d^{r-1}y}{dt^{r-1}}, \zeta_1, \ldots, \zeta_{n-r}\right) \\
u &= \frac{\frac{d^r y}{dt^r} - \alpha\left(y, \dot{y}, \ldots, \frac{d^{r-1}y}{dt^{r-1}}, \zeta_1, \ldots, \zeta_{n-r}\right)}{\beta\left(y, \dot{y}, \ldots, \frac{d^{r-1}y}{dt^{r-1}}, \zeta_1, \ldots, \zeta_{n-r}\right)}
\end{aligned} \tag{3.33}
$$

where the following change of variables has been introduced: $\zeta = \eta$. Note that the inverse operator is driven by the output y, and its first r derivatives and reconstructs the manipulated input u that produced y.

3.3.4 Example – Exothermic Chemical Reactor

Consider the problem of regulating the temperature in a nonisothermal CSTR in which an exothermic, irreversible, first-order reaction takes place. The mass and energy balances are given, in dimensionless form, as [30]:

$$
\begin{aligned}
\dot{x_1} &= -x_1 + \mathcal{D}a(1 - x_1)e^{\frac{x_2}{1+x_2/\gamma}} \\
\dot{x_2} &= -x_2 + B\mathcal{D}a(1 - x_1)e^{\frac{x_2}{1+x_2/\gamma}} + \beta(u - x_2) \\
y &= x_2
\end{aligned} \tag{3.34}
$$

where x_1 is the reactant concentration, x_2 is the reactor temperature, u is the cooling jacket temperature, and $\mathcal{D}a, B, \gamma$, and β are dimensionless parameters. Consider the problem of deriving the Hirschorn inverse for this system. The relative degree $r = 1$ since $L_g h(x) = \beta \neq 0$. Thus, the Hirschorn left inverse (equation 3.32) can be realized by the following nonlinear dynamic system:

$$
\begin{aligned}
\dot{z}_1 &= -z_1 + \mathcal{D}a(1 - z_1)e^{\frac{z_2}{1+z_2/\gamma}} \\
\dot{z}_2 &= \frac{dy}{dt} \\
u &= \frac{\frac{dy}{dt} + z_2 - B\mathcal{D}a(1 - z_1)e^{\frac{z_2}{1+z_2/\gamma}} + \beta z_2}{\beta}
\end{aligned} \tag{3.35}
$$

It is straightforward to verify that if the inverse system in equation 3.35 is driven by the output of the CSTR in equation 3.34, then the original input, u, from the CSTR is reconstructed.

3.3.5 Zero Dynamics

In the previous section, the concept of a nonlinear system inverse was introduced. It is apparent that there is no direct method for quantifying the behavior of the inverse dynamics. In other words, there are no simple quantities, such as the transmission zeros of a linear system, which characterize the stability of the inverse. Instead, one must analyze the dynamical system (3.33). Another complication which arises is that a nonlinear extension of transmission "zeros" yields multiple interpretations in the multivariable case [19]:

- The dynamics corresponding to the maximally unobservable system state variables.
- The invariant dynamics under which the system evolves when the output is constrained to be zero for all times.
- The dynamics corresponding a minimal realization of the process inverse.

Note that the three dynamic systems are equivalent for a linear system and a single-input single output, nonlinear system.

A question that arises naturally in the control context is whether the inverse of a dynamical system is stable. For example, in IMC design approach one utilizes the inverse of the process model directly in the controller synthesis [24]. In the linear context, this question is readily resolved through an analysis of the system zeros: systems with all zeros lying in the left-half plane are called *minimum-phase*, while systems with some zeros lying in the right-half plane are called *nonminimum-phase*. Referring to the system inverse derived in (3.33), it is clear that the nonlinear analog of the minimum-phase property is related to the stability of the dynamic system:

$$\begin{aligned}
\dot{\zeta}_1 &= q_1\left(y, \dot{y}, \ldots, \frac{d^{r-1}y}{dt^{r-1}}, \zeta_1, \ldots, \zeta_{n-r}\right) \\
&\vdots \\
\dot{\zeta}_{n-r} &= q_{n-r}\left(y, \dot{y}, \ldots, \frac{d^{r-1}y}{dt^{r-1}}, \zeta_1, \ldots, \zeta_{n-r}\right)
\end{aligned} \tag{3.36}$$

One approach to determine the stability of the inverse system is to analyze the unforced inverse dynamics, which are usually called the *zero dynamics*:

$$\begin{aligned}
\dot{\zeta}_1 &= q_1\left(0, 0, \ldots, 0, \zeta_1, \ldots, \zeta_{n-r}\right) \\
&\vdots \\
\dot{\zeta}_{n-r} &= q_{n-r}\left(0, 0, \ldots, 0, \zeta_1, \ldots, \zeta_{n-r}\right)
\end{aligned} \tag{3.37}$$

For example, in the case where the system output is constrained to a constant value (setpoint) — which can be assumed to be zero without loss of generality — the stability of the closed-loop system in which the inverse is employed as the controller is completely determined by the stability of these internal dynamics. In the case where the system output must follow a trajectory, the inverse dynamics are driven by the system output and its first $r-1$ derivatives. In this case, the stability of the forced zero dynamics (3.36) must be evaluated to determine the internal stability.

In the general case, one requires stronger conditions than stability of the zero dynamics to guarantee internal stability in order to avoid, for example, the peaking phenomena described in [29]. However, this is only a limitation on global or semi-global stabilization, and will not preclude local stabilization which can be guaranteed from an analysis of the local properties of the zero dynamics.

3.3.6 Example – Cyclopentenol Synthesis

Consider the synthesis of cyclopentenol from cyclopentadien [10], which follows the same reaction sequence as the van de Vusse reaction [31]: there is a primary reaction, a further reaction of the desired product, and a side reaction of the initial reactant. The reaction scheme looks like:

$$A \rightarrow B \rightarrow C$$

$$2A \rightarrow D$$

In this case, A is cyclopentadien, B is cyclopentenol, C is cyclopentandiol, and D is Dicyclopentadien. Assuming constant volume isothermal conditions, and mass action kinetics, the mass balances for A and B can be written as:

$$\begin{aligned}\dot{c}_A &= \frac{q}{V_R}(c_{A0} - c_A) - k_1 c_A - k_3 c_A^2 \\ \dot{c}_B &= -\frac{q}{V_R} c_B + k_1 c_A - k_2 c_B\end{aligned}$$

where c_A and c_B are the concentrations of A and B, respectively, c_{A0} is the inlet concentration of A, q is the inlet flow rate, V_R is the reactor volume, and k_1, k_2, and k_3 are reaction rate constants.

We are interested in determining the stability of the zero dynamics when the manipulated variable is the dilution rate ($u = \frac{q}{V_R}$), and the desired output is the concentration of B ($y = c_B$) [20]. The nonlinear state-space equations become:

$$\begin{aligned}\dot{x}_1 &= -k_1 x_1 - k_3 x_1^2 + u(x_{10} - x_1) \\ \dot{x}_2 &= k_1 x_1 - k_2 x_2 + u(-x_2) \\ y &= x_2\end{aligned} \tag{3.38}$$

where $x_1 = c_A$, $x_2 = c_B$, and $x_{10} = c_{A0}$. The system has relative degree $r = 1$ since $L_g h(x) = -x_2 \neq 0$. Hence, a minimal realization of the process inverse is a first-order nonlinear system. The change of coordinates (equation 3.30):

$$\begin{pmatrix}\xi \\ \eta\end{pmatrix} = \begin{pmatrix}h(x) \\ t_1(x)\end{pmatrix} = \begin{pmatrix}x_2 \\ \frac{x_{10}-x_1}{x_2}\end{pmatrix} \tag{3.39}$$

and its inverse:

$$\begin{pmatrix}x_1 \\ x_2\end{pmatrix} = \begin{pmatrix}x_{10} - \xi\eta \\ \xi\end{pmatrix} \tag{3.40}$$

can be used to realize the Byrnes-Isidori normal form:

$$\begin{aligned}\dot{\xi} &= \alpha(\xi,\eta) + \beta(\xi,\eta)u \\ \dot{\eta} &= q(\xi,\eta) \\ y &= \xi\end{aligned} \tag{3.41}$$

where:

$$\begin{aligned}\alpha(\xi,\eta) &= k_1(x_{10} - \eta\xi) - k_2\xi \\ \beta(\xi,\eta) &= -\xi \\ q(\xi,\eta) &= \frac{k_1(1-\eta)(x_{10} - \eta\xi) + k_2\xi\eta + k_3(x_{10} - \xi\eta)^2}{\xi}\end{aligned}$$

Thus, the zero dynamics are:

$$\dot{\eta} = q(\xi_0,\eta) = \frac{k_1(1-\eta)(x_{10} - \eta\xi_0) + k_2\xi_0\eta + k_3(x_{10} - \xi_0\eta)^2}{\xi_0} \tag{3.42}$$

where ξ_0 is the (nonzero) equilibrium value of the output. If we convert the zero dynamics back to the original coordinates and analyze the Jacobian linearization, we find that the sign of the quantity:

$$\kappa = k_2 - k_1 - 2k_3x_1 + k_1\frac{x_{10} - 2x_1}{x_2}$$

will determine whether the nonlinear system is locally minimum phase ($\kappa < 0$) or nonminimum phase ($\kappa > 0$). In fact, at the point where $\kappa = 0$, a singularity occurs and the system relative degree is not well defined. It is interesting to note that the locus of points where $\kappa = 0$ cross the conversion locus at the point of maximum conversion.

3.3.7 Example – Isothermal CSTR

This example involves the following multicomponent isothermal kinetic sequence carried out in a CSTR [5]:

$$A \rightleftharpoons B \rightarrow C$$

The desired product is component C, and the manipulated input is the feed flowrate of component B. The dimensionless mass balances for A, B, and C

are given below:

$$\begin{aligned}
\dot{x_1} &= 1 - x_1 - Da_1 x_1 + Da_2 x_2^2 \\
\dot{x_2} &= -x_2 + Da_1 x_1 - Da_2 x_2^2 - Da_3 x_2^2 + u \\
\dot{x_3} &= -x_3 + Da_3 x_2^2 \\
y &= x_3
\end{aligned} \tag{3.43}$$

where the Da_i terms are the respective Damköhler terms for the reactions.

The system has relative degree $r = 2$ since $L_g h(x) = 0$ and $L_g L_f h(x) = 2Da_3 x_2 \neq 0$. Hence, a minimal realization of the process inverse is a second-order nonlinear system. The change of coordinates (equation 3.30):

$$\begin{pmatrix} \xi_1 \\ \xi_2 \\ \eta \end{pmatrix} = \begin{pmatrix} x_3 - x_{3,ss} \\ -x_3 + Da_3 x_2^2 \\ x_1 - x_{1,ss} \end{pmatrix} \tag{3.44}$$

and its inverse:

$$\begin{pmatrix} x_1 \\ x_2 \\ x_3 \end{pmatrix} = \begin{pmatrix} \eta + x_{1,ss} \\ \sqrt{\frac{\xi_2 + \xi_1 + x_{3,ss}}{Da_3}} \\ \xi_1 + x_{3,ss} \end{pmatrix} \tag{3.45}$$

can be used to realize the Byrnes-Isidori normal form. The objective in this example is a derivation of the zero dynamics for this reactor. They can be constructed from the dynamics for η:

$$\dot{\eta} = -(1 + Da_1)\eta + \frac{Da_2}{Da_3}(\xi_1 + \xi_2) \tag{3.46}$$

Thus, the unforced zero dynamics are:

$$\dot{\eta} = -(1 + Da_1)\eta \tag{3.47}$$

Therefore, the unforced zero dynamics are globally stable, and this CSTR is globally minimum phase.

3.4 Controllability and Observability

3.4.1 Linear Systems

We briefly discuss the controllability and observability of the single-input, single-output linear system (3.2). A more complete presentation of these concepts is available elsewhere [4]. A dynamic system is said to be *controllable* if there exists an input $u(t)$ such that any initial state $x(t_0)$ can be driven to any other state $x(t_1)$ in finite time $t_1 > t_0$. A necessary and sufficient condition for the linear system (3.2) to be controllable is that the n x n controllability matrix:

$$W_c = \begin{bmatrix} b & Ab & \cdots & A^{n-1}b \end{bmatrix} \tag{3.48}$$

has rank n. The system is said to be *uncontrollable* if the rank$(W_c) < n$.

A dynamic system is said to be *observable* if there exists a time t_1 such that any initial state $x(t_0)$ can be distinguished from any other state x_0 using the input $u(t)$ and output $y(t)$ over the time interval $t_0 \leq t \leq t_1$. A necessary and sufficient condition for the linear system (3.2) to be observable is that the n x n observability matrix:

$$W_o = \begin{bmatrix} c \\ cA \\ \vdots \\ cA^{n-1} \end{bmatrix} \tag{3.49}$$

has rank n. If the rank$(W_c) < n$, the system is said to be *unobservable.*

The *Kalman decomposition* allows a linear system to be decomposed into controllable/uncontrollable and observable/unobservable parts. More specifically, there exists a linear change of coordinates $z = Tx$ such that (3.2) is transformed into:

$$\begin{aligned} \begin{bmatrix} \dot{z}_1 \\ \dot{z}_2 \\ \dot{z}_3 \\ \dot{z}_4 \end{bmatrix} &= \begin{bmatrix} A_{11} & 0 & A_{13} & 0 \\ A_{21} & A_{22} & A_{23} & A_{24} \\ 0 & 0 & A_{33} & 0 \\ 0 & 0 & A_{43} & A_{44} \end{bmatrix} \begin{bmatrix} z_1 \\ z_2 \\ z_3 \\ z_4 \end{bmatrix} + \begin{bmatrix} b_1 \\ b_2 \\ 0 \\ 0 \end{bmatrix} u \\ y &= \begin{bmatrix} c_1 & 0 & c_3 & 0 \end{bmatrix} \begin{bmatrix} z_1 \\ z_2 \\ z_3 \\ z_4 \end{bmatrix} \end{aligned} \tag{3.50}$$

It follows directly from (3.50) that: (1) z_1 represents the controllable and observable modes; (2) z_2 represents the controllable and unobservable modes;

(3) z_3 represents the uncontrollable and observable modes; and (4) z_4 represents the uncontrollable and unobservable modes.

3.4.2 Local Controllability

We now consider the controllability of the single-input, single-output nonlinear system (3.1). For simplicity, the subsequent analysis is *local*; *i.e.* the results are valid only in a neighborhood of the operating point. Global results are available elsewhere [15, 18, 25]. The nonlinear system (3.1) is said to be *weakly controllable* at x_0 if there exists an input $u(t)$ such that any initial state $x(t_0)$ in a neighborhood X_0 of x_0 can be driven to any other state $x(t_1) \in X_0$ in finite time $t_1 > t_0$ [15, 28]. It is important to note that this definition does not ensure that the state trajectory $x(t)$ will remain near x_0. For this reason, a local version of weak controllability is introduced. The nonlinear system (3.1) is said to be *locally weakly controllable* at x_0 if it is weakly controllable and there exists a neighborhood $X_1 \subset X_0$ such that $x(t) \in X_1$ for $t_0 \leq t \leq t_1$ [15, 28].

An important advantage of local controllability is that it can be determined by examining the rank of a matrix which is analogous to the linear controllability matrix. In particular, the nonlinear system (3.1) is locally weakly controllable at x_0 if the n x n controllability matrix [15, 28],

$$W_c(x) = \left[\begin{array}{cccc} g(x) & ad_f g(x) & \cdots & ad_f^{n-1} g(x) \end{array} \right] \tag{3.51}$$

has rank n at x_0. In this case, the system is said to satisfy the *controllability rank condition*. The system is said to be *uncontrollable* if this condition does not hold. It is illustrative to evaluate the controllability matrix $W_c(x)$ for the linear system (3.2), for which the Lie brackets can be expressed as:

$$ad_f^{k-1} g(x) = (-1)^{k-1} A^{k-1} b, \quad 1 \leq k \leq n \tag{3.52}$$

Hence, for a linear system the nonlinear controllability matrix $W_c(x)$ is identical to the linear controllability matrix W_c, modulo the $(-1)^{k-1}$ term which appears in $W_c(x)$.

In analogy to the linear case, the nonlinear system (3.1) can be locally decomposed into controllable/uncontrollable parts. If the system satisfies certain technical assumptions [18, 22], there exists a local change of coordinates $z = \Phi(x)$ defined in a neighborhood of x_0 such that the system is transformed into:

$$\begin{aligned} \dot{z}_1 &= f_1(z_1, z_2) + g_1(z_1, z_2)u \\ \dot{z}_2 &= f_2(z_2) \end{aligned} \tag{3.53}$$

The state z_1 represents the locally weakly controllable dynamics and has dimension $d_c = \text{rank}(W_c(x_0))$. By contrast, the state z_2, which has dimension $n - d_c$, represents the uncontrollable dynamics.

3.4.3 Local Observability

We now consider the observability of the nonlinear system (3.1). As before, only local results will be presented; the interested reader is referred elsewhere [15, 18, 25] for global considerations. The nonlinear system (3.1) is said to be *weakly observable* at x_0 if there exists a time t_1 such that any initial state $x(t_0)$ in a neighborhood X_0 of x_0 can be distinguished from any other state $x_1 \in X_0$ using the input $u(t)$ and output $y(t)$ over the time interval $t_0 \leq t \leq t_1$ [15, 18]. It is useful to introduce a local version of weak observability to ensure that the state trajectory $x(t)$ will remain near x_0 and, therefore, that t_1 is reasonably small. The nonlinear system (3.1) is said to be *locally weakly observable* if it is weakly observable and there exists a neighborhood $X_1 \subset X_0$ of such that $x(t) \in X_1$ for $t_0 \leq t \leq t_1$ [15, 28].

Local observability can be checked using a matrix which is analogous to the linear observability matrix. It can be shown [15, 18] that the nonlinear system (3.1) is locally weakly observable at x_0 if the n x n observability matrix:

$$W_o(x) = \begin{bmatrix} dh(x) \\ dL_f h(x) \\ \vdots \\ dL_f^{n-1} h(x) \end{bmatrix} \tag{3.54}$$

has rank n at x_0. In this case, the system is said to satisfy the *observability rank condition.* Otherwise, the system is said to be *unobservable.* It is illustrative to evaluate the nonlinear observability matrix $W_o(x)$ for the linear system (3.2). In this case, the Lie derivatives have the form:

$$dL_f^{k-1} h(x) = dCA^{k-1}x = CA^{k-1}, \quad 1 \leq k \leq n \tag{3.55}$$

Thus, $W_o(x)$ is identical to the linear observability matrix W_o when the system is linear.

In analogy to the linear case, the nonlinear system (3.1) can be locally decomposed into observable/unobservable parts. Assuming the system satisfies certain technical assumptions [18, 22], there exists a local change of coordinates $z = \Phi(x)$ defined in a neighborhood of x_0 such that the system is transformed into:

$$\begin{aligned} \dot{z}_1 &= f_1(z_1, z_2) + g_1(z_1, z_2)u \\ \dot{z}_2 &= f_2(z_2) + g_2(z_2)u \\ y &= h(z_2) \end{aligned} \tag{3.56}$$

The state z_2 represents the locally weakly observable dynamics and has dimension $d_o = \text{rank } W_o(x_0)$. By contrast, the state z_1 represents the unobservable dynamics and has dimension $n - d_o$.

3.4.4 Example – Cyclopentenol Synthesis

Consider again the cyclopentenol synthesis example from Section 3.3.6:

$$\begin{aligned} \dot{x}_1 &= -k_1x_1 - k_3x_1^2 + (x_{10} - x_1)u = f_1(x) + g_1(x)u \\ \dot{x}_2 &= k_1x_1 - k_2x_2 - x_2u = f_2(x) + g_2(x)u \\ y &= x_2 = h(x) \end{aligned} \tag{3.57}$$

In order for the system to be locally weakly controllable, the matrix $W_c(x) = \begin{bmatrix} g(x) & ad_f g(x) \end{bmatrix}$ must have rank two at the operating point x_0. It is easy to show that:

$$\begin{aligned} g(x) &= \begin{bmatrix} x_{10} - x_1 \\ -x_2 \end{bmatrix} \\ ad_f g(x) &= \begin{bmatrix} -f_1(x) - (k_1 + 2k_3x_1)g_1(x) \\ -f_2(x) - k_1g_1(x) + k_2g_2(x) \end{bmatrix} \end{aligned} \tag{3.58}$$

Thus, the model is locally controllable if:

$$\det W_c(x) = [-f_2(x) - k_1 g_1(x) + k_2 g_2(x)]g_1(x) + [f_1(x) + (k_1 + 2k_3 x_1)g_1(x)]g_2(x) \neq 0 \qquad (3.59)$$

at x_0. Although the controllability matrix has a generic rank of two, there may exist points x_0 for which $W_c(x)$ is singular. This possibility must be checked for particular x_0.

The model (3.57) is locally weakly observable if the matrix:

$$W_o(x) = \begin{bmatrix} dh(x) \\ dL_f h(x) \end{bmatrix} \qquad (3.60)$$

has rank two at the operating point x_0. It is easy to show that:

$$\begin{aligned} h(x) &= x_2 \\ L_f h(x) &= k_1 x_1 - k_2 x_2 \end{aligned} \qquad (3.61)$$

hence,

$$W_o(x) = \begin{pmatrix} 0 & 1 \\ k_1 & -k_2 \end{pmatrix} \qquad (3.62)$$

Thus, the model is locally observable for any x_0 because:

$$\det W_o(x) = -k_1 \neq 0 \qquad (3.63)$$

3.5 Input-Output Representations

3.5.1 Linear Systems

The general problem of input-output modeling for nonlinear systems involves the determination of a nonlinear functional which maps the entire past input history to the value of the output at the present time. For linear systems, the linear convolution integral:

$$y(t) = \int_{-\infty}^{\infty} h_1(\sigma)u(t-\sigma)d\sigma \qquad (3.64)$$

provides a functional which accomplishes this task. Here $h_1(\sigma)$, the "kernel" of the transformation, is the system's impulse response function. Causality requires that the lower limit on this integral be zero; it is therefore customary to write (3.64) as:

$$y(t) = \int_0^\infty h_1(\sigma)u(t-\sigma)d\sigma \tag{3.65}$$

3.5.2 Volterra Series

Volterra generalized this functional representation for nonlinear systems in the form of a power series [26]:

$$y(t) = y_1(t) + y_2(t) + y_3(t) + \cdots \tag{3.66}$$

where $y_1(t)$, the first order term, is defined as in (3.65), and the other terms in the series are defined as follows:

$$\begin{aligned}
y_2(t) &= \int_0^\infty \int_0^\infty h_2(\sigma_1,\sigma_2)\ u(t-\sigma_1)u(t-\sigma_2)\ d\sigma_1 d\sigma_2 \\
y_3(t) &= \int_0^\infty \int_0^\infty \int_0^\infty h_3(\sigma_1,\sigma_2,\sigma_3)\ u(t-\sigma_1)u(t-\sigma_2)u(t-\sigma_3)\ d\sigma_1 d\sigma_2 d\sigma_3 \\
y_i(t) &= \int_0^\infty \cdots \int_0^\infty h_i(\sigma_1,\ldots,\sigma_i)\ u(t-\sigma_1)\cdots u(t-\sigma_i)\ d\sigma_1 \cdots d\sigma_i
\end{aligned}$$

A discrete time representation of the Volterra operator takes the form of the power series:

$$y(k) = y_1(k) + y_2(k) + y_3(k) + \cdots \tag{3.67}$$

where the first term given by:

$$y_1(k) = \sum_{i=1}^{\infty} h_1(i)u(k-i) \tag{3.68}$$

is the convolution model employed in linear MPC approaches. The higher order terms are given by:

$$y_2(k) = \sum_{i=1}^{\infty}\sum_{j=1}^{\infty} h_2(i,j)\ u(k-i)u(k-j) \tag{3.69}$$

$$y_3(k) = \sum_{i=1}^{\infty}\sum_{j=1}^{\infty}\sum_{l=1}^{\infty} h_3(i,j,l)\ u(k-i)u(k-j)u(k-l) \tag{3.70}$$

Just as Taylor series are limited to approximating *analytic* functions [26], Volterra series — which are in fact temporal equivalents of Taylor series — are limited to approximating systems with fading memory [2]. Nevertheless, a wide variety of nonlinear chemical processes exhibit behavior which cannot be approximated adequately by linear models but for which a Volterra series models (even of only second order) provides a reasonable representation [8]. Such problems include chemical reactors which exhibit optima as well as high purity distillation columns which exhibit asymmetric behavior.

3.5.3 Fliess Canonical Form

For nonlinear systems, the input-output representation corresponding to the Byrnes-Isidori normal form is the Fliess canonical form [11] or Observability canonical form [23, 34]. A useful property of this normal form is that the pole dynamics and zero dynamics are structurally separated in the Jacobian approximation of the nonlinear system. As with the Byrnes-Isidori normal form, successive differentiations of the output are considered; however, in this case, the differentiations continue past the rth derivative until the nth derivative is obtained. Thus, the coordinate transformations are a function of both the original state, x, *and* the manipulated input, u. The resulting normal form is:

$$\begin{aligned}
\dot{z}_1 &= z_2 \\
&\vdots \\
\dot{z}_{n-1} &= z_n \\
\dot{z}_n &= F(y, \dot{y}, \ddot{y}, \ldots, y^{(n-1)}, u, \dot{u}, \ldots, u^{(n-r)}) \\
y &= z_1
\end{aligned} \tag{3.71}$$

Consequently, the nonlinear system can be represented by a single nth order ordinary differential equation:

$$\frac{d^n y}{dt^n} = F(y, \dot{y}, \ddot{y}, \ldots, y^{(n-1)}, u, \dot{u}, \ldots, u^{(n-r)}) \tag{3.72}$$

The system in (3.72) is also referred to as the *external differential form*, and a general algorithm for constructing it from a state-space representation is presented in [25].

By performing a Jacobian linearization of the system (3.71) one obtains:

$$\begin{aligned}\dot{z}_1 &= z_2\\ &\vdots\\ \dot{z}_{n-1} &= z_n\\ \dot{z}_n &= \alpha_1 z_1 + \alpha_2 z_2 + \ldots + \alpha_n z_n + \beta_0 u + \beta_1 \dot{u} + \ldots + \beta_{n-r} u^{(n-r)}\end{aligned} \tag{3.73}$$

where:

$$\alpha_i = \left.\frac{\partial F}{\partial z_i}\right|_0, \qquad \beta_i = \left.\frac{\partial F}{\partial u^{(i)}}\right|_0$$

The index 0 stands for evaluation at the nominal operating point, which can be taken as the origin without loss of generality. It is straightforward to show that the transfer function corresponding to (3.73) is given by:

$$G(s) = \frac{y(s)}{u(s)} = \frac{\beta_0 + \beta_1 s + \ldots + \beta_{n-r} s^{n-r}}{\alpha_1 + \alpha_2 s + \ldots + \alpha_n s^n} \tag{3.74}$$

Thus, the poles and zeros of the linear system are fully characterized by the final state equation. Returning to the original nonlinear system, one can easily show that the zero dynamics can be represented by the $(n-r)$th order differential equation:

$$0 = F(0, 0, 0, \ldots, 0, u, \dot{u}, \ldots, u^{(n-r)}) \tag{3.75}$$

3.5.4 Example – Exothermic Chemical Reactor

Consider the CSTR example from Section 3.3.4:

$$\dot{x_1} = -x_1 + \mathcal{D}a(1 - x_1)e^{\frac{x_2}{1+x_2/\gamma}} \tag{3.76}$$

$$\dot{x_2} = -x_2 + B\mathcal{D}a(1 - x_1)e^{\frac{x_2}{1+x_2/\gamma}} + \beta(u - x_2) \tag{3.77}$$

$$y = x_2 \tag{3.78}$$

To obtain the Fliess canonical form, one differentiates the output two times:

$$\dot{y} = -x_2 + B\mathcal{D}a(1 - x_1)e^{\frac{x_2}{1+x_2/\gamma}} + \beta(u - x_2) \tag{3.79}$$

$$\begin{aligned}\ddot{y} = {} & -\dot{x}_2 - B\mathcal{D}a\dot{x}_1 e^{\frac{x_2}{1+x_2/\gamma}} + B\mathcal{D}a(1 - x_1)e^{\frac{x_2}{1+x_2/\gamma}} \frac{1}{(1 + x/\gamma)^2}\\ & + \beta(\dot{u} - \dot{x}_2)\end{aligned} \tag{3.80}$$

Using the inverse transformation:

$$x_1 = 1 - \frac{\dot{y} - \beta(u-y) + y}{B\mathcal{D}ae^{\frac{y}{1+y/\gamma}}} \quad (3.81)$$

$$x_2 = y \quad (3.82)$$

one can obtain an external differential representation for the CSTR problem:

$$\begin{aligned} \ddot{y} &= (-2-\beta)\dot{y} + (-\beta-1)y + \beta\dot{u} + \beta u \\ &+ \frac{1}{(1+y/\gamma)^2}(\dot{y} - \beta(u-y) + y) + \\ &Dae^{\frac{y}{1+y/\gamma}}(B - \dot{y} + \beta(u-y) + y) \end{aligned} \quad (3.83)$$

(Note that for complex, higher-order state dynamics, the analytical derivation of the transformation from the inputs, outputs and their derivatives to the states may require symbolic manipulation software, or may not be globally solvable.)

The resultant structure is useful for several reasons: (i) the expression has an affine dependence on both u and $\dot{u}$, indicating that a dynamic feedback linearizing controller is easily constructed; and (ii) the structural form of the y nonlinearities suggests terms that might be included in a nonlinear input/output semi-empirical model structure for identification (*e.g.* terms such as $\frac{1}{(1+y/\gamma)^2}$ which are not immediately evident from the fundamental model.)

3.5.5 Realization Theory

The general problem of realizing a state-space representation of a nonlinear system from an external differential form is quite difficult. In particular, it often is not possible to completely eliminate derivatives of the input in the final structure, and it is even more difficult to obtain a control affine realization as in (3.1). Representative results in this area are given, for example, in the papers by Freedman and Willems [12], Glad [13], and Delaleau and Respondek [6]. They show that a system given in external differential form (3.72) can be represented, under certain technical conditions, by a general state space model:

$$\begin{aligned} \dot{z} &= g(z,u) \\ y &= h(z,u) \end{aligned} \quad (3.84)$$

A limited class of problems which admit such a solution include single-input, single-output systems with $n - r = 1$ [12].

3.5.6 Output Invariance

In developing input-output structures for multivariable dynamic systems, an important issue for controller synthesis is the case of individual inputs which do not affect a given output variable. In the linear case, this is straightforward to analyze – the corresponding transfer function element is identically zero. Consider the general (*i.e.*, possibly nonsquare) multivariable, state-space system:

$$\begin{aligned} \dot{x} &= f(x) + \sum_{i=1}^{m} g_i(x)u_i(t) \\ y_j(x) &= h_j(x) \qquad 1 \leq j \leq p \end{aligned} \tag{3.85}$$

In this section, we describe the conditions under which a given output channel, y_j, is unaffected by a given input, u_i.

Lemma 1 (Output Invariance [25]) *For the system (3.85), y_j is unaffected by u_i on a open set $U \subset \mathcal{R}^n$ if and only if for all $\gamma \geq 1$ and for any $\tau_1, \cdots, \tau_\gamma$ in the set $\{f, g_1, \cdots, g_m\}$, the following relations hold:*

$$\begin{aligned} L_{g_i} h_j(x) &= 0 \\ L_{g_i} L_{\tau_1} \cdots L_{\tau_\gamma} h_j(x) &= 0 \qquad \forall\, x \in U \end{aligned}$$

Note that this result only holds for analytic systems as there exists counterexamples for non-analytic systems [25]. If this result is applied to a linear system, the conditions become:

$$c_j A^k b_i = 0 \qquad k = 0, 1, \ldots$$

which indicates that output invariance is obtained if and only if b_i lies in the unobservable subspace of the output c_j. This suggests an alternative formulation of output invariance using the unobservability properties of the nonlinear system:

Proposition 1 *[25] For the system (3.85), y_j is unaffected by u_i on a open set $U \subset \mathcal{R}^n$ if there exists an involutive distribution D of constant dimension defined on U such that:*

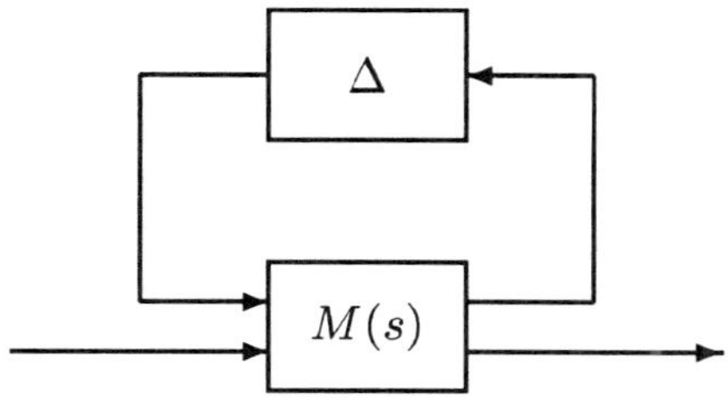

Figure 3.3: Uncertainty structure formulation

1. *D is invariant for (3.85)*

2. *$g_i \in D$*

3. *$D \in ker\ \ dh_j$*

where *ker* denotes the kernel (null-space).

3.5.7 Input-Output Stability

The results in this section rely on an input-output approach to nonlinear control system design and analysis. The question of closed-loop stability is analyzed for these systems in much the same manner as for linear systems; namely using the small gain theorem. We will briefly review this result, as well as the class of nonlinear systems which can be analyzed in this framework. Finally, a connection to linear robustness theory (μ analysis) is indicated.

The main result for stability of an interconnected block diagram such as that depicted in Figure 3.3 is given by the small gain theorem [33]. In this diagram, the uncertainty, Δ, is formulated as a separate element in a feedback loop about the nominal closed-loop system operator, M, which contains both the plant and the controller. This uncertainty can represent the effect of unmodeled dynamics and uncertain parameters in a process model. It is straightforward to show [24] that problems in which the uncertainty is of an additive, multiplicative, or inverse multiplicative nature can be cast in the structure in Figure 3.3.

Before stating the small gain result, we need to introduce the notion of an operator gain:

Definition 6 (Gain) *The incremental gain of an operator M, denoted by $g(M)$, is:*

$$g(M) = \sup \frac{\|M(x_1) - M(x_2)\|}{\|x_1 - x_2\|} \tag{3.86}$$

where the supremum is taken over all x_i in the domain of M, all $M(x_i)$ in the range of M, and all time for which $x \neq 0$.

This induced norm can be defined for various L_p norms (see for example, [32]). Having defined the gain of an operator, we can state the main result in [33]:

Theorem 4 (Small Gain Theorem) *If $g(M) \cdot g(\Delta) < 1$, then the closed-loop is internally input-output stable.*

Clearly the burden in this analysis is the calculation of the nonlinear operator norm. This remains an area of active research, and trends in computing algorithms may lead to more tractable solutions in the next few years.

The so-called M-Δ diagram in Figure 3.3 also represents a general framework for analysis of linear uncertain systems. In [9], there were extensions presented to include classes of nonlinear systems. It was shown that for norm bounded nonlinear operators, constant D-scalings could be used in the structured singular value analysis (see [7]) to give a conservative small gain condition for robust stability; an application to a CSTR was also presented. Furthermore, it is possible to calculate a Lyapunov function from the D-scalings, thus linking the input-output stability results to the Lyapunov stability.

A class of nonlinear systems is now described which fits into the proposed M-Δ structure framework. A conic sector is defined as:

$$\text{Cone}(C, R) \equiv \{(u, y) | \|y - Cu\| \leq \|Ru\|\} \tag{3.87}$$

where (u, y) is the input/output pair for the operator. A nonlinear operator enveloped tightly by a conic sector is most accurately approximated linearly by the cone center C. In general, the cone center will not coincide with the plant described by the Jacobian of the nonlinear model evaluated at the operating point. Note that because we have replaced a potentially highly nonlinear function by two linear time-invariant operators, this simplification is likely to be conservative. The Cone(C, R) describes many input-output pairs, some of which may yield poorer performance than the original operator.

There is a direct correspondence between a nonlinear cone-bounded operator and a time-varying gain. From the conic sector definition, the plant can be interpreted as being equal to the nominal value (C) which is perturbed by a time varying gain of magnitude R. R and C can be absorbed into the system to arrive at the general uncertainty structure in Figure 3.3, where Δ is a time-varying gain of norm one. The construction of interconnection structures for general uncertainty descriptions can be found in [24].

References

[1] B.W. Bequette. Nonlinear control of chemical processes: A review. *Ind. Eng. Chem. Res.*, 30:1391–1413, 1991.

[2] S. Boyd and L.O. Chua. Fading memory and the problem of approximating nonlinear operators with Volterra series. *IEEE Trans. Cir. Sys.*, CAS-32:1150–1161, 1985.

[3] C.I. Byrnes and A. Isidori. Global feedback stabilization of nonlinear systems. In *24th IEEE CDC*, pages 1031–1037, 1985.

[4] F. M. Callier and C. A. Desoer. *Linear System Theory.* Springer-Verlag, New York, 1991.

[5] J.P. Calvet. *A Differential Geometric Approach for the Nominal and Robust Control of Nonlinear Chemical Processes.* PhD thesis, Georgia Institute of Technology, Atlanta, GA, 1989.

[6] E. Delaleau and W. Respondek. Lowering the orders of derivatives of controls in generalized state space systems. In *Proceedings of 31st Conference on Decision and Control*, pages 3663–3668, 1992.

[7] J.C. Doyle. Analysis of feedback systems with structured uncertainties. *IEEE Proc., Part D*, 6:242–250, 1982.

[8] F.J. Doyle III, B.A. Ogunnaike, and R.K. Pearson. Nonlinear model-based control using second-order volterra models. *Automatica*, 31:697–714, 1995.

[9] F.J. Doyle III, A.P. Packard, and M. Morari. Robust controller design for a nonlinear CSTR. *Chem. Eng. Sci.*, 44:1929–1947, 1989.

[10] S. Engell and K.-U. Klatt. Gain Scheduling of a Non-Minimum Phase CSTR. In *Proc. of European Control Conference*, pages 2323–2328, 1993.

[11] M. Fliess. Generalized controller canonical forms for linear and nonlinear dynamics. *IEEE Trans. Autom. Control*, 35:994–1001, 1990.

[12] M.I. Freedman and J.C. Willems. Smooth representations of systems with differentiated inputs. *IEEE Trans. Autom. Control*, 23(1):16–21, 1978.

[13] S.T. Glad. Nonlinear state space and input output descriptions using differential polynomials. In *New Trends in Nonlinear Control Theory*, pages 182–189. Springer-Verlag, 1988.

[14] M.A. Henson and D.E. Seborg. Nonlinear control strategies for continuous fermentors. *Chem. Eng. Sci.*, 47:821–835, 1992.

[15] R. Hermann and A. J. Krener. Nonlinear controllability and observability. *IEEE Trans. Autom. Control*, AC-22:728–740, 1977.

[16] R. Hirschorn. Invertibility of multivariable nonlinear control systems. *IEEE Trans. Autom. Control*, 24:855–865, 1979.

[17] R. Hirschorn and J. Davis. Output tracking for nonlinear systems with singular points. *SIAM J. Control Optim.*, 25(3):547–557, 1987.

[18] A. Isidori. *Nonlinear Control Systems - An Introduction.* Springer-Verlag, New York, NY, 1989.

[19] A. Isidori and C.H. Moog. On the nonlinear equivalent of transmission zeros. In *Modelling and Adaptive Control, Proc. IIASA Conf., Sopron, Hungary.* Springer-Verlag, 1989.

[20] C. Kravaris and P. Daoutidis. Nonlinear state feedback control of second-order nonminimum phase systems. *Comput. Chem. Eng.*, 14(4/5):439–449, 1990.

[21] C. Kravaris and J.C. Kantor. Geometric methods for nonlinear process control. 1. Background 2. Controller synthesis. *Ind. Eng. Chem. Res.*, 29:2295–2323, 1990.

[22] A. J. Krener. A decomposition theory for differentiable systems. *SIAM J. Control Optimiz.*, 15:289–297, 1977.

[23] A.J. Krener. Normal forms for linear and nonlinear systems. *Contemp. Math.*, 68:157–189, 1987.

[24] M. Morari and E. Zafiriou. *Robust Process Control.* Prentice-Hall, Englewood Cliffs, NJ, 1989.

[25] H. Nijmeijer and A.J. van der Schaft. *Nonlinear Dynamical Control Systems.* Springer-Verlag, New York, NY, 1990.

[26] M. Schetzen. *The Volterra and Wiener Theories of Nonlinear Systems.* John Wiley & Sons, New York, NY, 1980.

[27] M. Spivak. *A Comprehensive Introduction to Differential Geometry.* Publish or Perish, Inc., Houston, TX, second edition, 1991.

[28] H. Sussmann and V. Jurdjevic. Controllability of nonlinear systems. *J. Diff. Eqs.*, 12:95–116, 1972.

[29] H.J. Sussmann. Limitations on the stabilizability of globally minimum phase systems. *IEEE Trans. Autom. Control*, AC-35:117–119, 1990.

[30] A. Uppal, W.H. Ray, and A.B. Poore. On the dynamic behavior of continuous stirred tanks. *Chem. Eng. Sci.*, 29:967–985, 1974.

[31] J.G. van de Vusse. Plug-flow type reactor versus tank reactor. *Chem. Eng. Sci.*, 19:994–997, 1964.

[32] M. Vidyasagar. *Nonlinear Systems Analysis.* Prentice Hall, Englewood Cliffs, NJ, second edition, 1993.

[33] G. Zames. On the input-output stability of nonlinear time-varying feedback systems, Parts I and II. *IEEE Trans. Autom. Control*, AC-11:228–239, 465–476, 1966.

[34] M. Zeitz. Canonical forms for nonlinear systems. In *Nonlinear Control Systems Design. Selected Papers from the IFAC Symposium*, pages 33–38. Oxford, 1989.

Chapter 4

Feedback Linearizing Control

MICHAEL A. HENSON
Department of Chemical Engineering
Louisiana State University
Baton Rouge, LA 70803-7303

DALE E. SEBORG
Department of Chemical Engineering
University of California
Santa Barbara, CA 93106-5080

4.1 Introduction

In this chapter, the basic theory of feedback linearization is presented and issues of particular relevance to process control applications are discussed. Two fundamental nonlinear controller design techniques — input-output linearization and state-space linearization — are discussed in detail. The theory also is presented for linear systems to facilitate understanding of the nonlinear results. Extensions are presented for disturbances and multivariable processes. Advanced topics such as dynamic feedback linearization, time delay compensation, constraint handling, robustness, and sampled-data systems are also discussed.

A survey of process control strategies and applications shows that: (1) a variety of nonlinear controller design techniques are based on input-output linearization; (2) few experimental studies of these techniques have been presented; and (3) many important problems remain unsolved. To illustrate design and implementation issues, feedback linearizing controllers are developed for three representative processes: a continuous stirred tank reactor, a continuous fermentor, and a pH neutralization system. This chapter differs from existing reviews [68, 82, 89, 94, 98, 99, 126, 134] of feedback linearization by providing a balanced discussion of theoretical and practical issues of interest to process control engineers.

Nonlinear Process Model

As discussed in previous chapters, there are several types of finite-dimensional, nonlinear process models. In this chapter, we will focus on continuous-time, state-space models of the form,

$$\begin{aligned} \dot{x} &= f(x) + g(x)u \\ y &= h(x) \end{aligned} \tag{4.1}$$

where: x is an n-dimensional vector of state variables; u is an m-dimensional vector of manipulated input variables; y is an m-dimensional vector of controlled output variables; $f(x)$ is an n-dimensional vector of nonlinear functions; $g(x)$ is an (n x m)-dimensional matrix of nonlinear functions; and $h(x)$ is an m-dimensional vector of nonlinear functions. The single-input, single-output (SISO) case where $m = 1$ will be emphasized to facilitate understanding of the basic concepts. The model (4.1) will be modified as necessary to describe more complex nonlinear processes, such as those with measured disturbances or time delays.

Feedback Linearization vs. Jacobian Linearization

Consider the Jacobian linearization of the nonlinear model (4.1) about an equilibrium point (u_0, x_0, y_0):

$$\begin{aligned} \dot{x} &= \left[\frac{\partial f(x_0)}{\partial x} + \frac{\partial g(x_0)}{\partial x}u_0\right](x - x_0) + g(x_0)(u - u_0) \\ y - y_0 &= \frac{\partial h(x_0)}{\partial x}(x - x_0) \end{aligned} \tag{4.2}$$

Using deviation variables, the Jacobian model can be written as a linear state-space system,

$$\begin{aligned} \dot{x} &= Ax + Bu \\ y &= Cx \end{aligned} \tag{4.3}$$

with obvious definitions for the matrices A, B, and C. It is important to note that the Jacobian model is an exact representation of the nonlinear model only at the point (x_0, u_0). As a result, a control strategy based on a linearized model may yield unsatisfactoy performance and robustness at other operating points.

In this chapter, we present a class of nonlinear control techniques that can produce a linear model that is an *exact* representation of the original nonlinear model over a large set of operating conditions. The general approach — typically called *feedback linearization* — is based on two operations: (1) nonlinear change of coordinates; and (2) nonlinear state feedback. We focus on *local* feedback linearization (*i.e.* the coordinate transformation and control law may be only locally defined) to avoid complications associated with the global problem.

After feedback linearization, the input-output model is linear,

$$\begin{aligned} \dot{\xi} &= A\xi + Bv \\ w &= C\xi \end{aligned} \tag{4.4}$$

where: ξ is an r-dimensional vector of transformed state variables; v is an m-dimensional vector of transformed input variables; w is an m-dimensional vector of transformed output variables; and the matrices A, B, and C have a very simple canonical structure. If $r < n$, an additional $n-r$ state variables must be introduced to complete the coordinate transformation. The integer r is called the *relative degree* and is a fundamental characteristic of a nonlinear system.

Feedback Linearization Approaches

Most feedback linearization approaches are based on *input-output linearization* or *state-space linearization*. In the input-output linearization approach, the objective is to linearize the map between the transformed inputs (v) and the *actual* outputs (y). A linear controller is then designed for the linearized input-output model, which can be represented by (4.4) with $r \leq n$ and $w = y$. However, there is a subsystem that typically is not linearized,

$$\dot{\eta} = q(\eta, \xi) \tag{4.5}$$

where η is an $(n-r)$-dimensional vector of transformed state variables and q is a $(n-r)$-dimensional vector of nonlinear functions. Input-output linearization techniques are restricted to processes in which these so-called *zero dynamics* are stable.

In the state-space linearization approach, the goal is to linearize the map between the transformed inputs and the *entire* vector of transformed state variables. This objective is achieved by deriving *artificial* outputs (w) that yield a feedback linearized model with state dimension $r = n$. A linear controller is then synthesized for the linear input-state model. However, this approach may fail to simplify the controller design task because the map between the transformed inputs and the original outputs (y) generally is nonlinear. As a result, input-output linearization is preferable to state-space linearization for most process control applications. For some processes, it is possible to simultaneously linearize the input-state and input-output maps because the original outputs yield a linear model with dimension $r = n$.

Static and Dynamic State Feedback

Feedback linearization produces a linear model by the use of nonlinear coordinate transformations and nonlinear state feedback. Coordinate transformations are described in Chapter 3; nonlinear state feedback is discussed below. In some applications, the control objectives can be achieved with a nonlinear static state feedback control law of the form,

$$u = \alpha(x) + \beta(x)v \tag{4.6}$$

where α is an m-dimensional vector of nonlinear functions and β is an m x m matrix of nonlinear functions. For some processes, it is not possible to satisfy the control objective with a static controller and a dynamic state feedback control law must be employed,

$$\begin{aligned} \dot{\zeta} &= \gamma(x,\zeta) + \delta(x,\zeta)v \\ u &= \alpha(x,\zeta) + \beta(x,\zeta)v \end{aligned} \tag{4.7}$$

where: ζ is a q-dimensional vector of controller state variables; γ is a q-dimensional vector of nonlinear functions; and δ is a q x m matrix of nonlinear functions. Specific forms for the nonlinear controller functions $(\alpha, \beta, \gamma, \delta)$ will be presented thoughout the chapter.

4.2 Input-Output Linearization

4.2.1 Linear System

An input-output linearizing controller is designed for the SISO version of the linear system (4.3). This exercise illustrates the basic controller design procedure employed in the nonlinear case. Recall from Chapter 3 that the system can be transformed into the following normal form via a linear change of coordinates $[\xi^T, \eta^T]^T = Tx$,

$$\begin{aligned} \dot{\xi}_1 &= \xi_2 \\ \dot{\xi}_2 &= \xi_3 \\ &\vdots \\ \dot{\xi}_r &= R\xi + S\eta + ku \\ \dot{\eta} &= P\xi + Q\eta \\ y &= \xi_1 \end{aligned} \tag{4.8}$$

where r is the *relative degree*. The static state feedback control law,

$$u = \frac{v - R\xi - S\eta}{k} \tag{4.9}$$

changes the r-th equation in (4.8) to: $\dot{\xi}_r = v$. The transformed input v is designed to stabilize the ξ subsystem,

$$v = -\alpha_r \xi_r - \alpha_{r-1}\xi_{r-1} - \cdots - \alpha_1 \xi_1 \tag{4.10}$$

In the original coordinates, the complete control law has the form,

$$u = \frac{-CA^r x - \alpha_r CA^{r-1}x - \cdots - \alpha_1 Cx}{CA^{r-1}b} \tag{4.11}$$

The proposed control law yields the following characteristic equation for the ξ subsystem:

$$s^r + \alpha_r s^{r-1} + \cdots + \alpha_2 s + \alpha_1 = 0 \tag{4.12}$$

Nominal stability of the ξ subsystem, and therefore boundedness of the output, is ensured if the controller tuning parameters α_i are chosen such that

(4.12) is a Hurwitz polynomial. As discussed in Chapter 3, the closed-loop system is internally stable (*i.e.* the η variables are bounded) if and only if the eigenvalues of the matrix Q are in the open left-half plane. Hence, the proposed controller design technique is restricted to linear systems which are *minimum phase*.

4.2.2 Controller Design

In this section, an input-output linearizing controller is designed for the SISO version of the nonlinear system (4.1). Extensions of the basic controller design procedure for disturbances and multivariable processes are discussed in subsequent sections.

Illustrative Example

In order to illustrate the basic concepts, we first consider the following two-dimensional nonlinear system:

$$\begin{aligned} \dot{x}_1 &= f_1(x_1, x_2) + g_1(x_1, x_2)u \\ \dot{x}_2 &= f_2(x_1, x_2) \\ y &= x_1 \end{aligned} \tag{4.13}$$

This model form can describe, for example, an irreversible reaction occuring in a constant volume, stirred tank reactor where x_1 is the reactor temperature, x_2 is the reactor concentration, and u is the coolant temperature [162].

If the nonlinear function g_1 is non-zero in the operating region of interest, the static state feedback control law,

$$u = \frac{v - f_1(x_1, x_2)}{g_1(x_1, x_2)} \tag{4.14}$$

changes the first equation of (4.13) to: $\dot{x}_1 = v$. Thus, the control law *exactly* linearizes the map between the transformed input v and the output y. Consequently, a *linear* controller can be designed to satisfy control objectives such as setpoint tracking. It is important to note that the x_2 dynamics remain nonlinear. As discussed in the next section, asymptotic stability of these *zero dynamics* is a necessary condition for nominal closed-loop stability.

General Design Procedure

We now consider the design of an input-output linearizing controller for the n-th order nonlinear system (4.1). This problem was originally posed and solved by Isidori and Krener [84, 105]. Recall from Chapter 3 that the system (4.1) can be transformed into normal form via a diffeomorphism $[\xi^T, \eta^T]^T = \Phi(x)$ if the relative degree r is well defined. The ξ coordinates are defined as,

$$\xi_k = \Phi_k(x) = L_f^{k-1} h(x), \qquad 1 \leq k \leq r \tag{4.15}$$

and $\eta_k = \Phi_{r+k}(x)$, $1 \leq k \leq n-r$, where $L_g \Phi_k(x) = 0$. The normal form can be written as:

$$\begin{aligned}
\dot{\xi}_1 &= \xi_2 \\
\dot{\xi}_2 &= \xi_3 \\
&\vdots \\
\dot{\xi}_r &= b(\xi,\eta) + a(\xi,\eta)u \\
\dot{\eta} &= q(\xi,\eta) \\
y &= \xi_1
\end{aligned} \tag{4.16}$$

The static state feedback control law,

$$u = \frac{v - b(\xi,\eta)}{a(\xi,\eta)} \tag{4.17}$$

changes the r-th equation of (4.16) to: $\dot{\xi}_r = v$. As a result, the map between the transformed input v and the output y is *exactly* linear. Thus, a linear state feedback controller can be synthesized to stabilize the ξ subsystem. For instance, the pole placement design (4.10) yields the characteristic polynomial (4.12) for the linear subsystem. When expressed in the original coordinates, the two control laws have the following form:

$$u = \frac{v - L_f^r h(x)}{L_g L_f^{r-1} h(x)} \tag{4.18}$$

$$v = -\alpha_r L_f^{r-1} h(x) - \alpha_{r-1} L_f^{r-2} h(x) - \cdots - \alpha_1 h(x) \tag{4.19}$$

Hence, the complete static state feedback control law can be written as:

$$u = \frac{-L_f^r h(x) - \alpha_r L_f^{r-1} h(x) - \cdots - \alpha_1 h(x)}{L_g L_f^{r-1} h(x)} \tag{4.20}$$

Integral Control

In most process control applications, the objective is to maintain the output at a non-zero setpoint despite unmeasured disturbances and plant/model mismatch. Consequently, the nonlinear controller (4.20) should contain an integral term that penalizes deviations between the output (y) and its setpoint (y_{sp}). The modified input-output linearizing controller can be written as,

$$u = \frac{-L_f^r h(x) - \alpha_r L_f^{r-1} h(x) - \cdots + \alpha_1 \left[y_{sp} - h(x)\right] + \alpha_0 \int_0^t \left[y_{sp} - h(x)\right] d\tau}{L_g L_f^{r-1} h(x)} \tag{4.21}$$

where α_0 is an additional controller tuning parameter associated with the integral term. The integral control law (4.21) yields the following characteristic equation for the ξ subsystem:

$$s^{r+1} + \alpha_r s^r + \cdots + \alpha_1 s + \alpha_0 = 0 \tag{4.22}$$

By choosing the controller parameters α_i in terms of a single tuning parameter ϵ, the following closed-loop transfer function is obtained for setpoint changes if $y(0) = y_{sp}(0)$ [67]:

$$\frac{y(s)}{y_{sp}(s)} = \frac{\frac{r+1}{\epsilon^r} s + 1}{(\epsilon s + 1)^{r+1}} \tag{4.23}$$

4.2.3 Nominal Stability

In this section, nominal stability of the closed-loop system resulting from input-output linearization is discussed. For simplicity, we focus on asymptotic stabilization rather than the more difficult problem of asymptotic setpoint tracking [82]. We will assume, without loss of generality, that $x_0 = 0$ is an equilibrium point. The objective is to find conditions which ensure that the origin is a locally or globally asymptotically stable equilbrium point of the input-output linearized system.

Consider the closed-loop system comprised of the nonlinear system (4.1) and the input-output linearizing controller (4.20). The closed-loop system has the following representation in the transformed coordinates,

$$\begin{aligned} \dot{\xi} &= A\xi \\ \dot{\eta} &= q(\xi, \eta) \\ y &= \xi_1 \end{aligned} \tag{4.24}$$

If the controller tuning parameters α_i are chosen such that the characteristic polynomial (4.12) is Hurwitz, then the linear state variables ξ converge exponentially to the origin for any initial state $\xi(0)$ for which the control law (4.17) remains well defined (*i.e.* $a \neq 0$).

For the moment, assume the nonlinear state variables η converge asymptotically to the origin. Recall that the original state variables are related to the transformed state variables as $[\xi^T, \eta^T]^T = \Phi(x)$. Further assume that the diffeomorphism Φ is well defined in the region of interest. In this case, Φ is smooth and invertible, and can be chosen such that $\Phi(0) = 0$. As a result, asymptotic convergence of ξ and η to the origin implies asymptotic convergence of x to the origin. Thus, the origin is a locally asymptotically stable equilibrium point if: (1) the η state variables are locally asymptotically stable; and (2) the diffeomorphism and input-output linearizing control law are locally defined. The origin is a globally asymptotically stable equilibrium point if these conditions hold globally. Hence, the stabilization problem is effectively reduced to finding conditions which guarantee that the η state variables converge asymptotically to the origin.

Local Stability

We first present a necessary and sufficient condition for the input-output linearized system to be locally asymptotically stable. Because the ξ state variables converge to zero, the second equation of (4.24) becomes,

$$\dot{\eta} = q(0, \eta) \tag{4.25}$$

in the limit as $t \to \infty$. Equation (4.25) is known as the *zero dynamics* and is the nonlinear analog of linear system zeros. Nonlinear systems with asymptotically stable zero dynamics are said to be *minimum phase*. Local asymptotic stability of the zero dynamics is clearly a necessary condition for the feedback linearized system (4.24) to be locally asymptotically stable. It has been shown that this condition also is sufficient [27, 123].

Global Stability

It is tempting to conjecture that global asymptotic stability of the zero dynamics is a sufficient condition for the feedback linearized system (4.24) to be globally asymptotically stable. The argument for this proposition proceeds as follows. The ξ state variables can be forced to zero arbitrarily fast by approriate selection of the controller tuning parameters α_i. Once the ξ variables converge to zero, the closed-loop trajectories are described by the zero dynamics (4.25). Because the zero dynamics are globally asymptotically stable by assumption, the η state variables converge to zero and the closed-loop system is globally asymptotically stable.

In fact, this argument is correct if the relative degree $r = 1$ [26, 27]. However, the argument does not hold in general if $r \geq 2$ due to the so-called "peaking phenomenon" [159, 160]. A high gain linear feedback can cause the linear state variables ξ to become very large before they decay to zero. These "peaking" variables act as destabilizing inputs to the zero dynamics. As a result, considerably more restrictive sufficient conditions are required to ensure that the system (4.24) is globally asymptotically stable if $r \geq 2$. Peaking cannot occur — and the closed-loop system therefore is globally asymptotically stable — if either of the following conditions are satisfied:

1. The only linear state variable entering the zero dynamics is the output $y = \xi_1$ [118, 148].

2. The zero dynamic satisfy a Lipschitz growth condition [27, 160].

Although neither condition may hold in practice, the authors have not encountered any process models with stable zero dynamics that cannot be asymptotically stabilized via input-output linearization.

4.2.4 Disturbance Decoupling

The input-output linearization technique presented above does not explicitly address process disturbances. In this section, conditions under which the output can be completely *decoupled* from disturbance variables are presented. This problem was originally posed and solved by several investigators using differential geometric analysis tools [37, 75, 85]. For simplicity, we employ an alternative formulation based on relative degrees [40]. Consider an SISO nonlinear system with a single disturbance d,

$$\begin{aligned} \dot{x} &= f(x) + g(x)u + p(x)d \\ y &= h(x) \end{aligned} \tag{4.26}$$

where $p(x)$ is an n-dimensional vector of nonlinear functions. The controller design technique presented below is easily extended to processes with multiple disturbances [40].

The *disturbance decoupling problem* is to find (if possible) a diffeomorphism and a nonlinear static state feedback control law such that: (1) the map between the transformed input and the output is linear; and (2) the output is completely unaffected by the disturbance. It is useful to define a relative degree for the disturbance that is analogous to the relative degree r associated with the manipulated input u. The disturbance d is said to have relative degree ρ at the point x_0 if:

1. $L_p L_f^k h(x) = 0$ for all x in a neighborhood of x_0 and all $k < \rho - 1$.

2. $L_p L_f^{\rho-1} h(x_0) \neq 0$.

If $\rho < r$, the disturbance affects the output more directly than does the manipulated input and disturbance decoupling cannot be achieved with a static state feedback control law. Consequently, a necessary condition for the solution of the disturbance decoupling problem is that $\rho \geq r$. Under this condition, the diffeomorphism in Section 4.2.2 transforms (4.26) into the normal form,

$$\begin{aligned} \dot{\xi}_1 &= \xi_2 \\ \dot{\xi}_2 &= \xi_3 \\ &\vdots \\ \dot{\xi}_r &= b(\xi,\eta) + a(\xi,\eta)u + s(\xi,\eta)d \\ \dot{\eta} &= q(\xi,\eta) + t(\xi,\eta)d \\ y &= \xi_1 \end{aligned} \tag{4.27}$$

where: $s(\xi,\eta) = L_p L_f^{\rho-1} h\left[\Phi^{-1}(\xi,\eta)\right]$ and $t_i(\xi,\eta) = L_p \Phi_{r+k}\left[\Phi^{-1}(\xi,\eta)\right]$, $1 \leq k \leq n-r$.

Assume that an on-line measurement of the disturbance is *not* available. If $\rho > r$, the function $s = 0$ and the linearizing control law (4.17) completely decouples the output from the disturbance. The transformed input v can be designed as usual. The requirement that $\rho > r$ is often called the ***disturbance matching condition***. Because the disturbance acts as an input to the zero dynamics, stability analysis is more difficult than in the disturbance-free case. Sufficient conditions for local asymptotic stabilization have been presented [25, 164].

Note that the control law (4.17) does not decouple the output from the disturbance if $\rho = r$. It is possible to achieve disturbance decoupling in this case if the disturbance d is measured and a feedforward/feedback control law is employed. The so-called ***disturbance decoupling problem with measurement*** is solved by the following nonlinear control law:

$$u = \frac{v - b(\xi,\eta) - s(\xi,\eta)d}{a(\xi,\eta)} \tag{4.28}$$

This feedforward/feedback control law changes the r-th equation in (4.27) to: $\dot{\xi}_r = v$. As a result, the input-output map is linear and the output is decoupled from the disturbance. When expressed in the original coordinates, the decoupling control law has the form:

$$u = \frac{v - L_f^r h(x) - L_p L_f^{r-1} h(x) d}{L_g L_f^{r-1} h(x)} \tag{4.29}$$

The transformed input v is designed as before. A more general solution to the disturbance decoupling problem with measurement has been proposed for the case $\rho < r$ [40]. However, the resulting control law contains derivatives of the disturbance up to order $r - \rho$. Because process measurements usually are corrupted with high frequency noise, this approach is difficult to successfully implement in practice.

4.2.5 Input-Output Decoupling

We now extend the input-output linearization approach to multiple-input, multiple-output (MIMO) processes. This extension is often called ***input-output decoupling*** because the input-output response is both linearized and decoupled. More precisely, the input-output decoupling problem is to find (if possible) a diffeomorphism and a state feedback control law such that: (1) the map between the transformed inputs v and controlled outputs y is linear; and (2) the i-th output y_i is decoupled from all inputs v_j for $i \neq j$. In this section, we will only consider static state feedback control laws [53, 85, 133, 138]; input-output decoupling based on dynamic state feedback is discussed elsewhere [44, 132].

The Decoupling Matrix

It is useful to represent the MIMO nonlinear system (4.1) as,

$$\begin{aligned} \dot{x} &= f(x) + \sum_{j=1}^{m} g_j(x) u_j \\ y_i &= h_i(x) \quad i = 1, 2, \ldots, m \end{aligned} \tag{4.30}$$

The nonlinear system is said to have *vector relative degree* $\{r_1, r_2, \ldots, r_m\}$ at the point x_0 if:

1. $L_{g_j} L_f^k h_i(x) = 0$ for all $1 \leq i, j \leq m$, for all $k < r_i - 1$, and for all x in a neighborhood of x_0.

2. The m x m *decoupling matrix*

$$A(x) = \begin{bmatrix} L_{g_1} L_f^{r_1 - 1} h_1(x) & \cdots & L_{g_m} L_f^{r_1 - 1} h_1(x) \\ \vdots & \ddots & \vdots \\ L_{g_1} L_f^{r_m - 1} h_m(x) & \cdots & L_{g_m} L_f^{r_m - 1} h_m(x) \end{bmatrix} \tag{4.31}$$

   is nonsingular at the point x_0.

The integer r_i represents the smallest relative degree of the i-th output with respect to any of the m inputs. Nonsingularity of the decoupling matrix $A(x)$ may be viewed as the MIMO generalization of the condition that $L_g L_f^{r-1} h(x_0) \neq 0$. It is a necessary and sufficient condition for the solution of the input-output decoupling control problem with static state feedback. A constructive proof of sufficiency is provided below; necessity is proven elsewhere [82, 133].

Normal Form

We develop a MIMO generalization of the normal form introduced in Chapter 3. The sum of the individual relative degrees is defined to be: $r = r_1 + \cdots + r_m$. Because $A(x)$ is nonsingular, the diffeomorphism $[\xi^T, \eta^T]^T = \Phi(x)$ can be constructed by choosing the first r coordinates as,

$$\xi_k^i = \Phi_k^i(x) = L_f^{k-1} h_i(x) \tag{4.32}$$

where $1 \leq k \leq r_i$ and $1 \leq i \leq m$. It is always possible to find $n-r$ additional coordinates $\eta^T = [\Phi_{r+1}(x), \ldots, \Phi_n(x)]$ such that,

$$\Phi^T(x) = \left[\Phi_1^1(x), \ldots, \Phi_{r_1}^1(x), \ldots, \Phi_1^m(x), \ldots, \Phi_{r_m}^m(x), \Phi_{r+1}(x), \ldots, \Phi_n(x)\right] \tag{4.33}$$

is invertible at the point x_0 [82]. If the vector fields $g_1(x), g_2(x), \ldots, g_m(x)$ are involutive (see Section 3.2.6), it is possible to choose the additional coordinates such that $L_{g_j}\Phi_{r+i}(x) = 0$ for all $1 \leq i \leq n-r$ and for all $1 \leq j \leq m$.

In general, the involutivity condition is not satisfied and the normal form is,

$$\begin{aligned}
\dot{\xi}_1^i &= \xi_2^i \\
\dot{\xi}_2^i &= \xi_3^i \\
&\vdots \\
\dot{\xi}_{r_i}^i &= b_i(\xi, \eta) + \sum_{j=1}^{m} a_{ij}(\xi, \eta) u_j \\
\dot{\eta} &= q(\xi, \eta) + \sum_{j=1}^{m} p_j(\xi, \eta) u_j \\
y_i &= \xi_1^i
\end{aligned} \tag{4.34}$$

where $1 \leq i \leq m$. The functions $a_{ij}(\xi, \eta)$ are elements of the decoupling matrix A expressed in terms of the transformed coordinates; the functions b_i are defined as,

$$b_i(\xi, \eta) = L_f^{r_i} h_i \left[\Phi^{-1}(\xi, \eta)\right] \tag{4.35}$$

the vector q is defined as in the SISO case; and the vectors $p_j(\xi, \eta)$ are defined as,

$$p_j(\xi, \eta) = \begin{bmatrix} L_{g_j}\Phi_{r+1}\left[\Phi^{-1}(\xi, \eta)\right] \\ \vdots \\ L_{g_j}\Phi_n\left[\Phi^{-1}(\xi, \eta)\right] \end{bmatrix} \tag{4.36}$$

The r_i-th equations in the normal form can be collected and written as:

$$\begin{bmatrix} \dot{\xi}^1_{r_1} \\ \vdots \\ \dot{\xi}^m_{r_m} \end{bmatrix} = b(\xi, \eta) + A(\xi, \eta)u \tag{4.37}$$

Controller Design

Because the decoupling matrix is nonsingular by assumption, the static state feedback control law which achieves input-output decoupling can be derived directly from (4.37),

$$u = A^{-1}(\xi, \eta)\left[v - b(\xi, \eta)\right] \tag{4.38}$$

where v is an m-dimensional vector of transformed input variables. This control law — which can be interpreted as the MIMO generalization of the input-output linearizing control law (4.17) — changes (4.37) to:

$$\begin{bmatrix} \dot{\xi}^1_{r_1} \\ \vdots \\ \dot{\xi}^m_{r_m} \end{bmatrix} = \begin{bmatrix} v_1 \\ \vdots \\ v_m \end{bmatrix} \tag{4.39}$$

As a result, the input-output response is both *linear* and *decoupled*,

$$\begin{aligned} \dot{\xi}^i_1 &= \xi^i_2 \\ \dot{\xi}^i_2 &= \xi^i_3 \\ &\vdots \\ \dot{\xi}^i_{r_i} &= v_i \\ y_i &= \xi^i_1 \end{aligned} \tag{4.40}$$

where $1 \leq i \leq m$. Each input v_i can be chosen as in the SISO case:

$$v_i = -\alpha_{r_i}^i \xi_{r_i}^i - \alpha_{r_i-1}^i \xi_{r_i-1}^i - \cdots - \alpha_1^i \xi_1^i \tag{4.41}$$

When expressed in the original coordinates, the decoupling control law (4.38) has the form,

$$u = -A^{-1}(x)b(x) + A^{-1}(x)v \tag{4.42}$$

where $A(x)$ is the decoupling matrix (4.31) and the m-dimensional vector $b(x)$ has elements $L_f^{r_i} h_i(x)$.

Nominal Stability

The input-output decoupling control law generally provides only partial linearization of the closed-loop system. An expression for the remaining $(n-r)$-dimensional nonlinear subsystem can be derived from (4.34) and (4.38),

$$\dot{\eta} = q(\xi, \eta) - P(\xi, \eta)A^{-1}(\xi, \eta)b(\xi, \eta) + P(\xi, \eta)A^{-1}(\xi, \eta)v \tag{4.43}$$

where the definition of the matrix P follows directly from (4.34). If each input v_i is designed as (4.41), the nonlinear subsystem can be written as $\dot{\eta} = \tilde{q}(\xi, \eta)$. The zero dynamics are obtained by setting $\xi(t) = 0$ for all $t \geq 0$:

$$\dot{\eta} = \tilde{q}(0, \eta) \tag{4.44}$$

As in the SISO case, local asymptotic stability of the zero dynamics is a sufficient condition for local asymptotic stability of the closed-loop system [27]. By contrast, this condition is not necessary for MIMO systems [83]. Sufficient conditions for an input-output decoupled system to be globally asymptotically stable are presented elsewhere [27, 148, 160].

4.3 State-Space Linearization

4.3.1 Linear System

The state-space linearization approach is applied to the SISO linear system (4.3). Although feedback linearization obviously is unnecessary in the linear case, this exercise illustrates the basic concepts of the nonlinear design procedure. First, we attempt to determined an output function,

$$\hat{y} = \hat{C}x \tag{4.45}$$

such that the relative degree with respect to $\hat{y}$ is equal to the system order: $r = n$. The variable $\hat{y}$ is called an ***artificial output*** because it generally is different than the controlled output y in (4.3). The artificial output can be derived by solving the following set of n linear equations for the vector $\hat{C}$,

$$\hat{C}\begin{bmatrix} B & AB & \cdots & A^{n-2}B & A^{n-1}B \end{bmatrix} = \begin{bmatrix} 0 & 0 & \cdots & 0 & \beta \end{bmatrix} \tag{4.46}$$

where β is a non-zero value. A solution exists if and only if the matrix $[B\ AB\ \cdots\ A^{n-1}B]$ is full rank; *i.e.* if and only if (A, B) are a controllable pair [28]. In this case, the solution is not unique because $\hat{C}A^{n-1}B$ can be any non-zero value.

We now derive the control law assuming that an artificial output has been determined. Consider the following change of coordinates:

$$\xi = Tx = \begin{bmatrix} \hat{C} \\ \hat{C}A \\ \vdots \\ \hat{C}A^{n-1} \end{bmatrix} x \tag{4.47}$$

It is easy to show that the linear system has the following normal form representation in the transformed coordinates,

$$\begin{aligned} \dot{\xi}_1 &= \xi_2 \\ \dot{\xi}_2 &= \xi_3 \\ &\vdots \\ \dot{\xi}_n &= R\xi + ku \\ y &= \xi_1 \end{aligned} \tag{4.48}$$

where $R = \hat{C}A^nT^{-1}$ and $k = \hat{C}A^{n-1}B$. The static state feedback control law,

$$u = \frac{v - R\xi}{k} \tag{4.49}$$

changes the last equation in (4.48) to: $\dot{\xi}_n = v$. The input v is designed to stabilize the transformed system:

$$v = -\alpha_n \xi_n - \alpha_{n-1}\xi_{n-1} - \cdots - \alpha_1 \xi_1 \tag{4.50}$$

In the original coordinates, the controller has the following form,

$$u = \frac{-\hat{C}A^n x - \alpha_n \hat{C}A^{n-1}x - \cdots - \alpha_1 \hat{C}x}{\hat{C}A^{n-1}B} \tag{4.51}$$

The transformed system has the following closed-loop characteristic equation:

$$s^n + \alpha_n s^{n-1} + \cdots + \alpha_2 s + \alpha_1 = 0 \tag{4.52}$$

Nominal stability of the transformed system — and therefore the original system — is guaranteed by choosing the controller tuning parameters α_i such that the characteristic polynomial is Hurwitz. Recall that nominal stability for the input-output linearization approach is ensured only if the linear system is minimum phase. On the other hand, the state-space linearization approach has the following disadvantages: (1) the linear system must be controllable; and (2) it may be difficult to satisfy output tracking objectives since the relationship between the *actual* output and the transformed state variables is: $y = CT^{-1}\xi$. We will see that these distinctions also hold in the nonlinear case.

4.3.2 Controller Design

The *state-space linearization* problem is to find (if possible) a diffeomorphism and a static state feedback control law such that the map between the transformed input and *entire* vector of transformed state variable is linear. This problem was orginally posed by Korobov [92]; a complete solution for the single-input case was provided by Brockett [24]. An alternative solution that facilitates the construction of the linearizing transformations also has been developed [81, 157]. In this section, the solution of the state-space linearization problem for single-input systems is presented; the multivariable case is discussed elsewhere [80, 86, 87]. An extension of the controller design procedure for disturbances is discussed in a later section.

Illustrative Example

For simplicity, we first consider the two-dimensional nonlinear system (4.13). In this case, an artificial output which yields a relative degree $r = 2$ can be determined by inspection: $\hat{y} = \hat{h}(x) = x_2$. If the diffeomorphism $\xi = \Phi(x)$ is chosen as,

$$\begin{aligned} \xi_1 &= \hat{h}(x) = x_2 \\ \xi_2 &= L_f\hat{h}(x) = f_2(x_1, x_2) \end{aligned} \tag{4.53}$$

the system has the following normal form representation:

$$\begin{aligned} \dot{\xi}_1 &= \xi_2 \\ \dot{\xi}_2 &= L_f^2\hat{h}[\Phi^{-1}(\xi)] + L_gL_f\hat{h}[\Phi^{-1}(\xi)]u \\ \hat{y} &= \xi_1 \end{aligned} \tag{4.54}$$

Assuming $L_gL_f\hat{h}$ is non-zero in the region of interest, the static state feedback control law,

$$u = \frac{v - L_f^2\hat{h}[\Phi^{-1}(\xi)]}{L_gL_f\hat{h}[\Phi^{-1}(\xi)]} \tag{4.55}$$

yields: $\dot{\xi}_2 = v$. Thus, the map between the transformed input v and the transformed state vector ξ is linear. The input v can be used to design a linear controller for the feedback linearized system. It is important to note that the map between ξ and the actual output y generally is *nonlinear*.

General Design Procedure

We now consider the n-dimensional nonlinear system (4.1). For the moment, assume that an artificial output $\hat{y} = \hat{h}(x)$ that yields $r = n$ has been determined. The existence and construction of such an output are discussed below. Consider the coordinate transformation:

$$\xi_k = \Phi_k(x) = L_f^{k-1}\hat{h}(x), \quad 1 \leq k \leq n \tag{4.56}$$

In the new coordinates, the system has the following normal form representation:

$$\begin{aligned}
\dot{\xi}_1 &= \xi_2 \\
\dot{\xi}_2 &= \xi_3 \\
&\vdots \\
\dot{\xi}_n &= b(\xi) + a(\xi)u \\
\hat{y} &= \xi_1
\end{aligned} \tag{4.57}$$

where $a(\xi) = L_g L_f^{n-1}\hat{h}[\Phi^{-1}(\xi)]$ and $b(\xi) = L_f^n\hat{h}[\Phi^{-1}(\xi)]$. If the function $a \neq 0$ throughout the region of operation, the static state feedback control law,

$$u = \frac{v - b(\xi)}{a(\xi)} \tag{4.58}$$

changes the n-th equation of the normal form to: $\dot{\xi}_n = v$. As a result, the map between the transformed input and each of the transformed state variables is linear. A linear state feedback controller can be synthesized for the state-space linearized system. For instance, the pole-placement design (4.50) yields the closed-loop characteristic polynomial (4.52). When expressed in the original coordinates, the complete control law is:

$$u = \frac{-L_f^n\hat{h}(x) - \alpha_n L_f^{n-1}\hat{h}(x) - \cdots - \alpha_1\hat{h}(x)}{L_g L_f^{n-1}\hat{h}(x)} \tag{4.59}$$

Artificial Output Determination

The controller design procedure presented above demonstrates that the existence of an artificial output that yields a maximal relative degree ($r = n$) is a sufficient condition for the solution of the state-space linearization problem. Although not shown here, this condition is also necessary [82]. We now present necessary and sufficient conditions for the existence of such an output. The proof of sufficiency provides a constructive procedure for determining a suitable output function.

Because the relative degree must equal the system order, it follows that the output function $\hat{h}(x)$ must satisfy:

$$L_g L_f^{k-1}\hat{h}(x) = 0, \qquad 1 \leq k \leq n-1 \tag{4.60}$$
$$L_g L_f^{n-1}\hat{h}(x) \neq 0$$

The determination of $\hat{h}(x)$ from these conditions requires the solution of a set of partial differential equations that include derivatives up to order $n-1$. However, the conditions can be rewritten as a set of first-order partial differential equations using the Lie bracket operator introduced in Chapter 3 [82]:

$$L_{ad_f^{k-1}g}\hat{h}(x) = 0, \qquad 1 \leq k \leq n-1 \tag{4.61}$$
$$L_{ad_f^{n-1}g}\hat{h}(x) \neq 0$$

Using the Frobenius theorem (see Chapter 3), it can be shown that a solution to these equations exists — and therefore the state-space linearization problem is solvable — if and only if the nonlinear systems is [157]:

1. Controllable – the matrix $\left[g(x)\ ad_f g(x)\ \cdots\ ad_f^{n-1}g(x)\right]$ has rank n.

2. Integrable – the vector fields $g(x), ad_f g(x), \ldots, ad_f^{n-2}g(x)$ are involutive.

Note that a set of vector fields $\{X_1(x), \ldots, X_p(x)\}$ is involutive if there exists scalar functions $\delta_{ijk}(x)$ such that:

$$ad_{X_i}X_j(x) = \sum_{k=1}^{p} \delta_{ijk}(x)X_k(x), \qquad 1 \leq i,j \leq p,\ i \neq j \tag{4.62}$$

If the system is two-dimensional, the integrability condition is always satisfied [82] and only the controllability condition must be checked. Using the theory discussed in Chapter 3, it is easy to show that the two-dimensional system (4.13) is controllable if $L_g L_f \hat{h}(x) \neq 0$.

Practical Issues

A potential disadvantage of the state-space linearization approach is that the artificial output $\hat{y}$ generally is different than the controlled output y. As a result, y usually is a nonlinear function of the transformed state variables: $y = h[\Phi^{-1}(\xi)]$. In this case, it is difficult to design a state-space linearizing controller to satisfy output tracking objectives because desirable $\hat{y}$ behavior does not necessarily imply desirable y behavior. State-space linearization therefore is most appropriate for stabilization problems in which the controlled output is not specified *a priori*.

For instance, consider the incorporation of integral action into the state-space linearizing control law (4.59). By analogy to the input-output linearization approach, the following control law is proposed:

$$u = \frac{-L_f^n \hat{h}(x) - \alpha_n L_f^{n-1}\hat{h}(x) - \cdots + \alpha_1 \left[y_{sp} - \hat{h}(x)\right] + \alpha_0 \int_0^t \left[y_{sp} - \hat{h}(x)\right] d\tau}{L_g L_f^{n-1}\hat{h}(x)} \tag{4.63}$$

This control law guarantees that $\hat{y} \to y_{sp}$ as long as the closed-loop system is asymptotically stable. However, offset-free tracking of y will be achieved if and only if $h(x_0) = \hat{h}(x_0)$, where x_0 is the equilibrium point corresponding to y_{sp}. A linear state/output map can be ensured by reformulating the control objectives in terms of an output that is a linear combination of the ξ state variables. However, this approach is difficult or even impossible to employ in practice. In general, the state-space linearizing controller must provide simultaneous linearization of the input/state and state/output maps to have output tracking capabilities. Unfortunately, the necessary and sufficient conditions for complete linearization are considerably more restrictive than those for state-space linearization [35, 121].

Additional disadvantages of the state-space linearization approach are that the existence conditions may be difficult to verify and the partial differential equations (4.61) may be difficult to solve analytically. It is not possible to achieve complete linearization of the input/state map if either of the existence conditions is not satisfied. However, some degree of partial linearization usually can be achieved. For example, in the input-output linearization approach an r-dimensional subsystem is linearized. In some applications, it may be desirable to maximize the dimension of the linear subsystem. Maximal linearization can be achieved by constructing an output function which yields the highest possible relative degree. Necessary and sufficient conditions for the solution of this problem are available [105, 122].

4.3.3 Nominal Stability

Nominal stability analysis for the input-output linearization approach is complex due to the presence of a nonlinear subsystem (the zero dynamics) in the closed-loop system. By contrast, the state-space linearization approach yields complete linearization of the state equations. As a result, zero dynamics are not present and nominal stability analysis is much simpler. This is the single most important advantage of state-space linearization as compared to input-output linearization.

Consider the closed-loop system comprised of the nonlinear system (4.1) and the state-space linearizing controller (4.59). In the transformed coordinates, the closed-loop system can be written as,

$$\begin{aligned} \dot{\xi} &= A\xi \\ \hat{y} &= \xi_1 \end{aligned} \tag{4.64}$$

where the matrix A has the characteristic equation (4.52). If the control law is well defined (*i.e.* $L_gL_f^{n-1}\hat{h} \neq 0$), then the state variables ξ converge exponentially to the origin for any initial condition $\xi(0)$ as long as the controller tuning parameters α_i are chosen such that (4.52) is a Hurwitz polynomial. Because the diffeomorphism $\Phi(x)$ is invertible and satisfies $\Phi(0) = 0$, exponential convergence of ξ to the origin implies asymptotic convergence of the actual state variables x to the origin.

4.3.4 Disturbances

An extension of the state-space linearization approach for nonlinear systems with disturbances is presented below. This problem is discussed in a series of papers by Calvet and Arkun [29, 30, 32, 33]. For simplicity, we consider the SISO nonlinear system (4.26) with a single disturbance. However, similar results are available for processes with multiple disturbances [30].

As in the disturbance-free case, we attempt to find an artificial output $\hat{y} = \hat{h}(x)$ which yields a relative degree r equal to the system order n. Such an output exists if and only if the controllability and integrability conditions in Section 4.3.2 are satisfied. If an artificial output can be constructed, the relative degree ρ for the disturbance d can be defined as in the input-output linearization approach (Section 4.2.4). Note that the relative degrees will always satisfy $\rho \leq r$. The artificial output function $\hat{h}(x)$ is used to construct the diffeomorphism (4.56). For a general value of ρ, the nonlinear system (4.26) will have the following normal form,

$$\begin{aligned}
\dot{\xi}_1 &= \xi_2 \\
&\vdots \\
\dot{\xi}_{\rho-1} &= \xi_\rho \\
\dot{\xi}_\rho &= \xi_{\rho+1} + \zeta_\rho(\xi)d \\
&\vdots \\
\dot{\xi}_{n-1} &= \xi_n + \zeta_{n-1}(\xi)d \\
\dot{\xi}_n &= b(\xi) + a(\xi)u + \zeta_n(\xi)d \\
\hat{y} &= \xi_1
\end{aligned} \tag{4.65}$$

where $a(\xi)$ and $b(\xi)$ are defined as in the disturbance-free case, and $\zeta_k(\xi) = L_p\Phi_k\left[\Phi^{-1}(\xi)\right]$.

Disturbance Decoupling

The normal form shows that the disturbance cannot be completely decoupled from the artificial output $\hat{y}$ with a state feedback control law. In fact, it is easy to prove that the disturbance decoupling problem (without measurement) cannot be solved for *any* output that is a linear combination of the transformed state variables. Most importantly, the state-space approach does not provide a systematic framework for deriving decoupling control laws for the *actual* output y. This is a significant disadvantage of state-space linearization as compared to input-output linearization, which yields disturbance decoupling if the relative degree of the actual output is less than ρ.

However, the disturbance decoupling problem with measurement can be solved if the matching condition $\rho = n$ holds. In this case, the following feedforward/state feedback control law is employed:

$$u = \frac{v - b(\xi) - \zeta_n(\xi)d}{a(\xi)} \tag{4.66}$$

This control law changes the final equation in the normal form (4.65) to: $\dot{\xi}_n = v$. Hence, the transformed system is completely linear and the input v

can be designed as usual. When expressed in the original coordinates, the feedforward/feedback controller is:

$$u = \frac{v - L_f^n \hat{h}(x) - L_p L_f^{n-1} \hat{h}(x) d}{L_g L_f^{n-1} \hat{h}(x)} \tag{4.67}$$

However, this control law generally does not provide disturbance decoupling with respect to the actual output.

Stabilization of the Quasi-Linear System

In most process applications, the disturbance is unmeasured and/or the matching assumption $\rho = n$ does not hold. In this case, the state-space linearizing control law (4.58) yields a *quasi-linear* system,

$$\begin{aligned} \dot{\xi} &= A\xi + Bv + \zeta(\xi)d \\ \hat{y} &= C\xi \end{aligned} \tag{4.68}$$

where the definitions of A, B, C, and ζ follow directly from the feedback linearized version of the normal form (4.65). The objective is to design a *linear* controller that stabilizes the quasi-linear system (and therefore stabilizes the original nonlinear system). A controller synthesis procedure based on Lyapunov stability theory has been proposed [32, 33]. The technique is based on two rather restrictive assumptions: (1) the disturbance matching condition $\rho = n$ holds; and (2) the nonlinear vector function ζ satisfies a nonlinear growth condition. The first assumption can be relaxed, but this generally results in poor performance. The interested reader is referred to the original papers [32, 33] for further details.

4.4 Advanced Topics

In this section, several advanced topics on feedback linearizing controller design are discussed. We focus on the input-output linearization approach because it usually is more useful for process applications. Results are presented for the state-space linearization approach as appropriate. For the sake of brevity, we omit several important topics such as approximate feedback linearization [63, 104], adaptive feedback linearization [88, 149], and non-minimum phase compensation [64, 161].

4.4.1 General Nonlinear Systems

We extend the input-output linearization approach to *general nonlinear systems* that are not necessarily affine in the manipulated input:

$$\begin{aligned} \dot{x} &= f(x,u) \\ y &= h(x) \end{aligned} \tag{4.69}$$

This problem is important because some processes are naturally described by non-affine models. Two alternative controller design strategies are presented [67, 163]. The first technique employs the non-affine nonlinear system (4.69) directly, while the second approach is based on an *extended system* that is control affine. Although not discussed here, similar results are available for the state-space linearization approach [134, 157].

Controller Design Based on the Original System

In this case, the input-output linearizing controller design is based directly on the non-affine system (4.69). We define the Lie derivative of the scalar function $h(x)$ with respect to the vector function $f(x, u)$ as:

$$L_f h(x,u) = \frac{\partial h(x)}{\partial x} f(x,u) \tag{4.70}$$

Higher-order Lie derivatives are defined as,

$$L_f^k h(x,u) = \frac{\partial L_f^{k-1} h(x,u)}{\partial x} f(x,u) \tag{4.71}$$

where: $L_f^0 h(x,u) = h(x)$. The system is said to have relative degree r at the point (x_0, u_0) if:

1. $\frac{\partial}{\partial u} L_f^k h(x,u) = 0$ for all x in a neighborhood of x_0, all u in a neighborhood of u_0, and all $k < r$.

2. $\frac{\partial}{\partial u} L_f^r h(x_0, u_0) \neq 0$.

It follows that the first r Lie derivatives do not depend explicitly on the input: $L_f^k h(x,u) = L_f^k h(x), \quad 0 \leq k \leq r-1$.

The diffeormorphism $[\xi^T, \eta^T]^T = \Phi(x)$ that places the nonlinear system (4.69) in normal form is constructed as follows. The ξ variables are chosen as,

$$\xi_k = \Phi_k(x) = L_f^{k-1}h(x) \tag{4.72}$$

where $1 \le k \le r$. The η variables can be chosen as,

$$\eta_k = \Phi_{r+k}(x), \quad 1 \le k \le n-r \tag{4.73}$$

such that their time derivatives are independent of u: $L_f\Phi_{r+k}(x,u) = L_f\Phi_{r+k}(x)$. When expressed in the new coordinates, the nonlinear system has the following normal form representation,

$$\begin{aligned}
\dot{\xi}_1 &= \xi_2 \\
\dot{\xi}_2 &= \xi_3 \\
&\vdots \\
\dot{\xi}_r &= r(\xi, \eta, u) \\
\dot{\eta} &= q(\xi, \eta) \\
y &= \xi_1
\end{aligned} \tag{4.74}$$

where $r(\xi,\eta,u) = L_f^r h\left[\Phi^{-1}(\xi,\eta), u\right]$ and the $(n-r)$-dimensional vector q has components $q_k(\xi,\eta) = L_f\Phi_{r+k}\left[\Phi^{-1}(\xi,\eta)\right]$.

The input-output linearizing control law is obtained by solving the following nonlinear algebraic equation for the input u:

$$r(\xi, \eta, u) = v \tag{4.75}$$

If this equation is solvable, the input-output response is linearized as the r-th equation in the normal form (4.74) becomes: $\dot{\xi}_r = v$. The transformed input v is designed as in the control affine case. When expressed in terms of the original state variables, the controller equation is:

$$L_f^r h(x, u) = v \tag{4.76}$$

In most cases, this equation will not have an analytical solution and therefore it must be solved on-line at each sampling point. For a solution to exist at the point (x_0, v_0), the following two conditions must be satisfied:

1. There exists a u_0 such that $L_f^r h(x_0, u_0) = v_0$.

2. $\frac{\partial}{\partial u} L_f^r h(x_0, u_0) \neq 0$.

The second condition holds since the system is assumed to have a well defined relative degree; the first condition may not hold. If both conditions are satisfied, the implicit function theorem (Chapter 3) guarantees the existence of a locally defined static state feedback control law,

$$u = \Psi(x, v) \tag{4.77}$$

where the implicit function Ψ satisfies $L_f^r h\,[x, \Psi(x, v)] = v$.

Controller Design Based on an Affine System

In this case, the input-output linearizing controller design is based on an *extended system* that is control affine. A new manipulated input w is defined as $w = \dot{u}$, and u is viewed as a state variable. By defining the extended state vector as $\bar{x} = [x^T \; u]^T$, the nonlinear system (4.69) can be represented as,

$$\begin{aligned} \dot{\bar{x}} &= \bar{f}(\bar{x}) + \bar{g}(\bar{x})w \\ y &= \bar{h}(\bar{x}) \end{aligned} \tag{4.78}$$

where:

$$\hat{f}(\bar{x}) = \begin{bmatrix} f(x, u) \\ 0 \end{bmatrix}, \quad g(\bar{x}) = \begin{bmatrix} 0 \\ 1 \end{bmatrix}, \quad \bar{h}(\bar{x}) = h(x) \tag{4.79}$$

Due to the introduction of the integrator, the relative degree of the extended system is equal to $r + 1$, where r is the relative degree of the original non-affine system.

Because the extended system (4.78) is control affine, the input-output linearizing controller can be designed in the usual manner. The resulting control law has the following form when expressed in terms of the actual input u:

$$\dot{u} = \frac{v - L_{\bar{f}}^{r+1} \bar{h}(x, u)}{L_{\bar{g}} L_{\bar{f}}^r \bar{h}(x, u)} \tag{4.80}$$

A *dynamic* state feedback control law of the form (4.7) can be obtained by defining the controller state variable as $\zeta = u$:

$$\begin{aligned}\dot{\zeta} &= -\frac{L_{\bar{f}}^{r+1}\bar{h}(x,\zeta)}{L_{\bar{f}}^{r+1}\bar{h}(x,\zeta)} + \frac{1}{L_{\bar{f}}^{r+1}\bar{h}(x,\zeta)}v \\ u &= \zeta\end{aligned} \tag{4.81}$$

The transformed input v can be designed to place the closed-loop poles:

$$v = -\alpha_{r+1}L_{\bar{f}}^{r}\bar{h}(x,u) - \alpha_r L_{\bar{f}}^{r-1}\bar{h}(x) - \cdots - \alpha_1\bar{h}(x) \tag{4.82}$$

4.4.2 Time Delay Compensation

Many nonlinear processes contain time delays due to transportation lags and measurement delays. In this section, an extension of the basic input-output linearizing design strategy for time delay systems is presented [73, 102]. A similar technique has been developed for the state-space linearization approach [79]. The SISO nonlinear system is assumed to have the form,

$$\begin{aligned}\dot{x} &= f(x) + g(x)u(t-\theta) \\ y &= h(x)\end{aligned} \tag{4.83}$$

where θ is a known, constant time delay associated with the manipulated input. Note that a time delay in the output can be handled simply by combining an input-output linearizing controller and a linear output predictor (*e.g.* Smith predictor) [73].

Motivation

We demonstrate that an input-output linearizing controller for the time delay system (4.83) requires future values of the process state variables. The system can be placed in the following normal form using the standard change of coordinates:

$$\begin{aligned}\dot{\xi}_1 &= \xi_2 \\ \dot{\xi}_2 &= \xi_3 \\ &\vdots\end{aligned} \tag{4.84}$$

$$\begin{aligned}\dot{\xi}_r &= b(\xi,\eta) + a(\xi,\eta)u(t-\theta)\\ \dot{\eta} &= q(\xi,\eta)\\ y &= \xi_1\end{aligned}$$

The input-output linearizing control law is:

$$u(t-\theta) = \frac{v(t) - b(\xi,\eta)}{a(\xi,\eta)} \tag{4.85}$$

Hence, the current input $u(t)$ must be computed as:

$$u(t) = \frac{v(t+\theta) - b[\xi(t+\theta),\eta(t+\theta)]}{a[\xi(t+\theta),\eta(t+\theta)]} \tag{4.86}$$

When expressed in the original coordinates, the linearizing control law has the form:

$$u(t) = \frac{v(t+\theta) - L_f^r h[x(t+\theta)]}{L_g L_f^{r-1} h[x(t+\theta)]} \tag{4.87}$$

This control law cannot be implemented directly because $x(t+\theta)$ and $v(t+\theta)$ are not known at time t.

Controller Design

The following strategy is employed to make the linearizing control law (4.87) implementable. First, the future state vector $x(t+\theta)$ is replaced by the estimate $\hat{x}(t+\theta|t)$. The calculation of $\hat{x}(t+\theta|t)$ is discussed below. In addition, the future input $v(t+\theta)$ is replaced by the estimate $\hat{v}(t+\theta|t)$, which is computed as:

$$\hat{v}(t+\theta|t) = -\alpha_r L_f^{r-1} h[\hat{x}(t+\theta|t)] - \cdots - \alpha_1 h[\hat{x}(t+\theta|t)] + w(t) \tag{4.88}$$

Note that (4.88) is identical to (4.19) except that $\hat{x}(t+\theta|t)$ has been substituted for $x(t)$ and the signal $w(t)$ has been introduced. As discussed below, $w(t)$ is the output of a linear time delay compensator. The overall control law can be written as,

$$u(t) = \frac{w(t) - \sum_{k=1}^{r+1} \alpha_k L_f^{k-1} h[\hat{x}(t+\theta|t)]}{L_g L_f^{r-1} h[\hat{x}(t+\theta|t)]} \tag{4.89}$$

where $\alpha_{r+1} = 1$.

Under the assumption that $\hat{x}(t+\theta|t) = x(t+\theta)$ for all $t \geq 0$, it can be shown that the controller tuning parameters α_k can be chosen to yield the following closed-loop transfer function in the absence of plant/model mismatch [73]:

$$\frac{y(s)}{w(s)} = \frac{\epsilon^r e^{-\theta s}}{(\epsilon s+1)^r} \tag{4.90}$$

A linear controller is designed for the transfer function (4.90) by considering w as the manipulated input. The following control law provides time delay compensation and integral action:

$$\frac{w(s)}{y_{sp}(s) - y(s)} = \frac{\epsilon^{-r}(\epsilon s+1)^r}{(\epsilon s+1)^r e^{-\theta s}} \tag{4.91}$$

This controller can be implemented as a linear Smith predictor [150]. By combining (4.90) and (4.91), it is easy to show that the overall transfer function for setpoint changes is,

$$\frac{y(s)}{y_{sp}(s)} = \frac{e^{-\theta s}}{(\epsilon s+1)^r} \tag{4.92}$$

where ϵ is the controller tuning parameter.

Predictor Design

The remaining task is to compute the one-time-delay-ahead estimate of the process state vector $\hat{x}(t+\theta|t)$. The following prediction formula has been proposed [73],

$$\begin{aligned} \hat{x}(t+\theta|t) &= x(t) + \int_t^{t+\theta} [f[\hat{x}(\tau|t)] + g[\hat{x}(\tau|t)]u(\tau-\theta)]\, d\tau + \\ & \quad \lambda\,[x(t) - \hat{x}(t|t-T_d)] \end{aligned} \tag{4.93}$$

where λ and T_d are tuning parameters and the predictor is initialized with the plant state $\hat{x}(t|t) = x(t)$. The filter parameter λ accounts for modeling errors; it should be maintained in the range $0 \leq \lambda < \frac{\theta}{T_d}$. It is chosen near the upper limit if there is little plant/model mismatch and near zero if significant modeling errors are present. The term $\hat{x}(t|t-T_d)$ in the prediction (4.93) is calculated as,

$$\hat{x}(t|t-T_d) = x(t-T_d) + \int_{t-T_d}^{t} [f[\hat{x}(\tau|t-T_d)] + g[\hat{x}(\tau|t-T_d)]u(\tau-\theta)]\,d\tau \tag{4.94}$$

where $\hat{x}(t-T_d|t-T_d) = x(t-T_d)$. If the model is perfect and the predictor is initialized as $\hat{x}(0) = x(\theta)$, it can be shown that perfect state predictions are obtained when $\lambda = 1$ [73].

4.4.3 Constrained Nonlinear Systems

Many processes have constraints on input and output variables. In this section, we present an input-output linearization strategy for constrained nonlinear systems [109]. The basic idea is to map the original constraints into constraints on the feedback linearized system. The resulting linear system is controlled with a linear model predictive controller with explicit constraint handling capability. For the sake of brevity, we do not discuss alternative constraint compensation schemes based on feedback linearization [30, 91, 131, 152].

Constraint Mapping

The controller design is based on the SISO nonlinear model (4.1) with the following constraints:

$$u_{min} \leq u \leq u_{max}, \quad \Delta u_{min} \leq \Delta u \leq \Delta u_{max}, \quad y_{min} \leq y \leq y_{max} \tag{4.95}$$

The objective is to transform these constraints into constraints on the feedback linearized system. First, the linear subsystem is discretized to facilitate the linear model predictive controller design,

$$\begin{aligned} \xi(k+1) &= A_d\xi(k) + B_d v(k) \\ y(k) &= C\xi(k) \end{aligned} \tag{4.96}$$

where A_d is an r x r matrix, B_d is an r x 1 vector, and C is an 1 x r vector. The eigenvalues of A_d are on the unit circle, but the pair (A_d,B_d) is controllable.

The next step is to map the constraints on the original nonlinear system into constraints on the discretized linear system (4.96). The output constraints for the two systems are identical since the output is not transformed as part of the controller design. By contrast, the constraints on the actual input u must be mapped into constraints on the transformed input v. This transformation must be performed at each sampling instant because the mapping is state dependent (shown below). Moreover, the transformation must be performed for the entire control horizon N of the predictive controller. The constraint mapping is performed using the feedback linearizing control law (4.18) and the current measurement of the process state vector $x(k)$:

$$v(k) = L_f^r h[x(k)] + L_g L_f^{r-1} h[x(k)]u(k) = b[x(k)] + a[x(k)]u(k) \tag{4.97}$$

The constraints on the first input in the control horizon, $v(k|k)$, are calculated as,

$$\begin{aligned} v_{min}(k|k) &= \min_{u(k)} \quad b[x(k)] + a[x(k)]u(k) \\ v_{max}(k|k) &= \max_{u(k)} \quad b[x(k)] + a[x(k)]u(k) \end{aligned} \tag{4.98}$$

where:

$$\begin{aligned} u_{min} \leq \quad & u(k) \quad \leq u_{max} \\ \Delta u_{min} + u(k-1) \leq \quad & u(k) \quad \leq \Delta u_{max} + u(k-1) \end{aligned} \tag{4.99}$$

This optimization problem is trivial to solve since the objective function is affine in $u(k)$.

By contrast, computing the constraints for future inputs in the control horizon, $v(k+1|k), \ldots, v(k+N-1|k)$, is problematic. This difficulty occurs because future values of the input and state variables are needed to compute the future constraints, as indicated by the mapping (4.97). A simple way to overcome this problem is to extend the constraints calculated for $v(k|k)$ over the entire control horizon. An important property of this technique is that the *implemented* input,

$$u(k) = \frac{v(k|k) - L_f^r h[x(k)]}{L_g L_f^{r-1} h[x(k)]} \tag{4.100}$$

is guaranteed to satisfy the *actual* input constraints. On the other hand, the constraints calculated for the future inputs (which are not implemented) generally will not agree with the actual constraints. This is a serious disadvantage since incorrect future constraints may lead to implemented control moves that are overly conservative or aggressive. More accurate constraints can be obtained by utilizing the control moves calculated at the previous iteration of the predictive controller [109].

Linear Model Predictive Controller Design

The constrained linear model used for model predictive controller design has the form,

$$\begin{aligned} \xi(k+1) &= A_d\xi(k) + B_d v(k), \quad v_{min}(k) \le v(k) \le v_{max}(k) \\ y(k) &= C\xi(k), \quad y_{min} \le y(k) \le y_{max} \end{aligned} \tag{4.101}$$

It is important to note that the input constraints are *time varying*. The linear model is used to predict the effects of future control moves on future values of the output. To obtain improved predictions in the presence of plant/model mismatch, at each time step the linear model state is initialized with the current plant state:

$$\xi(k|k) = \begin{bmatrix} h[x(k)] \\ L_f h[x(k)] \\ \vdots \\ L_f^{r-1} h[x(k)] \end{bmatrix} \tag{4.102}$$

The matrix A_d is unstable because all its eigenvalues are located on the unit circle. Consequently, a model predictive controller design technique specifically developed for unstable linear system is employed [130]. At each time step, the manipulated input is generated by solving the following open-loop optimal control problem [130],

$$\begin{aligned} \min_{V(k)} \quad & s\left[v(k+N-1|k) - v_s\right]^2 + \sum_{j=0}^{N-1} \left[\xi(k+j|k) - \xi_s\right]^T Q \left[\xi(k+j|k) - \xi_s\right] + \\ & r\left[v(k+j|k) - v_s\right]^2 + s\left[v(k+j|k) - v(k+j-1|k)\right]^2 \end{aligned} \tag{4.103}$$

where: $\xi(k+j|k)$ is the predicted value of $\xi(k+j)$; ξ_s and v_s are target values for ξ and v, respectively; $r > 0$ and $s \ge 0$ are scalar tuning parameters; and Q is a positive semidefinite tuning matrix. The decision vector $V(k)$ contains N values of the manipulated input: $V(k) = [v(k|k), v(k+1|k), \ldots, v(k+N-1|k)]^T$. Inputs beyond the control horizon are set equal to the target value: $v(k+j|k) = v_s,\ j \geq N$.

The targets ξ_s and v_s are calculated from the steady-state form of (4.101) under the condition that $y = y_{sp}$. The target values must lie within the feasible region defined by the input and output constraints for the optimal control problem to have a solution. In practice, a disturbance model is used to shift the targets in order to eliminate offset [109]. Because all the eigenvalues of A_d are located on the unit circle, a necessary condition for the optimization problem to have a solution is that the state variables are driven to their target values by the end of the control horizon: $\xi(k+N|k) = \xi_s$. Thus, the optimization problem must be solved subject to the following constraints:

$$\begin{aligned} &v_{min}(k) \leq v(k+j|k) \leq v_{max}(k), \quad 0 \leq j \leq N-1 \\ &y_{min} \leq C\xi(k+j|k) \leq y_{max}, \quad j \geq 1 \\ &\xi(k+N|k) = \xi_s \end{aligned} \tag{4.104}$$

If necessary, input and /or output constraints can be relaxed to ensure that the optimization problem is feasible [109, 145]. The optimal control problem (4.103) can be manipulated to yield a quadratic program that can be solved with standard software [130].

4.4.4 Robust Controller Design

We discuss robust controller design techniques based on the input-output and state-space linearization approaches. This is a very important topic because modeling errors usually preclude exact cancellation of nonlinear terms. The SISO nonlinear system is assumed to be described by the nominal nonlinear model (4.1) with state dependent perturbations:

$$\begin{aligned} \dot{x} &= f(x) + \Delta f(x) + [g(x) + \Delta g(x)]\, u \\ y &= h(x) \end{aligned} \tag{4.105}$$

Note that the perturbations $\Delta f(x)$ and $\Delta g(x)$ are *not* known. Available controller design techniques differ according to the *structure matching conditions* and *growth conditions* imposed on the perturbations. Less restrictive matching conditions generally require more restrictive bounds to be placed on the perturbations.

Input-Output Linearization Techniques

First we define relative degrees for the state dependent perturbations. The perturbation $\Delta f(x)$ is said to have relative degree γ_f at the point x_0 if:

1. $L_{\Delta f}L_f^k h(x) = 0$ for all x in a neighborhood of x_0 and all $k < \gamma_f - 1$.
2. $L_{\Delta f}L_f^{\gamma_f - 1} h(x_0) \neq 0$.

The relative degree γ_g for the pertubation $\Delta g(x)$ is defined analogously. For the moment, assume that the structure matching conditions $\gamma_f \geq r$ and $\gamma_g \geq r$ are satisfied. Using the standard change of coordinates, the uncertain system (4.105) has the following normal form representation,

$$
\begin{aligned}
\dot{\xi}_1 &= \xi_2 \\
\dot{\xi}_2 &= \xi_3 \\
&\vdots \\
\dot{\xi}_r &= b(\xi,\eta) + \Delta b(\xi,\eta) + [a(\xi,\eta) + \Delta a(\xi,\eta)]\, u \\
\dot{\eta} &= q(\xi,\eta) + r(\xi,\eta) + s(\xi,\eta)u \\
y &= \xi_1
\end{aligned}
\tag{4.106}
$$

where the functions a, b, and q are defined as usual and:

$$
\begin{aligned}
\Delta a(\xi,\eta) &= L_{\Delta g}L_f^{r-1}h\left[\Phi^{-1}(\xi,\eta)\right] \\
\Delta b(\xi,\eta) &= L_{\Delta f}L_f^{r-1}h\left[\Phi^{-1}(\xi,\eta)\right] \\
r_k(\xi,\eta) &= L_{\Delta f}\Phi_{r+k}\left[\Phi^{-1}(\xi,\eta)\right], \quad 1 \leq k \leq n-r \\
s_k(\xi,\eta) &= L_{\Delta g}\Phi_{r+k}\left[\Phi^{-1}(\xi,\eta)\right], \quad 1 \leq k \leq n-r
\end{aligned}
\tag{4.107}
$$

Under the more restrictive matching conditions that $\gamma_f > r$ and $\gamma_g > r$, the perturbations do not appear in the r-th equation in the normal form. As a result, the standard input-output linearizing control law (4.20) yields a stable closed-loop system if the perturbed zero dynamics are asymptotically stable [119].

On the other hand, plant/model mismatch should be considered explicitly in the controller design if $\gamma_f \leq r$ and/or $\gamma_g \leq r$. Several Lyapunov-based design techniques have been proposed for this important case. Kravaris and Palanki [100] assume the existence of scalar functions $\Delta f^*(x)$ and $\Delta g^*(x)$ that satisfy the matching conditions:

$$L_{\Delta f} L_f^{r-1} h(x) = g(x)\Delta f^*(x), \qquad \Delta g(x) = g(x)\Delta g^*(x) \tag{4.108}$$

These conditions are rather restrictive as they require that the peturbations be expressed in terms of the input vector $g(x)$. Behtash [22] considers similar, but slightly less restrictive, matching conditions. Arkun and Calvet employ the following matching conditions,

$$\Delta f(x) = \Delta f_1(x) + \Delta f_2(x), \qquad \Delta g(x) = g(x)\Delta g^*(x) \tag{4.109}$$

where $\Delta f_1(x)$ is a matched uncertainty and $\Delta f_2(x)$ is an unmatched uncertainty with relative degree $\gamma_{f_2} = r$. A stabilizing input-output linearizing controller is design by solving an algebraic Riccati equation. Liao *et al.* decompose both perturbations into matched and unmatched parts:

$$\Delta f(x) = g(x)\Delta f_1(x) + \Delta f_2(x), \qquad \Delta g(x) = g(x)\Delta g_1(x) + \Delta g_2(x) \tag{4.110}$$

Restrictive bounds must be placed on the unmatched uncertainties $\Delta f_2(x)$ and $\Delta g_2(x)$ to guarantee closed-loop stability.

State-Space Linearization Techniques

Su *et al.* [158] have shown that the state-space linearization approach possesses some degree of robustness to modeling errors. The analysis does not require matching conditions, but it is assumed that the plant and model are "close" in a topological sense. Controller design techniques that explicitly address plant/model mismatch usually employ: (1) matching conditions on both $\Delta f(x)$ and $\Delta g(x)$; or (2) no matching condition on $\Delta f(x)$, but $\Delta g(x) = 0$. The first group of robust controller design techniques are based on the following matching conditions:

$$\Delta f(x) = g(x)\Delta f^*(x), \quad \Delta g(x) = g(x)\Delta g^*(x) \tag{4.111}$$

In this case, the standard change of coordinates places the nonlinear system (4.1) in the normal form,

$$\begin{aligned}
\dot{\xi}_1 &= \xi_2 \\
\dot{\xi}_2 &= \xi_3 \\
&\vdots \\
\dot{\xi}_n &= b(\xi,\eta) + \Delta b(\xi,\eta) + [a(\xi,\eta) + \Delta a(\xi,\eta)]\, u \\
\hat{y} &= \xi_1
\end{aligned} \tag{4.112}$$

where the nonlinear functions Δa and Δb are defined as in (4.107) with $r = n$. Spong [155] uses L_2-optimal control theory to synthesize a stabilizing controller for the feedback linearized system, while Ha and Gilbert [59] employ a Lyapunov-based design strategy.

Several robust controller design techniques are based on the alternative matching condition $\Delta g(x) = 0$. In this case, the perturbation $\Delta f(x)$ is allowed to be arbitrary at the expense of assuming exact knowledge of the input vector $g(x)$. Spong and Vidyasagar [156] have proposed a design strategy in which the stable factorization approach is applied to the feedback linearized system. A Lyapunov design technique based on the solution of a algebraic Riccati equation has been developed by Calvet and Arkun [31]. Doyle and Morari [49] have proposed a robust controller design strategy based on conic sector bounds and approximate state-space linearization.

4.4.5 Discrete-Time and Sampled-Data Systems

Feedback linearizing controller design for discrete-time and sampled-data nonlinear systems is discussed below. In many ways, the results parallel the continuous-time results presented throughout this chapter. However, the operations used to construct the discrete-time controllers are quite different. In addition, discrete-time and sampled-data systems offer unique possibilities (*e.g.* deadbeat control) and complexities (*e.g.* sampling) not encountered in the continuous-time case. For simplicity, we focus on SISO models of the form:

$$\begin{aligned} x(k+1) &= F[x(k), u(k)] \\ y(k) &= h[x(k)] \end{aligned} \tag{4.113}$$

Because most processes are inherently continous, the function F usually is obtained by discretizing a continuous-time model or identified directly from process data. The discrete-time model is not assumed to be control affine because: (1) exact sampling generally does not yield an affine model; and (2) an affine model usually does not simplify the controller design task.

Discrete-Time Systems

For the moment, we neglect the effects of sampling and discuss the feedback linearization of inherently discrete-time systems. Sampled-data systems are considered below. For the sake of brevity, we focus on the input-output linearization problem [110, 127]. Similar results are available for disturbance decoupling [55, 134], input-output decoupling [56, 134], and state-space linearization [110, 134].

The *composition* of the scalar function $h(x)$ and the vector function $F(x)$ is defined as: $h \circ F(x) = h[F(x)]$. Higher order compositions are defined recursively: $h \circ F^i(x) = h \circ F^{i-1}[F(x)]$, where $h \circ F^0(x) = h(x)$. The composition operator plays the same role as does the Lie derivative in the continuous-time case. The discrete-time system (4.113) is said to have relative degree r at the point (x_0, u_0) if:

1. $\frac{\partial}{\partial u(k)} h \circ F^i[x(k), u(k)] = 0$ for all (x, u) in a neighborhood of (x_0, u_0) and all $i \leq r-1$.
2. $\frac{\partial}{\partial u(k)} h \circ F^r[x_0, u_0] \neq 0$.

By definition of the relative degree we can write,

$$h \circ F^i[x(k), u(k)] = h \circ F_0^i[x(k)], \quad 1 \leq k \leq r-1 \tag{4.114}$$

since these functions do not depend explicitly on the input $u(k)$.

If the relative degree is well defined, a diffeomorphism $[\xi^T(k), \eta^T(k)]^T = \Phi[x(k)]$ that places the system in normal form is constructed as follows. The ξ vaiables are chosen as,

$$\xi_i(k) = h \circ F_0^{i-1}[x(k)], \quad 1 \leq i \leq r \tag{4.115}$$

The remaining variables $\eta_i(k) = \Phi_{r+i}[x(k)]$, $1 \leq i \leq n-r$, can be chosen such that Φ is invertible and $\frac{\partial}{\partial u(k)}\Phi_{r+i}\circ F[x(k), u(k)] = 0$. As a result, the normal form is,

$$\begin{aligned}
\xi_1(k+1) &= \xi_2(k) \\
\xi_2(k+1) &= \xi_3(k) \\
&\vdots \\
\xi_r(k+1) &= a[\xi(k), \eta(k), u(k)] \\
\eta(k+1) &= q[\xi(k), \eta(k)] \\
y(k) &= \xi_1(k)
\end{aligned} \tag{4.116}$$

where:

$$a[\xi(k), \eta(k), u(k)] = h\circ F^r[\Phi^{-1}(\xi(k), \eta(k)), u(k)] \tag{4.117}$$

$$q_i[\xi(k), \eta(k)] = \Phi_{r+i}\circ F[\Phi^{-1}(\xi(k), \eta(k))], \quad 1 \leq i \leq n-r$$

Note that the functions q_i do not depend on $u(k)$ by construction.

The input-output linearizing control law is obtained by solving the following nonlinear algebraic equation for $u(k)$,

$$a[\xi(k), \eta(k), u(k)] = v(k) \tag{4.118}$$

where $v(k)$ is the transformed input. This equation can be solved locally if the conditions of the implicit function theorem are satisfied (see Section 4.4.1). When expressed in terms of the original state variables, the linearizing control law is:

$$h\circ F^r[x(k), u(k)] = v(k) \tag{4.119}$$

If the transformed input is chosen as $v(k) = y_{sp}(k)$, output deadbeat control is obtained [120]: $y(k+r) = y_{sp}(k)$. Alternatively, the linear controller can be designed as,

$$v(z) = \frac{g_d(z)}{1 - z^{-r} g_d(z)}[y_{sp}(z) - y(z)] \tag{4.120}$$

where $g_d(z)$ is the desired closed-loop transfer function. A simple choice for g_d is,

$$g_d(z) = \frac{1-\alpha}{1-\alpha z^{-1}} \tag{4.121}$$

where the tuning parameter α represents the closed-loop pole. Note that the linear control law has integral action. For this design, the nominal closed-loop transfer function for setpoint changes is:

$$\frac{y(z)}{y_{sp}(z)} = z^{-r}\frac{1-\alpha}{1-\alpha z^{-1}} \tag{4.122}$$

As in the continuous-time case, closed-loop stability can be ensured only if the zero dynamics,

$$\eta(k+1) = q[0, \eta(k)] \tag{4.123}$$

are asymptotically stable.

Sampled-Data Systems

Feedback linearizing controller design becomes considerably more complex if the effects of sampling are considered. We briefly describe the potential difficulties and proposed solutions. Monaco and Normand-Cyrot [128] demonstrate that exact sampling always produces a discrete-time system of relative degree one. They also show that if the continuous-time system is minimum phase and has relative degree r, sufficiently fast sampling always yields a discrete-time system with: (1) stable zero dynamics if $r = 1$; and (2) unstable zero dynamics if $r \geq 3$. As a result, a discretized input-output linearizing controller is likely to provide satisfactory performance if $r = 1$, but it may produce unexpected results if $r \geq 2$. Glad [54] also considered sampling effects in developing output deadbeat controllers for sampled-data systems. The $r \geq 3$ case is handled by employing a multi-rate sampling strategy in which the control move is implemented each sampling period but the state is only measured every r sampling periods.

It has been shown that state-space linearizability of a continuous-time system can be destroyed by sampling [57]. Arapostathis *et al.* [14] demonstrate that sampling can impose severe restrictions — in addition to the standard controllability and involutivity conditions — on the structure of the continuous-time system. For two-dimensional systems, state-space linearizability of the sampled-data system implies that the continuous-time system can be completely linearized by coordinate transformations alone. To overcome these difficulties, several state-space linearization techniques based on multi-rate sampling have been proposed. Grizzle and Kokotovic [57] consider implementing the control move each sampling period while measuring the state every N sampling periods, where $N \geq 2$.

4.5 Process Control Strategies and Applications

A variety of nonlinear controller design strategies have been developed for process applications. Many of these techniques are based, either implicitly or explicitly, on exact linearization of the input-output response. In this section, these linearizing control strategies are reviewed and critically evaluated. Applications of feedback linearizing control techniques to process systems also are discussed. The applications are categorized in terms of the unit operations involved; a representative list of references is provided in each case. Additional applications are described in other review articles on feedback linearizing control [68, 89, 94, 99, 126].

4.5.1 Process Control Techniques

Techniques for Processes of Relative Degree One

Generic model control [111, 113], internal decoupling [16, 17] and reference system synthesis [19] are nonlinear controller design techniques developed specifically for process control applications. We outline the generic model control (GMC) design strategy and show that GMC is an input-output linearization technique for processes of relative degree one. Similar analyses of the internal decoupling and reference system synthesis techniques are omitted for the sake of brevity.

Consider the SISO nonlinear system (4.1) The rate-of-change of the output can be written as:

$$\dot{y} = L_f h(x) + L_g h(x) u \tag{4.124}$$

Assume that the desired rate-of-change is,

$$\dot{y}_d = k_1(y_{sp} - y) + k_2 \int_0^t (y_{sp} - y)\, d\tau \tag{4.125}$$

where k_1 and k_2 are controller tuning parameters. The goal is to determine a control law such that $\dot{y} = \dot{y}_d$ for all $t \geq 0$. If the relative degree $r = 1$, then the function $L_g h(x) \neq 0$ and the following state feedback control law achieves the control objective:

$$u = \frac{k_1(y_{sp} - y) + k_2 \int_0^t (y_{sp} - y)\, d\tau - L_f h(x)}{L_g h(x)} \tag{4.126}$$

The closed-loop system is input-output linearized,

$$\dot{y} = k_1(y_{sp} - y) + k_2 \int_0^t (y_{sp} - y)\, d\tau \tag{4.127}$$

and has the following transfer function for setpoint changes if $y(0) = y_{sp}(0)$:

$$\frac{y(s)}{y_{sp}(s)} = \frac{k_1 s + k_2}{s^2 + k_1 s + k_2} \tag{4.128}$$

The tuning parameters k_1 and k_2 are used to place the closed-loop poles. Note that the GMC control law (4.126) and the input-output linearizing control law (4.21) are *identical* when $r = 1$.

Globally Linearizing Control

Globally linearizing control (GLC) [95] is a controller design strategy developed for nonlinear process applications. We show that GLC is an input-output linearization technique for processes of arbitrary relative degree. Consider the SISO nonlinear system (4.1). By definition of the relative degree, the first r derviative of the output are:

$$\begin{aligned} y^{(k)} &= L_f^k h(x), \quad 1 \leq k \leq r-1 \\ y^{(r)} &= L_f^r h(x) + L_g L_f^{r-1} h(x) u \end{aligned} \tag{4.129}$$

The state feedback control law is chosen as,

$$u = \frac{\bar{v} - \beta_r L_f^r h(x) - \beta_{r-1} L_f^{r-1} h(x) - \cdots - \beta_0 h(x)}{\beta_r L_g L_f^{r-1} h(x)} \tag{4.130}$$

where $\bar{v}$ is a new input and β_k are controller tuning parameters.

It follows from (4.129) that the proposed control law yields a linearized input-output response:

$$\beta_r y^{(r)} + \beta_{r-1} y^{(r-1)} + \cdots + \beta_0 y = \bar{v} \tag{4.131}$$

A proportional-integral controller is designed for the feedback linearized system,

$$\bar{v} = k_c \left[(y_{sp} - y) + \frac{1}{\tau_I} \int_0^t (y_{sp} - y)\, d\tau \right] \tag{4.132}$$

where the gain k_c and integral time τ_I are additional controller tuning parameters. The complete GLC control law obtained by combining (4.130) and (4.132) yields the following closed-loop transfer function for setpoint changes if $y(0) = y_{sp}(0)$:

$$\frac{y(s)}{y_{sp}(s)} = \frac{k_c s + \frac{k_c}{\tau_I}}{\beta_r s^{r+1} + \beta_{r-1} s^r + \cdots + (\beta_0 + k_c)s + \frac{k_c}{\tau_I}} \tag{4.133}$$

The tuning parameters β_k, k_c, and τ_I are used to place the closed-loop poles.

The GLC control law is closely related to the input-output linearizing control law (4.21). In fact, the two control laws are *identical* if $\beta_r = 1$ and the tuning parameters of the input-output linearizing controller are chosen as:

$$\alpha_0 = \frac{k_c}{\tau_I}, \qquad \alpha_1 = \beta_0 + k_c, \qquad \alpha_k = \beta_{k-1},\ 2 \le k \le r \tag{4.134}$$

Although not discussed here, extensions of the GLC technique for measured disturbances [40], multivariable processes [101], and discrete-time systems [151] parallel those presented earlier for the input-output linearization approach.

Nonlinear Internal Model Control Techniques

Internal model control (IMC) is a powerful controller design strategy for linear process models [129]. Two distinctive characteristics of IMC are: (1) the controller design is based on the inverse of the process model; and (2) the error between the plant and model outputs is used as a feedback signal. Several nonlinear controller design techniques which include these two features have been proposed for process applications [51, 69, 137]. In this section, we discuss a particular nonlinear IMC technique [69] and provide comparisons with the input-output linearization approach.

As in the linear case, the nonlinear IMC control law is comprised of a model inverse controller C and a robustness filter F. For the moment, we assume $F = 1$ and design the controller C to optimize the nominal performance measure,

$$\min_{C} \| y_{sp}(t) - y(t) \| \tag{4.135}$$

where $\| \cdot \|$ represents a desired norm. The feedback signal to C is $e = y_{sp} - y + \tilde{y}$, where y is the plant output and $\tilde{y}$ is the model output. In the absence of modeling errors, $e = y_{sp}$ and the plant output can be written as $y = MCy_{sp}$, where M represents the process model. As a result, the controller design problem can be reformulated as:

$$\min_{C} \| (1 - MC)y_{sp}(t) \| \tag{4.136}$$

If the system is initially at rest and $y(0) = y_{sp}(0)$, the performance criterion is identically zero for any norm and setpoint trajectory when C is chosen to be the right-inverse of the model (see Chapter 3),

$$u = \frac{y_{sp}^{(r)} - L_f^r h(\tilde{x})}{L_g L_f^{r-1} h(\tilde{x})} \tag{4.137}$$

where $\tilde{x}$ are the model state variables.

The model inverse controller C is not suitable for implementation because: (1) "perfect" control requires unreasonably large control moves; (2) the controller is not "proper" in the sense that it requires derivatives of the setpoint; and (3) the perfect model assumption is not satisfied in practice. As a result, C is augmented with the following *nonlinear* filter F:

$$v^{(r)} = -\alpha_r L_f^{r-1} h(\tilde{x}) - \alpha_{r-1} L_f^{r-2} h(\tilde{x}) - \cdots - \alpha_1 h(\tilde{x}) + \alpha_1 e \tag{4.138}$$

The filter output $v^{(r)}$ replaces the signal $y_{sp}^{(r)}$ in the nonlinear control law (4.137). If the system is initially at rest and $y(0) = y_{sp}(0)$, the controller parameters α_k can be chosen to yield the nominal closed-loop transfer function:

$$\frac{y(s)}{y_{sp}(s)} = \frac{1}{(\epsilon s + 1)^r} \tag{4.139}$$

The controller tuning parameter ϵ can take values in the range $0 < \epsilon < \infty$. The nonlinear IMC technique possesses similar stability, perfect control, and zero offset properties as linear IMC [129]. In fact, the linear and nonlinear IMC design strategies yield identical controllers when applied to a stable, minimum-phase linear model.

The nonlinear IMC controller can be interpreted as a variant of the input-output linearizing controller (4.21). Despite the similarities, the nonlinear IMC technique offers several unique features including:

1. A control structure in which the difference between the plant and model outputs is used as a feedback signal for the nonlinear controller.
2. Implicit integral action that is a result of the IMC control structure.
3. An output feedback controller implementation because the nonlinear controller uses state variables generated by the model.

The third feature represents a significant advantage of the nonlinear IMC strategy as compared to the input-output linearization approach, which is based on full-state feedback. However, the nonlinear IMC technique employs the model as an open-loop observer and therefore it is restricted to open-loop stable processes.

4.5.2 Process Control Applications

Chemical Reactors

Feedback linearizing control strategies have been applied to a wide variety of chemical reactor models. The most commonly used model describes the irreversible, exothermic reaction $A \rightarrow B$, occurring in a constant volume, continuously stirred tank reactor [66]:

$$\begin{aligned} \dot{C}_A &= \frac{q}{V}(C_{Af} - C_A) - k_0 \exp\left(-\frac{E}{RT}\right) C_A \qquad (4.140) \\ \dot{T} &= \frac{q}{V}(T_f - T) + \frac{(-\Delta H)}{\rho C_p} k_0 \exp\left(-\frac{E}{RT}\right) C_A + \frac{UA}{V \rho C_p}(T_c - T) \end{aligned}$$

The severe static and dynamic nonlinear behavior of this model are well documented [162]. The coolant temperature T_c or feed flowrate q is usually employed as the manipulated input, while either the reactor concentration C_A or temperature T is chosen as the controlled output. Other commonly

Table 4.1: Applications to Chemical Reactors

References	Approach	Focus
[95]	IOL	basic controller design
[1]	IOL	global feedback stabilization
[19, 40]	IOL	disturbance decoupling
[42]	IOL	input-output decoupling
[67]	IOL	general nonlinear processes
[73, 102]	IOL	time delay compensation
[8, 109, 91]	IOL	constrained processes
[3, 12, 15, 100]	IOL	robust controller design
[151, 154]	IOL	discrete-time models
[23, 51, 71, 136]	IOL	internal model control
[50, 143]	IOL	approximate feedback linearization
[48, 96, 169]	IOL	non-minimum phase compensation
[13, 21]	IOL	adaptive feedback linearization
[107, 125]	IOL	differential/algebraic models
[41, 97, 108, 170]	IOL	output feedback control
[11, 76, 89]	SSL	basic controller design
[7]	SSL	global feedback stabilization
[30, 32, 33]	SSL	feedforward/feedback control
[79]	SSL	time delay compensation
[49, 90]	SSL	robust controller design
[50]	SSL	approximate feedback linearization
[117]	SSL	output feedback control

used reactors models differ from (4.140) by including coolant jacket dynamics and/or by considering more complex reactions. A summary of applications of feedback linearization to chemical reactors is presented in Table 4.1. For each reference, the table contains the controller design approach employed (input-output linearization (IOL) or state-space linearization (SSL)) and the major focus of the paper.

Biological Reactors

Feedback linearizing controllers have been developed for several types of biological reactor models. The most commonly used model describes the growth of a single cell population from a single, rate-limiting substrate in a continuous stirred tank reactor [4]:

$$\begin{aligned} \dot{X} &= \mu(X,S)X - DX \\ \dot{S} &= -\sigma(X,S)X - D(S_f - S) \end{aligned} \tag{4.141}$$

Typically, the specific growth rate μ is modeled by a Monod relation,

$$\mu(X,S) = \mu(S) = \frac{\mu_m}{K_m + S} \tag{4.142}$$

and a constant yield expression is used to describe the specific substrate consumption rate σ:

$$\sigma(X,S) = \frac{1}{Y_{X/S}}\mu(S) \tag{4.143}$$

Bioreactor models of this form can exhibit significant static and dynamic nonlinear behavior [69]. The dilution rate D or feed substrate concentration S_f is usually employed as the manipulated input, while the cell concentration X or substrate concentration S is often chosen as the controlled output. A summary of applications of feedback linearization to bioreactor models is presented in Table 4.2.

Other Processes

Several other types of processes can benefit from feedback linearizing control as a result of their strongly nonlinear behavior. Important examples of such processes include polymerization reactors, high purity distillation columns, and weakly buffered pH neutralization systems. A summary of applications of feedback linearization to models of these processes is shown in Table 4.3.

Experimental Studies

As compared to simulation studies, relatively few experimental applications of feedback linearizing control have been presented. However, the number of experimental studies has increased significantly over the past several years. As shown in Table 4.4, input-output linearization techniques have

Table 4.2: Applications to Biological Reactors

References	Approach	Focus
[20, 38, 70]	IOL	basic controller design
[140]	IOL	input-output decoupling
[69]	IOL	internal model control
[70, 116]	IOL	approximate feedback linearization
[45]	IOL	non-minimum phase compensation
[46]	IOL	adaptive feedback linearization
[52, 70, 78]	SSL	basic controller design
[77]	SSL	multiple-input processes

Table 4.3: Applications to Other Processes

References	Process	Approach
[34, 36, 61, 126, 144]	distillation column	IOL
[6]	distillation column	SSL
[62, 112]	evaporator	IOL
[147]	heat exchanger	IOL
[5]	heat exchanger	SSL
[68, 74, 106, 137, 166, 168]	pH neutralization	IOL
[2, 43, 93, 101, 103, 124, 165]	polymerization reactor	IOL
[9, 10]	polymerization reactor	SSL
[114, 135]	storage tank	IOL
[141]	supercritical extractor	IOL

been applied to a variety of processes including biological reactors, distillation columns, pH neutralization systems, and polymerization reactors. By contrast, we have found no experimental applications of the state-space linearization approach.

Table 4.4: Experimental Studies

Reference	Process	Focus
[139]	bioreactor	input-output decoupling
[142, 146]	bioreactor	adaptive input-output linearization
[18, 39]	distillation column	input-output linearization
[115]	distillation column	disturbance decoupling
[47]	distillation column	input-output decoupling
[167, 171]	pH neutralization	input-output linearization
[58, 72]	pH neutralization	adaptive input-output linearization
[152]	polymerization reactor	input-output linearization
[153]	polymerization reactor	input-output decoupling

4.6 Case Studies

4.6.1 Continuous Stirred Tank Reactors in Series

Process Model

The process consists of two constant volume reactors in which an irreversible, exothermic reaction A $\rightarrow$ B occurs. The effluent stream from the first reactor serves as the feed stream for the second reactor. The reactors are cooled by a single coolant stream flowing cocurrently with the reaction stream. Neglecting coolant dynamics, the process model consists of four nonlinear ordinary differential equations [67]:

$$\begin{aligned}
\dot{C}_{A1} &= \frac{q}{V_1}(C_{Af} - C_{A1}) - k_0 C_{A1}\exp\left(-\frac{E}{RT_1}\right) \\
\dot{T}_1 &= \frac{q}{V_1}(T_f - T_1) + \frac{(-\Delta H)k_0 C_{A1}}{\rho C_p}\exp\left(-\frac{E}{RT_1}\right) \\
&\quad + \frac{\rho_c C_{pc}}{\rho C_p V_1} q_c \left[1 - \exp\left(-\frac{UA_1}{q_c \rho_c C_{pc}}\right)\right](T_{cf} - T_1) \qquad (4.144) \\
\dot{C}_{A2} &= \frac{q}{V_2}(C_{A1} - C_{A2}) - k_0 C_{A2}\exp\left(-\frac{E}{RT_2}\right)
\end{aligned}$$

Table 4.5: Nominal Parameters for the CSTR Model

Variable	Value	Variable	Value
q	100 L/min	k_0	7.2×10^{10} min^{-1}
C_{Af}	1 mol/L	$\frac{E}{R}$	1×10^4 K
T_f	350 K	$-\Delta H$	4.78×10^4 J/mol
T_{cf}	350 K	ρ, ρ_c	1000 g/L
V_1, V_2	100 L	C_p, C_{pc}	0.239 J/g·K
hA_1, hA_2	1.67×10^5 J/min·K		

$$\begin{aligned}\dot{T}_2 &= \frac{q}{V_2}(T_1 - T_2) + \frac{(-\Delta H)k_0 C_{A2}}{\rho C_p}\exp\left(-\frac{E}{RT_2}\right) + \frac{\rho_c C_{pc}}{\rho C_p V_2} q_c \\ &\times \left[1 - \exp\left(-\frac{UA_2}{q_c \rho_c C_{pc}}\right)\right]\left[T_1 - T_2 + \exp\left(-\frac{UA_1}{q_c \rho_c C_{pc}}\right)(T_{cf} - T_1)\right]\end{aligned}$$

We have used standard notation [162] where the subscripts 1, 2, c, and f denote the first reactor, second reactor, coolant stream, and feed stream, respectively. Nominal parameters for the model are shown in Table 4.5. The objective is to control the effluent composition from the second tank (C_{A2}) by manipulating the coolant flow rate (q_c). Note that the model is not control affine when $u = q_c$. If the state variables and controlled output are defined as,

$$x^T = \begin{bmatrix} C_{A1} & T_1 & C_{A2} & T_2 \end{bmatrix}, \quad y = C_{A2} \tag{4.145}$$

the model has the form of the general nonlinear system (4.69). As a result, the nonlinear controller design is based on the input-output linearization techniques discussed in Section 4.4.1.

Controller Design

For the purpose of comparison, three controllers are designed: a conventional proportional-integral (PI) controller; a static input-output linearizing controller; and a dynamic input-output linearizing controller. The PI controller is tuned for setpoint changes, yielding $k_c = 350$ L^2/mol·min and $\tau_I = 0.25$ min. The nonlinear controllers are designed by assuming that all four state variables are available for feedback. The static nonlinear controller design

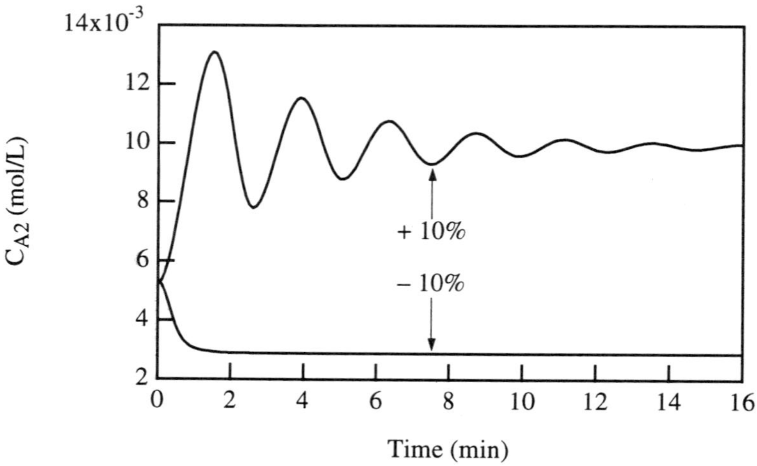

Figure 4.1: Open-loop response for ± 10% change in q_c [67].

is based on the original nonlinear model (4.144), which has relative degree $r = 2$. The control moves of the static controller are generated by solving a nonlinear algebraic equation of the form (4.76) at each sampling instant. The dynamic nonlinear controller is designed using an extended model of the form (4.78), which has relative degree $r = 3$ in this case. For both nonlinear controllers, the transformed input v is chosen to yield a nominal closed-loop transfer function of the form (4.23) where $\epsilon = 0.25$ min. The expressions for the two nonlinear controllers are rather complicated, and therefore they are not presented here.

Simulation Results

The open-loop composition responses in Figure 4.1 demonstrate that the model exhibits highly nonlinear behavior. The disturbance rejection performance of the three controllers for ± 5% unmeasured disturbances in the feed composition (C_{Af}) is shown in Figure 4.2. The nonlinear controllers cannot completely decouple the disturbance from the output because C_{Af} has relative degree $\rho = 1$. The static nonlinear controller provides excellent performance and is clearly superior to the other controllers for both disturbances. Note that the PI controller provides more effective rejection of the + 5% disturbance than does the dynamic nonlinear controller.

The controllers are compared for unmeasured feed temperature (T_f) dis-

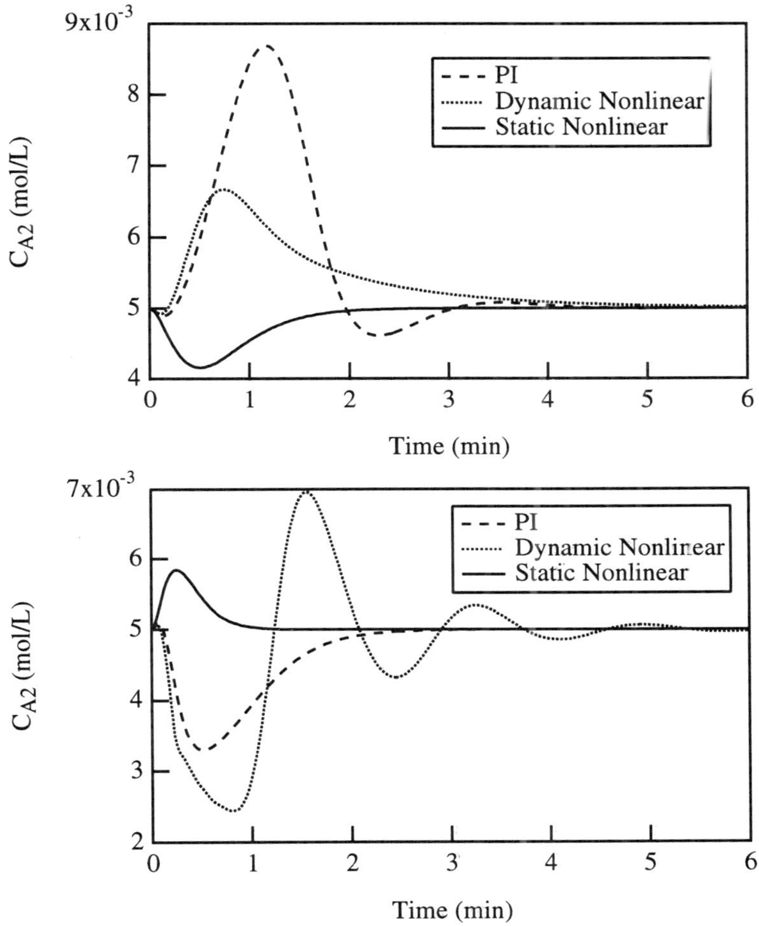

Figure 4.2: Closed-loop response for C_{Af} disturbances of -5 % (top) and $+5$ % (bottom) [67].

turbances of $\pm$ 25 K in Figure 4.3. Because T_f has relative degree $\rho = 3$, the static nonlinear controller provides perfect disturbance decoupling and therefore it outperforms the other controllers. The dynamic nonlinear controller provides vastly superior performance as compared to the PI controller for the $-$ 25 K disturbance. The responses of the two controllers are comparable for the $+$ 25 K disturbance. These results indicate that a disturbance can be rejected most effectively if the difference between the relative de-

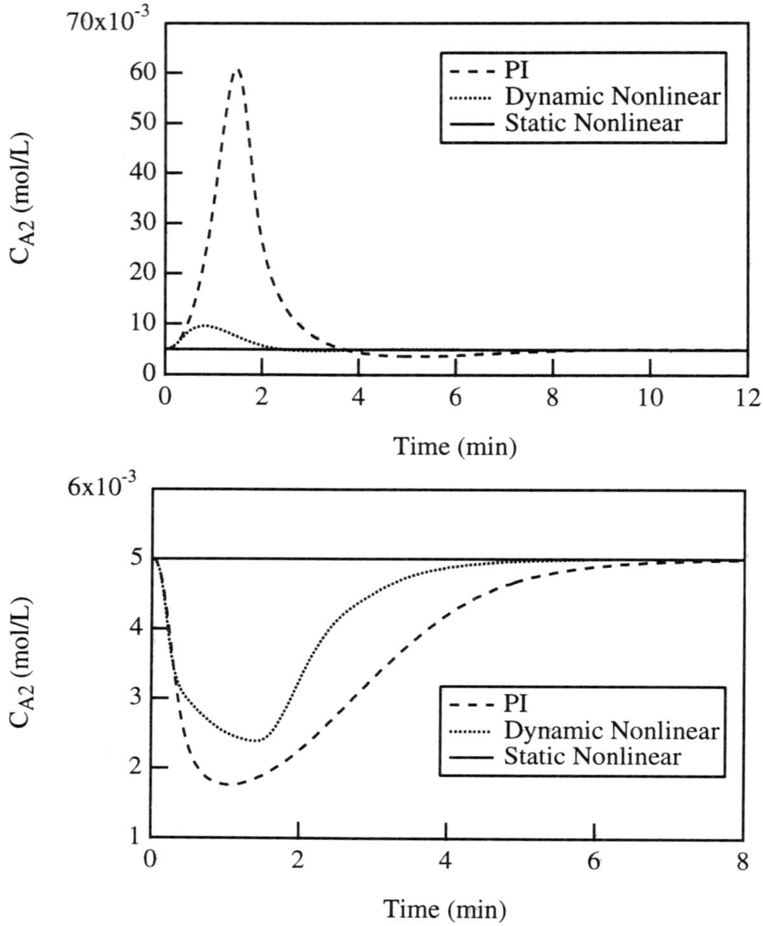

Figure 4.3: Closed-loop response for q disturbances of -25 % (top) and $+5$ % (bottom) [67].

grees of the manipulated input and the disturbance is small. This difference is always smaller for the static controller than for the dynamic controller. Consequently, we propose that the manipulated input and controlled output of an input-output linearizing controller should be chosen such that the relative degree r is *minimized*. This proposition provides another argument against the state-space linearization approach, which results in a *maximal* relative degree.

4.6.2 Continuous Fermentor

Process Model

The process consists of a constant volume reactor in which a single, rate limiting substrate promotes biomass growth and product formation. By assuming constant yields, a process model with three nonlinear ordinary differential equations can be obtained [4],

$$\begin{aligned} \dot{X} &= -DX + \mu(S,P)X \\ \dot{S} &= D(S_f - S) - \frac{1}{Y_{X/S}}\mu(S,P)X \\ \dot{P} &= -DP + [\alpha\mu(S,P) + \beta]\, X \end{aligned} \tag{4.146}$$

where: X, S, and P are the biomass, substrate, and product concentrations, respectively; D is the dilution rate; S_f is the feed substrate concentration; and $Y_{X/S}$, α, and β are yield parameters. The specific growth rate μ is modeled as,

$$\mu(S,P) = \frac{\mu_m\left(1 - \frac{P}{P_m}\right)S}{K_m + S + \frac{S^2}{K_i}} \tag{4.147}$$

where: μ_m is the maximum specific growth rate; and P_m, K_m, and K_i are constant parameters. Nominal operating conditions are shown in Table 4.6.

The control objective in many continuous fermentations is to maximize the steady-state biomass production. This can be a difficult task since parameters such as the maximum specific growth rate μ_m and cell-mass yield $Y_{X/S}$ may exhibit significant time-varying behavior. It can be shown [69] that near optimal steady-state performance can be achieved by manipulating the dilution rate D and regulating the biomass concentration X at a constant value. A nonlinear model of the form (4.1) can be obtained by choosing:

$$u = D, \quad x^T = \begin{bmatrix} X & S & P \end{bmatrix}, \quad y = X \tag{4.148}$$

Controller Design

Three controllers are designed: a PI controller; an input-output linearizing controller based on full state feedback; and a nonlinear IMC controller that

Table 4.6: Nominal Operating Conditions for the Fermentor Model

Variable	Value	Variable	Value
$Y_{X/S}$	0.4 g/g	α	2.2 g/g
β	0.2 h^{-1}	μ_m	0.48 h^{-1}
P_m	50 g/L	K_m	1.2 g/L
K_i	22 g/L	S_f	20 g/L
D	0.202 h^{-1}	X	6.0 g/L
S	5.0 g/L	P	19.14 g/L

requires only a biomass concentration measurement. The PI controller parameters are determined initially by applying IMC tuning rules [129] to a first-order linear model obtained from the open-loop responses in Figure 4.4. The IMC closed-loop time constant is chosen as one-third the open-loop time constant. The controller parameters are fine-tuned for setpoint responses, yielding $k_c = 0.07$ L/g·h and $\tau_I = 4.5$ h.

The nonlinear controllers are designed by noting that the relative degree $r = 1$ since $L_g h(x) = -x_1$. The input output linearizing controller is synthesized as described in Section 4.2.2:

$$u = \frac{v - \mu(x_2, x_3)x_1}{-x_1} \tag{4.149}$$

$$v = \frac{2}{\epsilon}[y_{sp} - x_1] + \frac{1}{\epsilon^2}\int_0^t [y_{sp} - x_1]\,d\tau$$

Using the design procedure in Section 4.5.1, the following nonlinear IMC controller is obtained,

$$u = \frac{\dot{v} - \mu(\tilde{x}_2, \tilde{x}_3)\tilde{x}_1}{-\tilde{x}_1} \tag{4.150}$$

$$\dot{v} = -\frac{1}{\epsilon}\tilde{x}_1 + \frac{1}{\epsilon}(y_{sp} - x_1 + \tilde{x}_1)$$

where the tilde represents a variable obtained from the process model. Both nonlinear controllers are tuned with $\epsilon = 1$ h, which is approximately one-third the open-loop time constant for the – 10% dilution rate change in Figure 4.4.

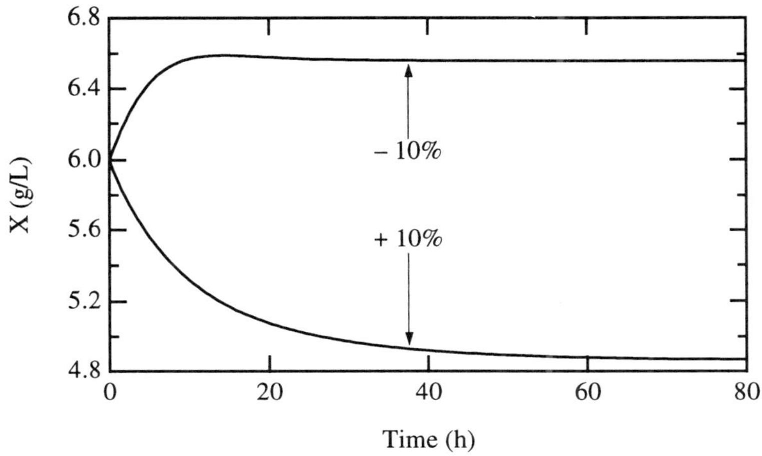

Figure 4.4: Open-loop response for ± 10% change in D [69].

Simulation Results

The open-loop biomass concentration response shown in Figure 4.4 demonstrate that the fermentor exhibits significant static and dynamic nonlinear behavior. In Figure 4.5, the three controllers are compared for an unmeasured step disturbance of – 12.5% in the maximum specific growth rate (μ_m). The input-output linearizing cannot provide complete decoupling because the disturbance relative degree $\rho = 1$ and μ_m is unmeasured. However, the linearizing controller yields superior regulatory performance as compared to the PI and nonlinear IMC controllers. This result is expected since the linearizing controller has access to the entire state vector. On the other hand, the nonlinear IMC controller yields vastly improved disturbance rejection as compared to the PI controller despite the fact that both controllers only use a measurement of the biomass concentration.

The controllers are compared for a – 20% step disturbance in the cell-mass yield ($Y_{X/S}$) in Figure 4.6. The input-output linearizing controller yields perfect disturbance decoupling because it has access to the entire state vector and $\rho = 2$. By contrast, the nonlinear IMC controller uses state estimates from the model and therefore it cannot provide exact decoupling. However, the IMC controller yields superior performance as compared to the PI controller. The estimated state variables used by the IMC controller also are shown in Figure 4.6. Despite the significant errors caused by the

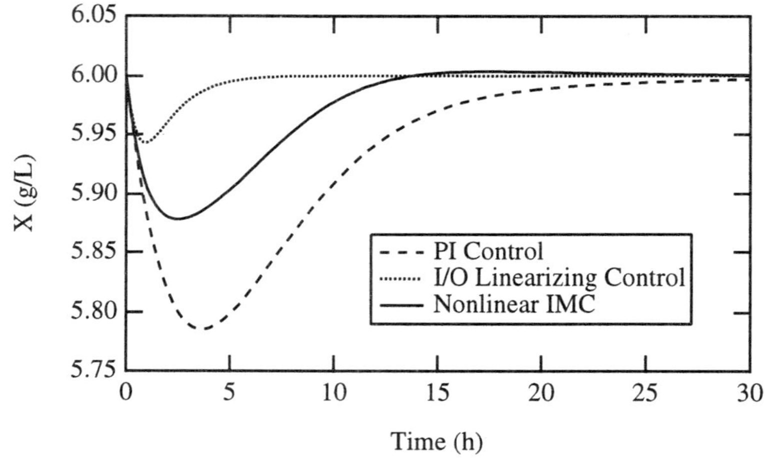

Figure 4.5: Closed-loop response for – 12.5% disturbance in μ_m [69].

unmeasured disturbance, the IMC controller provides satisfactory control.

Figure 4.7 shows the responses of the controllers responses for the same $Y_{X/S}$ disturbance when there is a stuctural error in the specific growth rate. The growth rate used for nonlinear controller design is a simple Monod expression:

$$\mu = \frac{\mu_m}{K_m + S} \tag{4.151}$$

The PI controller response is identical to that shown in Figure 4.6 since the modeling error does not affect the process. By contrast, the nonlinear controller responses are changed significantly. The response of the nonlinear IMC controller is actually improved, while the input-output linearizing controller is no longer able to provide complete decoupling. In fact, the IMC controller outperforms the linearizing controller in this case. Additional simulation results [69] show that the nonlinear IMC controller is superior to the PI controller and compares favorably to the input-output linearizing controller based on full-state feedback.

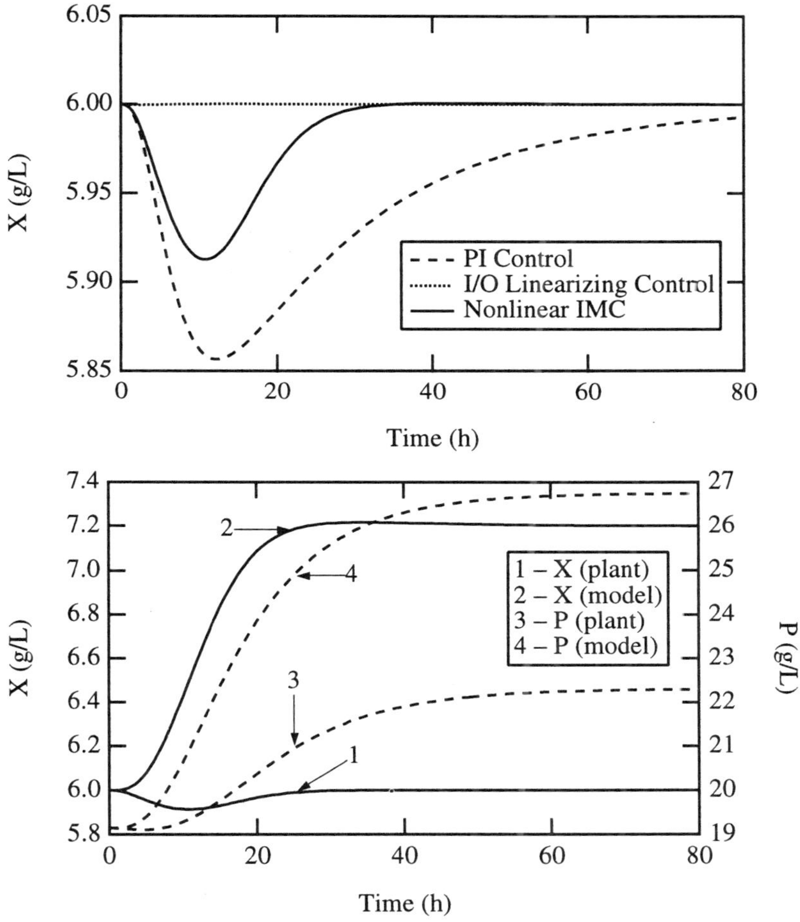

Figure 4.6: Closed-loop response for – 20% $Y_{X/S}$ disturbance [69].

4.6.3 pH Neutralization System

Process Model

The process consists of an acid (HNO_3) stream, buffer ($NaHCO_3$) stream, and base ($NaOH$) stream that are mixed in a stirred tank. The chemical equilibria is modeled by introducing two reaction invariants [58] for each inlet stream,

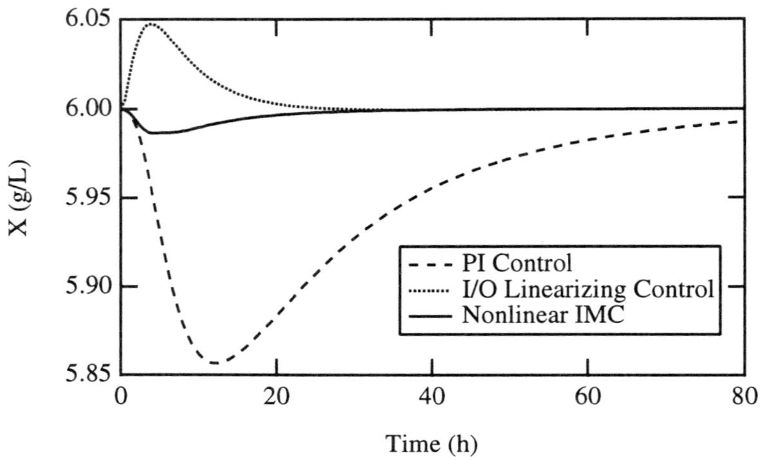

Figure 4.7: Closed-loop response with modeling error for $-$ 20% $Y_{X/S}$ disturbance [69].

$$\begin{aligned} W_{ai} &= [H^+]_i - [OH^-]_i - [HCO_3^-]_i - 2[CO_3^=]_i \\ W_{bi} &= [H_2CO_3]_i + [HCO_3^-]_i + [CO_3^=]_i \end{aligned} \tag{4.152}$$

where $i = 1$ for the acid stream, $i = 2$ for the buffer stream, and $i = 3$ for the base stream. By combining mass balances on each of the ionic species in the system, the following differential equations for the effluent reaction invariants can be derived [60]:

$$\begin{aligned} \dot{W}_{a4} &= \frac{1}{Ah}(W_{a1} - W_{a4})q_1 + \frac{1}{Ah}(W_{a2} - W_{a4})q_2 + \frac{1}{Ah}(W_{a3} - W_{a4})q_3 \\ \dot{W}_{b4} &= \frac{1}{Ah}(W_{b1} - W_{b4})q_1 + \frac{1}{Ah}(W_{b2} - W_{b4})q_2 + \frac{1}{Ah}(W_{b3} - W_{b4})q_3 \end{aligned} \tag{4.153}$$

where: q_1, q_2, and q_3 are the volumetric flow rates of the acid, buffer, and base streams, respectively; A is the cross-sectional area of the mixing tank; and h is the liquid level.

The effluent pH is determined from W_{a4} and W_{b4} using the following relation,

Table 4.7: Nominal Operating Conditions for the pH System

Variable	Value	Variable	Value
W_{a1}	0.003 M	W_{b1}	0 M
W_{a2}	– 0.03 M	W_{b2}	0.03 M
W_{a3}	– 3.05×10^{-3} M	W_{b3}	5×10^{-5} M
K_{a1}	4.47×10^{-7}	K_{a2}	5.62×10^{-11}
A	207 cm^2	n	0.607
z	11.5 cm	q_1	16.6 ml/s
q_2	0.55 ml/s	q_3	15.6 ml/s
W_{a4}	– 4.32×10^{-4} M	W_{b4}	5.28×10^{-4} M
h	14.0 cm	pH	7.0

$$W_{a4} + 10^{pH-14} - 10^{-pH} + W_{b4}\frac{1 + 2 \times 10^{pH-pK_2}}{1 + 10^{pK_1-pH} + 10^{pH-pH_2}} = 0 \qquad (4.154)$$

where pK_1 and pK_2 are the base-10 logarithms of the equilibrium constants associated with H_2CO_3 and HCO_3^- disassociation, respectively. Because the pH probe is located downstream from the mixing tank, there is an unmodeled time delay of approximately 10 s associated with the pH measurement. The liquid level is modeled as,

$$\dot{h} = \frac{1}{A}\left[q_1 + q_2 + q_3 - C_v(h+z)^n\right] \qquad (4.155)$$

where C_v is the valve coefficient, n is the valve exponent, and z is the vertical distance between the bottom of the mixing tank and the outlet of the effluent stream. On-line measurements of the liquid level and effluent pH are available, while the reaction invariants must be estimated. Nominal operating conditions are shown in Table 4.7.

The objective is to control the pH despite unmeasured acid and buffer flow rate disturbances by manipulating the base flow rate. It is important to have an accurate estimate of the buffer flow rate because it determines the buffering capacity of the systems. A nonlinear state-space model is obtained by defining:

$$x^T = \left[\begin{array}{ccc} W_{a4} & W_{b4} & h \end{array}\right], \quad u = q_3, \quad y = pH, \quad d = q_2 \qquad (4.156)$$

The resulting model has the form,

$$\begin{aligned} \dot{x} &= f(x) + g(x)u + p(x)d \\ c(x,y) &= 0 \end{aligned} \tag{4.157}$$

where the definitions of the functions $f(x)$, $g(x)$, $p(x)$, and $c(x,y)$ follow directly from the process model. Note that the output equation is an *implicit* function of the output (y); that is, a closed form representation $y = h(x)$ cannot be determined. As a result, a modified technique is proposed for the design of the input-output linearizing controller.

Input-Output Linearizing Controller Design

The input-output linearizing controller is designed by taking the time derivative of the output equation and rearranging,

$$\dot{y} = -c_y^{-1}(x,y)c_x(y)[f(x) + g(x)u + p(x)d] \tag{4.158}$$

where:

$$\begin{aligned} c_x^T(y) &= \begin{bmatrix} 1 & \frac{1+2\times10^{y-pK_2}}{1+10^{pK_1-y}+10^{y-pK_2}} & 0 \end{bmatrix} \\ c_y(x,y) &= \ln(10)\left[10^{y-14} + 10^{-y}\right] + \\ & \ln(10)\left[x_2 \frac{10^{pK_1-y} + 10^{y-pK_2} + 4\left(10^{pK_1-y}\right)\left(10^{y-pK_2}\right)}{(1+10^{pK_1-y}+10^{y-pK_2})^2}\right] \end{aligned} \tag{4.159}$$

Because the term $c_y^{-1}(x,y)c_x(y)g(x)$ that multiplies u is non-zero for all operating points of interest, the model is said to have relative degree $r = 1$. The input-output linearizing control law is:

$$u = \frac{v + c_y^{-1}(x,y)c_x(y)[f(x) + p(x)d]}{-c_y^{-1}(x,y)c_x(y)g(x)} \tag{4.160}$$

The transformed input v is chosen as,

$$v = 2\epsilon^{-1}[y_{sp} - y] + \epsilon^{-2}\int_0^t [y_{sp} - y]d\tau \tag{4.161}$$

where ϵ is the controller tuning parameter. In the absence of plant/model mismatch, the control law yields the closed-loop transfer function (4.23) with $r = 1$. The controller is tuned with $\epsilon = 1$ min, which is approximately one-half the time constant for the open-loop responses in Figure 4.8. To facilitate experimental implementation, the control law is discretized with $\Delta t = 15s$ [72].

State and Parameter Estimation

The input-output linearizing controller requires full state feedback. In practice, the reaction invariants cannot be measured and therefore they must be estimated from available on-line measurements. A closed-loop nonlinear observer with exponentially stable error dynamics (see Chapter 6) does not exist since the Jacobian linearization of the model is unobservable at every equilibrium point [72]. As an alternative, an open-loop nonlinear observer that does not require observability is employed. Because the differential equations for the reaction invariants are decoupled, the invariants can be estimated sequentially [72]. Invariant W_{b4} is estimated in an open-loop fashion. An estimate of W_{a4} is generated from the estimated W_{b4} and the measured pH using the output equation.

The proposed controller does not explicitly account for buffering changes. As shown below, unstable behavior is obtained if the buffer flow rate drops significantly below the nominal value. Under these conditions, the process gain increases dramatically and some type of on-line adaptation is necessary to achieve satisfactory performance. A straightforward approach is to treat the buffer flow rate as an unknown parameter that is estimated with a recursive least-squares algorithm. The parameter estimator receives measurements of the pH and level, as well as reaction invariant estimates from the open-loop observer. A detailed derivation of the estimator is presented elsewhere [72].

Experimental Results

Open-loop pH responses for the sequence of base and buffer flow rate changes in Table 4.8 are shown in Figure 4.8. Note that the actual run time is plotted and therefore the plots do not begin with zero time. The response in Figure 4.8(a) shows that the system exhibits significant static nonlinearities with respect to manipulated input changes. The pH response for buffer flow rate changes is shown in Figure 4.8(b). Significant static nonlinear behavior is

Table 4.8: Standard Inlet Flow Changes

Time (min)	q_1 (ml/s)	q_2 (ml/s)	q_3 (ml/s)
0	16.6	0.55	15.6
30	15.1	1.2	13.6
60	18.1	2.0	17.6
90	16.6	1.0	15.6
120	–	0	–
150	–	0.55	–

observed, especially when $q_2 \to 0$ ml/s. The pH response for base flow rate changes is similar to that obtained from the process model, while significant deviations are observed between the simulated and experimental responses for buffer flow rate changes [65].

As discussed in [72], non-adaptive and adaptive versions of the input-output linearizing controller easily outperform a PI controller. However, Figure 4.9 shows that the non-adaptive controller exhibits unstable behavior for low buffering conditions. The sequence of buffer flow rate disturbances is the same as that in Table 4.8 except that the flow rate is only reduced to 0.2 ml/s at $t = 120$ min. The non-adaptive controller produces a highly oscillatory response as a result of the $q_2 = 0.2$ ml/s disturbance at $t = 215$ min. This unstable behavior is attributable to poor estimates of the reaction invariants at low buffering conditions.

The performance of the adaptive nonlinear controller for the same sequence of buffer flow rate disturbances is shown in Figure 4.10. The controller is able to provide excellent control over a wide range of buffering conditions. Unlike the non-adaptive controller (Figure 4.9), the adaptive controller is able to maintain the system at the setpoint even when $q_2 \to 0.2$ ml/s. The estimated buffer flow rate produced by the adaptive controller also is shown in Figure 4.10. At steady-state conditions, the estimation error is less than 15% of the actual value. Improved knowledge of the buffering capacity results in more accurate estimates of the reaction invariants, which ultimately leads to superior closed-loop performance as compared to the non-adaptive case.

The performance of the adaptive nonlinear controller for the sequence of acid flow rate disturbances in Table 4.8 is shown in Figure 4.11. Acid flow rate disturbances represent a robustness test since the parameter estimator

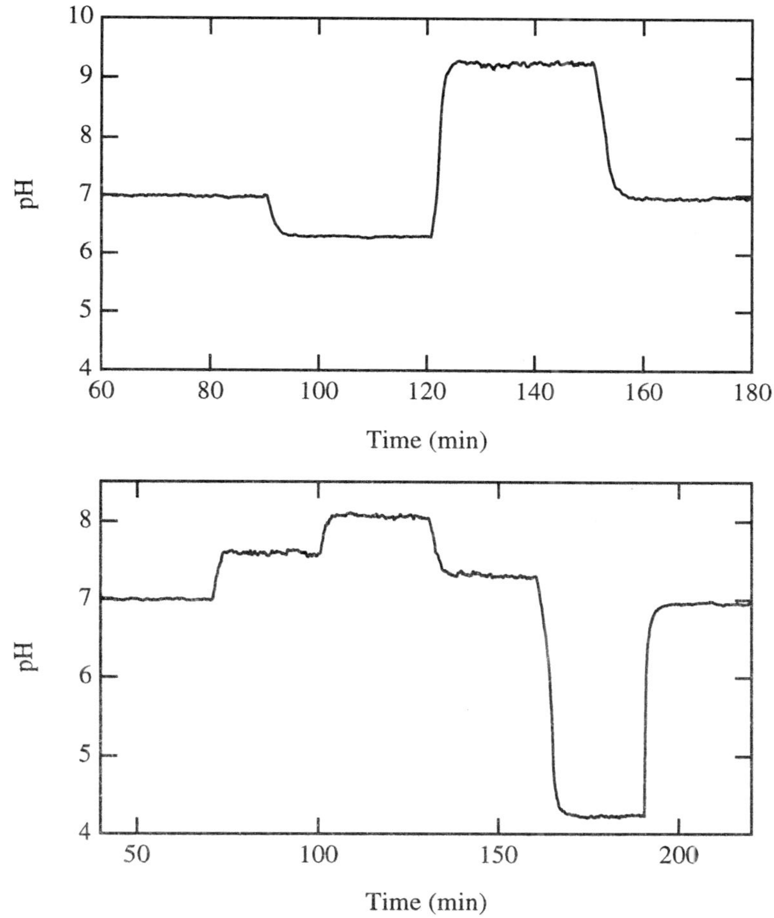

Figure 4.8: Open-loop response for base (top) and buffer (bottom) flow rate changes [72].

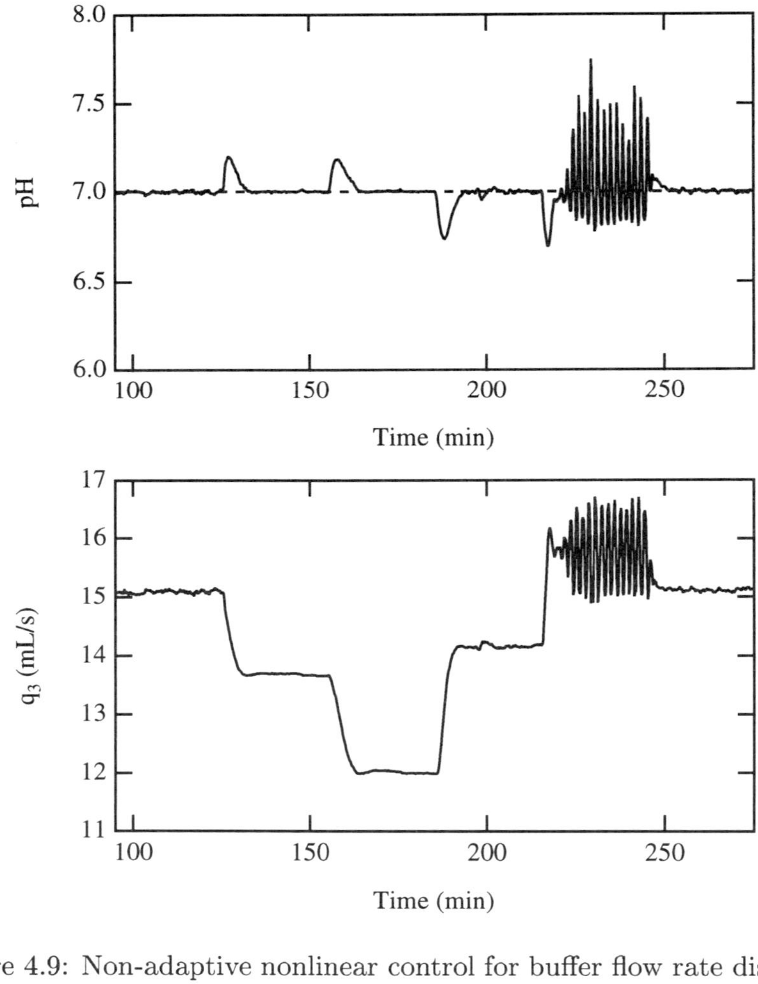

Figure 4.9: Non-adaptive nonlinear control for buffer flow rate disturbances [72].

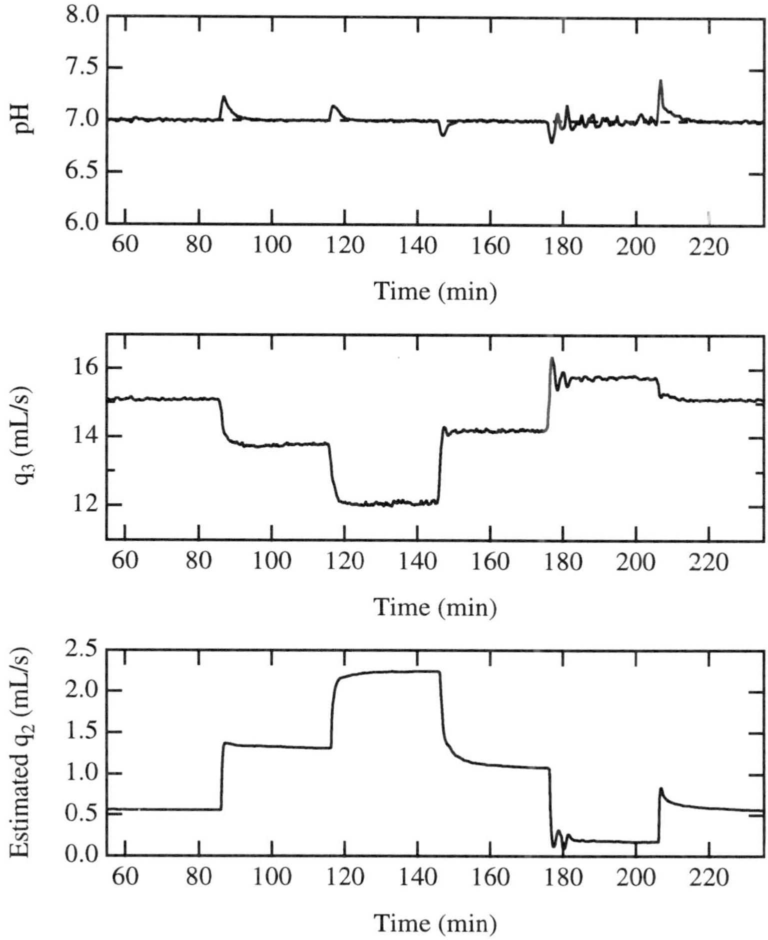

Figure 4.10: Adaptive nonlinear control for buffer flow rate disturbances [72].

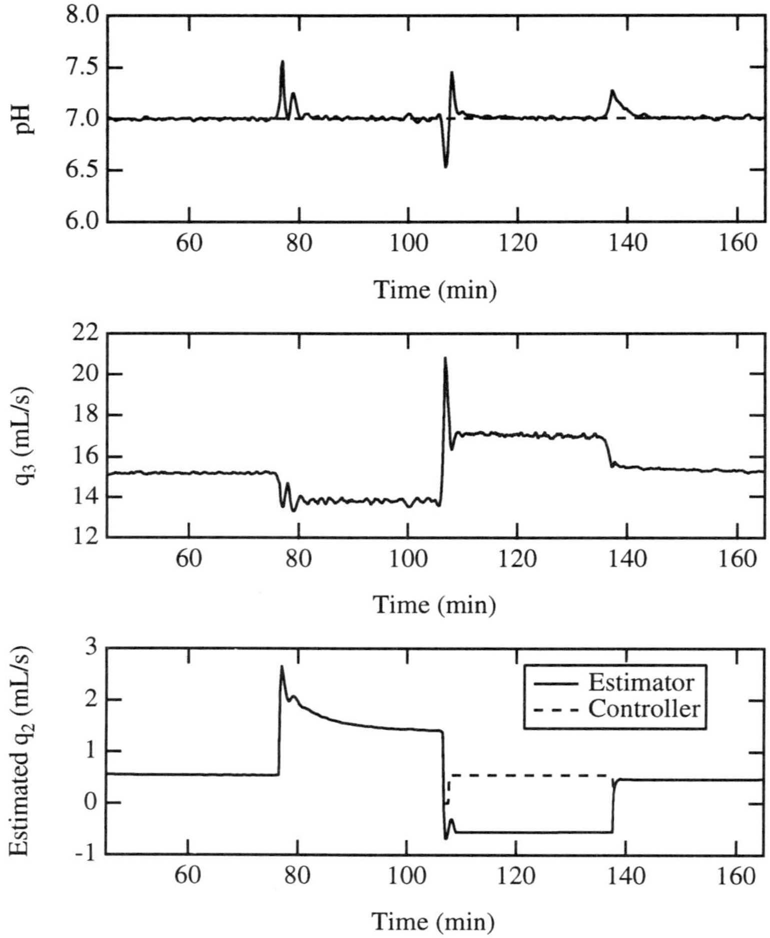

Figure 4.11: Adaptive nonlinear control for acid flow rate disturbances [72].

is designed only to account for buffering changes. Note that the estimated buffer flow rate used by the controller is set equal to the nominal value (0.55 ml/s) if the value produced by the estimator is sufficiently negative. Despite using poor estimates of the buffer flow rate and the reaction invariants, the adaptive controller provides superior pH responses as compared to the PI and non-adaptive nonlinear controllers (not shown).

References

[1] A. K. Adebekun and F. J. Schork. On the global stabilization of n-th order reactions. *Chem. Eng. Commun.*, 100:47–59, 1991.

[2] D. K. Adebekun and F. J. Schork. Continuous solution polymerization reactor control. 1. Nonlinear reference control of methyl methacrylate polymerization. *Ind. Eng. Chem. Res.*, 28:1308–1324, 1989.

[3] K. Adebekun. The robust global stabilization of a stirred tank reactor. *AIChE J.*, 38:651–659, 1992.

[4] P. Agrawal, G. Koshy, and M. Ramseier. An algorithm for operating a fed-batch reactor at optimum specific growth rate. *Biotech. Bioeng.*, 33:115, 1989.

[5] A. W. Alsop and T. F. Edgar. Nonlinear heat exchanger control through the use of partially linearized control variables. *Chem. Eng. Commun.*, 75:155–170, 1989.

[6] A. W. Alsop and T. F. Edgar. Nonlinear control of a high-purity distillation column by the use of partially linearizied control variables. *Computers. chem. Engng.*, 14:665–678, 1990.

[7] J. Alvarez, J. Alvarez, and E. Gonzalez. Global nonlinear control of a continuous stirred tank reactor. *Chem. Eng. Sci.*, 44:1147–1160, 1989.

[8] J. Alvarez, J. Alvarez, and R. Suarez. Nonlinear bounded control for a class of continuous agitated tank reactors. *Chem. Eng. Sci.*, 46:3235–3249, 1991.

[9] J. Alvarez, R. Suarez, and A. Sanchez. Nonlinear decoupling control of free-radical polymerization continuous stirred tank reactors. *Chem. Eng. Sci.*, 45:3341–3357, 1990.

[10] J. Alvarez, R. Suarez, and A. Sanchez. Semiglobal nonlinear control based on complete input-output linearization and its application to the start-up of a continuous polymerization reactor. *Chem. Eng. Sci.*, 49:3617–3630, 1994.

[11] J. Alvarez-Gallegos. Application of nonlinear system transformations to control design for a chemical reactor. *IEE Proc. - Pt. D*, 135:90–94, 1988.

[12] J. Alvarez-Ramirez. Stability of a class of uncertain continuous stirred chemical reactors with a nonlinear feedback. *Chem. Eng. Sci.*, 49:1743–1748, 1994.

[13] H. Aoufoussi, M. Perrier, J. Chaouki, C. Chavarie, and D. Dochain. Feedback linearizing control of a fluidized bed reactor. *Canadian J. Chem. Eng.*, 70:356–367, 1992.

[14] A. Arapostathis, B. Jakubczyk, H.-G. Lee, S. I. Marcus, and E. D. Sontag. The effect of sampling on linear equivalence and feedback equivalence. *System Control Lett.*, 13:373–381, 1989.

[15] Y. Arkun and J.-P. Calvet. Robust stabilization of input/output linearizable systems under uncertainty and disturbances. *AIChE J.*, 38:1145–1156, 1992.

[16] J. G. Balchen. Nonlinear decoupling in process control. *Model. Ident. Control*, 12:81–94, 1991.

[17] J. G. Balchen, B. Lie, and I. Solberg. Internal decoupling in non-linear process control. *Model. Ident. Control*, 9:137–148, 1988.

[18] M. Barolo, G. B. Guarise, S. A. Rienzi, and A. Trotta. Nonlinear model-based startup and operation control of a distillation column: An experimental study. *Ind. Eng. Chem. Res.*, 33:3160–3167, 1994.

[19] R. D. Bartusiak, C. Georgakis, and M. J. Reilly. Nonlinear feedforward/feedback control structures designed by reference synthesis. *Chem. Eng. Sci.*, 44:1837–1851, 1989.

[20] G. Bastin. Nonlinear and adaptive control in biotechnology: A tutorial. In *Proc. European Control Conf.*, pages 2001–2012, Grenoble, France, 1991.

[21] G. Bastin. Adaptive non-linear control of a fed-batch stirred tank reactors. *Int. J. Adaptive Control Signal Process.*, 6:273–284, 1992.

[22] S. Behtash. Robust output tracking for non-linear systems. *Int. J. Control*, 51:1381–1407, 1990.

[23] J. Bhat, K. P. Madhavan, and M. Chidambaram. Multivariable global input/output linearized internal model control of a semibatch reactor. *Ind. Eng. Chem. Res.*, 30:1541–1547, 1991.

[24] R. W. Brockett. Feedback invariants for nonlinear systems. In *Proc. IFAC World Congress*, pages 1115–1120, Helsinki, Finland, 1978.

[25] C. I. Byrnes and A. Isidori. Nonlinear distrurbance decoupling with stability. In *Proc. IEEE Conf. on Decision and Control*, pages 513–518, Los Angeles, 1987.

[26] C. I. Byrnes and A. Isidori. New results and examples in nonlinear feedback stabilization. *Systems Control Lett.*, 12:437–442, 1989.

[27] C. I. Byrnes and A. Isidori. Asymptotic stabilization of minimum phase nonlinear systems. *IEEE Trans. Autom. Control*, AC-36:1122–1137, 1991.

[28] F. M. Callier and C. A. Desoer. *Linear System Theory.* Springer-Verlag, New York, NY, 1991.

[29] J.-P. Calvet and Y. Arkun. Feedforward and feedback linearization of non-linear systems with disturbances. *Int. J. Control*, 48:1551–1559, 1988.

[30] J.-P. Calvet and Y. Arkun. Feedforward and feedback linearization of nonlinear systems and its implementation using internal model control (IMC). *Ind. Eng. Chem. Res.*, 27:1822–1831, 1988.

[31] J.-P. Calvet and Y. Arkun. Robust control design for uncertain nonlinear systems under feedback linearization. In *Proc. IEEE Conf. Decision and Control*, pages 102–106, Tampa, FL, 1989.

[32] J.-P. Calvet and Y. Arkun. Stabilization of feedback linearized nonlinear systems with bounded perturbations. In *Proc. American Control Conf.*, pages 747–752, Pittsburgh, 1989.

[33] J.-P. Calvet and Y. Arkun. Design of P and PI stabilizing controllers for quasi-linear systems. *Computers chem. Engng.*, 14:415–426, 1990.

[34] R. Castro, J. Alvarez, and J. Alvarez. Nonlinear disturbance decoupling control of a binary distillation column. *Automatica*, 26:567–572, 1990.

[35] D. Cheng, A. Isidori, W. Respondek, and T. J. Tarn. Exact linearization of nonlinear systems with outputs. *Math. Systems Theory*, pages 63–83, 1988.

[36] C.-B. Chung. Dynamic simulation and nonlinear-model-based product quality control of a crude tower. *AIChE J.*, 41:122–134, 1995.

[37] D. Claude. Decoupling of nonlinear systems. *System Control Lett.*, 1:242–248, 1982.

[38] D. J. Costello, P. L. Lee, and P. F. Greenfield. Control of anaerobic digestors using generic model control. *Bioproc. Eng.*, 4:119–122, 1989.

[39] B. J. Cott, R. G. Durham, P. L. Lee, and G. R. Sullivan. Process model-based engineering. *Computers chem. Engng.*, 13:973–984, 1989.

[40] P. Daoutidis and C. Kravaris. Synthesis of feedforward/state feedback controllers for nonlinear processes. *AIChE J.*, 35:1602–1616, 1989.

[41] P. Daoutidis and C. Kravaris. Dynamic output feedback control of minimum-phase nonlinear processes. *Chem. Eng. Sci.*, 47:837–849, 1992.

[42] P. Daoutidis and C. Kravaris. Dynamic output feedback control of minimum-phase multivariable nonlinear processes. *Chem. Eng. Sci.*, 49:433–447, 1994.

[43] P. Daoutidis, M. Soroush, and C. Kravaris. Feedforward/feedback control of multivariable nonlinear processes. *AIChE J.*, 36:1471–1484, 1990.

[44] J. Descusse and C. H. Moog. Decoupling with dynamic compensation for strong invertible affine nonlinear systems. *Int. J. Control*, 42:1387–1398, 1985.

[45] D. Dochain. Adaptive control algorithms for nonminimum phase bioreactors. *Computers chem. Engng.*, 16:449–462, 1992.

[46] D. Dochain, M. Perrier, and A. Pauss. Adaptive control of the hydrogen concentration in anaerobic digestions. *Ind. Eng. Chem. Res.*, 30:129–136, 1991.

[47] P. L. Douglas, P. S. Fountain, G. R. Sullivan, and W. Zhou. Model based control of a high purity distillation column. *Canadian J. Chem. Eng.*, 72:1055–1065, 1994.

[48] F. J. Doyle, F. Allgower, S. Oliveria, E. Gilles, and M. Morari. On nonlinear systems with poorly behaved zero dynamics. In *Proc. American Control Conf.*, pages 2571–2575, Chicago, 1992.

[49] F. J. Doyle and M. Morari. A conic sector-based methodology for nonlinear control design. In *Proc. American Control Conf.*, pages 2746–2751, San Diego, CA, 1990.

[50] F. J. Doyle and M. Morari. A survey of approximate linearization approaches for chemical reactor control. In *Proc. IFAC Symposium DYCORD+ '92*, pages 207–212, College Park, MD, 1992.

[51] C. G. Economou, M. Morari, and B. O. Palsson. Internal model control. 5. Extension to nonlinear systems. *Ind. Eng. Chem. Des. Dev.*, 25:403–411, 1986.

[52] J. El Moubaraki, G. Bastin, and L. Levine. Nonlinear control of biotechnological processes with growth-production decoupling. *Math. Biosciences*, 116:21–44, 1993.

[53] E. Freund. The structure of decoupled non-linear systems. *Int. J. Control*, 21:443–450, 1975.

[54] S. T. Glad. Output dead-beat control for nonlinear systems with one zero at infinity. *Systems Control Lett.*, 9:249–255, 1987.

[55] J. W. Grizzle. Controlled invariance for discrete-time nonlinear systems with an application to the disturbance decoupling problem. *IEEE Trans. Autom. Control*, AC-30:868–874, 1985.

[56] J. W. Grizzle. Local input-output decoupling of discrete-time nonlinear systems. *Int. J. Control*, 43:1517–1530, 1986.

[57] J. W. Grizzle and P. V. Kokotovic. Feedback linearization of sampled-data systems. *IEEE Trans. Autom. Control*, AC-33:857–859, 1988.

[58] T. K. Gustafsson and K. V. Waller. Nonlinear and adaptive control of pH. *Ind. Eng. Chem. Res.*, 31:2681–2693, 1992.

[59] I. J. Ha and E. G. Gilbert. Robust tracking in nonlinear systems. *IEEE Trans. Autom. Control*, AC-32:763–771, 1987.

[60] R. C. Hall and D. E. Seborg. Modelling and multi-loop control of a multivariable pH neutralization process: Part I: Modelling and multiloop control. In *Proc. American Control Conf.*, pages 1833–1827, Pittsburgh, 1989.

[61] M. Han and S. Park. Control of high-purity distillation column using a nonlinear wave theory. *AIChE J.*, 39:787–796, 1993.

[62] T. J. Harris and P. J. McLellan. Generic model control: A case study revisited. *Canadian J. Chem. Eng.*, 68:1066–1070, 1990.

[63] J. Hauser, S. Sastry, and P. Kokotovic. Nonlinear control via approximate input-output linearization: The ball and beam example. *IEEE Trans. Autom. Control*, AC-37:392–398, 1992.

[64] J. Hauser, S. Sastry, and G. Meyer. Nonlinear control design for slightly non-minimum phase systems: Application to V/STOL aircraft. *Automatica*, 28:665–679, 1992.

[65] M. A. Henson. *Feedback Linearization Strategies for Nonlinear Process Control.* PhD thesis, University of California, Santa Barbara, CA, 1992.

[66] M. A. Henson and M. J. Kurtz. Feedback linearizing controller design for chemical processes: Challenges and recent advances. In *Proc. IFAC Symposium on Nonlinear Control Systems Design*, Tahoe City, CA, in press.

[67] M. A. Henson and D. E. Seborg. Input-output linearization of general nonlinear processes. *AIChE J.*, 36:1753–1757, 1990.

[68] M. A. Henson and D. E. Seborg. Critique of exact linearization strategies for process control. *J. Process Control*, 1:122–139, 1991.

[69] M. A. Henson and D. E. Seborg. An internal model control strategy for nonlinear systems. *AIChE J.*, 37:1065–1081, 1991.

[70] M. A. Henson and D. E. Seborg. Nonlinear control strategies for continuous fermentors. *Chem. Eng. Sci.*, 47:821–835, 1992.

[71] M. A. Henson and D. E. Seborg. Theoretical analysis of unconstrained nonlinear model predictive control. *Int. J. Control*, 58:1053–1080, 1993.

[72] M. A. Henson and D. E. Seborg. Adaptive nonlinear control of a pH neutralization process. *IEEE Trans. Control Systems Technology*, 2:169–182, 1994.

[73] M. A. Henson and D. E. Seborg. Time delay compensation for nonlinear processes. *Ind. Eng. Chem. Res.*, 33:1493–1500, 1994.

[74] M. A. Henson and D. E. Seborg. Adaptive input-output linearization of a pH neutralization process. *Int. J. Adaptive Control Signal Process.*, submitted.

[75] R. M. Hirschorn. (A,B)-invariant distributions and disturbance decoupling of nonlinear systems. *SIAM J. Control Optimiz.*, 19:1–19, 1981.

[76] K. A. Hoo and J. C. Kantor. An exothermic continuous stirred tank reactor is feedback equivalent to a linear system. *Chem. Eng. Commun.*, 37:1–10, 1985.

[77] K. A. Hoo and J. C. Kantor. Global linearization and control of a mixed-culture bioreactor with competition and external inhibition. *Math. Biosciences*, 82:43–62, 1986.

[78] K. A. Hoo and J. C. Kantor. Linear feedback equivalence and control of an unstable biological reactor. *Chem. Eng. Commun.*, 46:385–399, 1986.

[79] H.-S. Huang and G.-B. Wang. Deadtime compensation for nonlinear processes with disturbances. *Int. J. Systems Sci.*, pages 1761–1776, 1992.

[80] L. R. Hunt, R. Su, and G. Meyer. Design for multi-input systems. In R. W. Brockett, R. S. Millman, and H. J. Sussmann, editors, *Differential Geometric Control Theory*, pages 268–297. Birkhauser, Boston, MA, 1983.

[81] L. R. Hunt, R. Su, and G. Meyer. Global transformations of nonlinear systems. *IEEE Trans. Autom. Control*, AC-28:24–31, 1983.

[82] A. Isidori. *Nonlinear Control Systems.* Springer-Verlag, New York, NY, 1989.

[83] A. Isidori and J. W. Grizzle. Fixed modes and nonlinear interacting control with stability. *IEEE Trans. Autom. Control*, AC-33:907–914, 1988.

[84] A. Isidori and A. J. Krener. On feedback equivalence of nonlinear systems. *Systems Control Lett.*, 2:118–121, 1982.

[85] A. Isidori, A. J. Krener, C. Gori-Giorgi, and S. Monaco. Nonlinear decoupling via feedback: A differential geometric approach. *IEEE Trans. Autom. Control*, AC-26:331–345, 1981.

[86] A. Isidori, C. H. Moog, and A. De Luca. A sufficient condition for full linearization via dynamic state feedback. In *Proc. IEEE Conf. on Decision and Control*, pages 203–208, 1986.

[87] B. Jakubczyk and W. Respondek. On linearization of control systems. *Bull. Acad. Polonaise Sci. Ser. Sci. Math.*, 28:517–522, 1980.

[88] I. Kanellakopoulos, P. V. Kokotovic, and R. Marino. An extended direct scheme for robust adaptive nonlinear control. *Automatica*, 27:247–255, 1991.

[89] J. C. Kantor. An overview of nonlinear geometrical methods for process control. In D. M. Prett and M. Morari, editors, *Shell Process Control Workshop*, pages 225–250. Butterworth, London, 1987.

[90] J. C. Kantor and M. R. Keenan. Static nonlinear control of chemical processes. In *Proc. IFAC World Congress*, pages 287–290, Munich, 1987.

[91] T. A. Kendi and F. J. Doyle. An anti-windup scheme for input-output linearization. In *Proc. European Control Conf.*, Rome, Italy, 1995.

[92] V. I. Korobov. Controllability and stability of certain nonlinear systems. *Diff. Eqs. (Russian)*, 9:614–619, 1973.

[93] D. J. Kozub and J. F. MacGregor. Feedback control of polymer quality in semi-batch copolymerization reactors. *Chem. Eng. Sci.*, 47:929–942, 1992.

[94] C. Kravaris and Y. Arkun. Geometric nonlinear control – An overview. In Y. Arkun and W. H. Ray, editors, *Chemical Process Control IV*. CACHE, AIChE, Austin, TX, 1991.

[95] C. Kravaris and C.-B. Chung. Nonlinear state feedback synthesis by global input-output linearization. *AIChE J.*, 33:592–603, 1987.

[96] C. Kravaris and P. Daoutidis. Nonlinear state feedback control of second-order non-minimum phase nonlinear systems. *Comput. chem. Engng.*, 14:439–449, 1990.

[97] C. Kravaris, P. Daoutidis, and R. A. Wright. Output feedback control of nonminimum-phase nonlinear processes. *Chem. Eng. Sci.*, 49:2107–2122, 1994.

[98] C. Kravaris and J. C. Kantor. Geometric methods for nonlinear process control: 1. Background. *Ind. Eng. Chem. Res.*, 29:2295–2310, 1990.

[99] C. Kravaris and J. C. Kantor. Geometric methods for nonlinear process control: 2. Controller synthesis. *Ind. Eng. Chem. Res.*, 29:2310–2323, 1990.

[100] C. Kravaris and S. Palanki. Robust nonlinear state feedback under structured uncertainty. *AIChE J.*, 34:1119–1127, 1988.

[101] C. Kravaris and M. Soroush. Synthesis of multivariable nonlinear controllers by input/output linearization. *AIChE J.*, 36:249–264, 1990.

[102] C. Kravaris and R. A. Wright. Deadtime compensation for nonlinear processes. *AIChE J.*, 35:1535–1542, 1989.

[103] C. Kravaris, R. A. Wright, and J. F. Carrier. Nonlinear controllers for trajectory tracking in batch processes. *Computers. chem. Engng.*, 13:73–82, 1989.

[104] A. J. Krener. Approximate linearization by state feedback and coordinate change. *System Control Lett.*, 5:181–185, 1984.

[105] A. J. Krener, A. Isidori, and W. Respondek. Partial and robust linearization by feedback. In *Proc. IEEE Conf. Decision and Control*, pages 126–130, San Antonio, 1983.

[106] B. D. Kulkarni, S. S. Tambe, N. V. Shukla, and P. B. Deshpande. Nonlinear pH control. *Chem. Eng. Sci.*, 46:995–1003, 1991.

[107] A. Kumar and P. Daoutidis. Feedbach control of nonlinear differential-algebraic-equation systems. *AIChE J.*, 41:619–636, 1995.

[108] M. J. Kurtz and M. A. Henson. Nonlinear output feedback control of chemical reactors. In *Proc. American Control Conf.*, pages 2667–2671, Seattle, WA, 1995.

[109] M. J. Kurtz and M. A. Henson. Input-output linearizing control of constrained nonlinear processes. *J. Process Control*, in press.

[110] H.-G. Lee, A. Arapostathis, and S. I. Marcus. Linearization of discrete-time systems. *Int. J. Control*, 45:1803–1822, 1987.

[111] P. L. Lee. Direct use of nonlinear models for process control. In Y. Arkun and W. H. Ray, editors, *Chemical Process Control IV*. CACHE, AIChE, Austin, TX, 1991.

[112] P. L. Lee, R. B. Newell, and G. R. Sullivan. Generic model control – A case study. *Canadian J. Chem. Eng.*, 67:478–484, 1989.

[113] P. L. Lee and G. R. Sullivan. Generic model control (GMC). *Comput. chem Engng.*, 12:573, 1988.

[114] P. L. Lee, W. Zhou, I. T. Cameron, R. B. Newell, and G. R. Sullivan. Constrained generic model control of a surge tank. *Computers. chem. Engng.*, 15:191–195, 1991.

[115] J. Levine and P. Rouchon. Quality control of binary distillation columns via nonlinear aggregated models. *Automatica*, 27:463–480, 1991.

[116] C.-Y. Lien and T.-W. Wang. Application of feedback linearization to bioreactor control. In *Proc. American Control Conf.*, pages 2161–2166, San Diego, CA, 1990.

[117] L. C. Limqueco and J. C. Kantor. Nonlinear output feedback control of an exothermic reactor. *Computers chem. Engng.*, 4/5:427–437, 1990.

[118] Z. Lin and A. Saberi. Semi-global stabilization of partially linear composite systems via feedback of the state of the linear part. *System Control Lett.*, 20:199–207, 1993.

[119] J.-S. Liu and K. Yuan. Characterization of allowable perturbations for robust decoupling of affine non-linear systems. *Int. J. Control*, 54:729–735, 1991.

[120] J.-S. Liu and K. Yuan. Deadbeat output regulators for decouplable recursive non-linear systems. *Int. J. Systems Sci.*, 22:2671–2678, 1991.

[121] L. R. Hunt M. Luksic and R. Su. Exact linearization of input-output systems. *Int. J. Control*, pages 247–255, 1986.

[122] R. Marino. On the largest feedback linearizable subsystem. *Systems Control Lett.*, pages 345–351, 1986.

[123] R. Marino. Feedback stabilization of single-input nonlinear systems. *System Control Lett.*, 10:201–206, 1988.

[124] K. B. McAuley and J. F. MacGregor. Nonlinear product property control in industrial gas-phase polyethylene reactors. *AIChE J.*, 39:855–866, 1993.

[125] P. J. McLellan. A differential-algebraic perspective on nonlinear controller design methodologies. *Chem. Eng. Sci.*, 49:1663–1679, 1994.

[126] P. J. McLellan, T. J. Harris, and D. W. Bacon. Error trajectory descriptions of nonlinear controller designs. *Chem. Eng. Sci*, 45:3017–3034, 1990.

[127] S. Monaco and D. Normand-Cyrot. Minimum-phase nonlinear discrete-time systems and feedback stabilization. In *Proc. IEEE Conf. on Decision and Control*, pages 979–986, Los Angeles, 1987.

[128] S. Monaco and D. Normand-Cyrot. Zero dynamics of sampled nonlinear systems. *System Control Lett.*, 11:229–234, 1988.

[129] M. Morari and E. Zafiriou. *Robust Process Control.* Prentice-Hall, Englewood Cliffs, NJ, 1989.

[130] K. R. Muske and J. B. Rawlings. Model predictive control with linear models. *AIChE J.*, 39:262–287, 1993.

[131] V. Nevistic and L. Del Re. Feasible suboptimal model predictive control for linear plants with state dependent constraints. In *Proc. American Control Conf.*, pages 2862–2866, 1994.

[132] H. Nijmeijer and W. Respondek. Dynamic input-output decoupling of nonlinear control systems. *IEEE Trans. Autom. Control*, AC-33:1065–1070, 1988.

[133] H. Nijmeijer and J. M. Schumacher. The regular local noninteracting control problem for nonlinear control systems. *SIAM J. Control Optimiz.*, 24:1232–1245, 1986.

[134] H. Nijmeijer and A. J. van der Schaft. *Nonlinear Dynamical Control Systems*. Springer-Verlag, New York, NY, 1990.

[135] B. A. Ogunnaike. Controller design for nonlinear process systems via variable transformations. *Ind. Eng. Chem. Proc. Des. Dev.*, 25:241–248, 1986.

[136] J. Parrish and C. Brosilow. Nonlinear inferential control of reactor effluent concentration from temperature and flow measurements. In *Proc. American Control Conf.*, pages 1027–1033, 1986.

[137] J. R. Parrish and C. B. Brosilow. Nonlinear inferential control. *AIChE J.*, 34:633, 1988.

[138] W. A. Porter. Diagonalization and inverses for non-linear systems. *Int. J. Control*, 11:67–76, 1970.

[139] T. Proell, A. Hilaly, M. N. Karim, and D. Guyre. Comparison of different optimization and control schemes in an industrial scale microalgae fermentation. In *Proc. American Control Conf.*, pages 1323–1328, San Francisco, CA, 1991.

[140] T. Proell and N. M. Karim. Nonlinear control of a bioreactor model using exact and I/O linearization. *Int. J. Control*, 60:499–519, 1994.

[141] B. Ramchandran, J. B. Riggs, and H. R. Heichelheim. Nonlinear plant-wide control: Application to a supercritical fluid extraction process. *Ind. Eng. Chem. Res.*, 31:290–300, 1992.

[142] M. Ramseier, P. Agrawal, and D. A. Mellichamp. Nonlinear adaptive control of fermentation processes utilizing *a priori* knowledge. In *Proc. American Control Conf.*, pages 1305–1310, San Francisco, CA, 1991.

[143] D. Rangel, C. Y. Lien, and T. W. Wang. Application of feedback linearization to the control of nonlinear chemical reactors. In *AIChE Annual Mtg.*, Chicago, 1990.

[144] K. Y. Rani and K. Gangiah. Adaptive generic model control: Dual composition control of distillation. *AIChE J.*, 37:1634–1644, 1991.

[145] J. B. Rawlings and K. R. Muske. The stability of constrained receding horizon control. *IEEE Trans. Autom. Control*, AC-38:1512–1516, 1993.

[146] P. Renard, D. Dochain, G. Bastin, H. Naveau, and E.-J. Nyns. Adaptive control of anaerobic digestion processes — A pilot-scale application. *Biotech. Bioeng.*, 31:287–294, 1988.

[147] J. B. Riggs and R. R. Rhinehart. Comparison between two nonlinear process-model based controllers. *Computers. chem. Engng.*, 14:1075–1081, 1990.

[148] A. Saberi, P. V. Kokotovic, and H. J. Sussmann. Global stabilization of partially linear composite systems. *SIAM J. Control Optimiz.*, 28:1491–1503, 1990.

[149] S. Sastry and A. Isidori. Adaptive control of linearizable systems. *IEEE Trans. Autom. Control*, AC-34:1123–1131, 1989.

[150] D. E. Seborg, T. F. Edgar, and D. A. Mellichamp. *Process Dynamics and Control.* Wiley, New York, NY, 1989.

[151] M. Soroush and C. Kravaris. Discrete-time nonlinear controller synthesis by input-output linearization. *AIChE J.*, 38:1923–1945, 1992.

[152] M. Soroush and C. Kravaris. Nonlinear control of a batch polymerization reactor: An experimental study. *AIChE J.*, 38:1429–1448, 1992.

[153] M. Soroush and C. Kravaris. Multivariable nonlinear control of a continuous polymerization reactor: An experimental study. *AIChE J.*, 39:1920–1937, 1993.

[154] M. Soroush and C. Kravaris. Synthesis of discrete-time nonlinear feedforward/feedback controllers. *AIChE J.*, 40:473–495, 1994.

[155] M. W. Spong. Robust stabilization for a class of nonlinear systems. In C. I. Byrnes and A. Lindquist, editors, *Theory and Applications of*

Nonlinear Control Systems, pages 155–166. Elsevier Science Publishers, 1986.

[156] M. W. Spong and M. Vidyasagar. Robust linear compensator design for nonlinear robotic control. *IEEE J. Robotics Autom.*, RA-3:345–351, 1987.

[157] R. Su. On the linear equivalents of nonlinear systems. *Systems Control Lett.*, pages 48–52, 1982.

[158] R. Su, G. Meyer, and L. R. Hunt. Robustness in nonlinear control. In R. W. Brockett, R. S. Millman, and H. J. Sussmann, editors, *Differential Geometric Control Theory*, pages 316–337. Birkhauser, Boston, MA, 1983.

[159] H. J. Sussmann. Limitations on the stabilizability of globally minimum phase systems. *IEEE Trans. Autom. Control*, AC-35:117–119, 1990.

[160] H. J. Sussmann and P. V. Kokotovic. The peaking phenomenon and the global stabilization of nonlinear systems. *IEEE Trans. Autom. Control*, AC-36:424–440, 1991.

[161] A. Tornambe. Asymptotic inverse dynamics of feedback linearizable systems. *Systems Control Lett.*, 16:145–153, 1991.

[162] A. Uppal, W. H. Ray, and A. B. Poore. On the dynamic behavior of continuous stirred tank reactors. *Chem. Eng. Sci.*, 29:967, 1974.

[163] A. J. van der Schaft. Linearization and input-output decoupling for general nonlinear systems. *Systems Control Lett.*, 5:27–33, 1984.

[164] L. van der Wegen and H. Nijmeijer. The local disturbance decoupling problem with stability for nonlinear systems. *System Control Lett.*, 12:139–149, 1989.

[165] Z. L. Wang, F. Pla, and J. P. Corriou. Nonlinear adaptive control of batch styrene polymerization. *Chem. Eng. Sci.*, 50:2081–2091, 1995.

[166] G. L. Williams, R. R. Rhinehart, and J. R. Riggs. In-line process-model-based control of wastewater pH using dual base injection. *Ind. Eng. Chem. Res.*, 29:1254–1259, 1990.

[167] Y. H. Wong, P. R. Krishnaswamy, W. K. Teo, B. D. Kulkarni, and P. B. Deshpande. Experimental application of robust nonlinear control law to pH control. *Chem. Eng. Sci.*, 49:199–207, 1994.

[168] R. A. Wright and C. Kravaris. Nonlinear control of pH processes using the strong acid equivalent. *Ind. Eng. Chem. Res.*, 30:1561–1572, 1991.

[169] R. A. Wright and C. Kravaris. Nonminimum-phase compensation for nonlinear processes. *AIChE J.*, 38:26–40, 1992.

[170] R. A. Wright and C. Kravaris. Dynamically equivalent outputs and their use in nonlinear controller synthesis. *Chem. Eng. Sci.*, 48:3207–3223, 1993.

[171] R. A. Wright, M. Soroush, and C. Kravaris. Strong acid equivalent control of pH processes: An experimental study. *Ind. Eng. Chem. Res.*, 30:2437–2444, 1991.

Chapter 5

Model Predictive Control

EDWARD S. MEADOWS
Departamento de Eletrônica
Pontifícia Universidade Católica do Rio Grande do Sul
Porto Alegre, RS BRAZIL

JAMES B. RAWLINGS
Department of Chemical Engineering
The University of Wisconsin
Madison, WI USA

5.1 Introduction

Model predictive control (MPC) is an optimal-control based method to select control inputs by minimizing an objective function. The objective function is defined in terms of both present and predicted system variables and is evaluated using an explicit model to predict future process outputs. This chapter discusses the current practice in MPC, computational and design issues, performance and stability and future research directions. Model predictive control incorporates ideas from systems theory, system identification and optimization. The application of these ideas is illustrated using simulation results from two chemical reactor models.

Because MPC is implemented with digital computers, much of this chapter deals with discrete-time formulations. In discrete time, a nonlinear, deterministic process model can be represented by

$$x_{k+1} = f(x_k, u_k) \tag{5.1}$$

$$y_k = g(x_k) \tag{5.2}$$

The states x_k, controls u_k and outputs y_k are vectors.

The MPC control problem is as follows: With knowledge of the current output y_k, we seek a control that minimizes the objective function

$$J = \phi(y_{k+N|k}) + \sum_{j=0}^{N-1} L(y_{k+j|k}, u_{k+j|k}, \Delta u_{k+j|k}) \tag{5.3}$$

Of the N-move control sequence that minimizes the above objective, only the first is implemented. When another measurement becomes available, the parameters of the problem are updated and a new optimization problem is formulated whose solution provides the next control. This repeated optimization using an objective function that is modified through process feedback is one of the principal defining features of MPC.

The objective function that appears in Equation 5.3 illustrates some of the commonly used terminology in the literature of MPC. States, outputs and controls are usually doubly indexed as $x_{k+j|k}$, $y_{k+j|k}$ and $u_{k+j|k}$ to indicate values at time $k+j$ given information up to and including time k. Since MPC requires prediction, the double subscript also carries the connotation of prediction when the first index is larger than the second. Variables that are singly indexed may be used to represent controls, states or outputs that have already been implemented, computed or measured, up to the current time k. An optimal sequence of controls is often indicated using an asterisk ($u^*_{k+j|k}$). The corresponding state and output values then become $x^*_{k+j|k}$ and $y^*_{k+j|k}$, respectively.

Another feature of MPC illustrated by Equation 5.3 is the presence of $\Delta u_{k+j|k}$ in the objective, in which $\Delta u_{k+j|k} = u_{k+j|k} - u_{k+j-1|k}$. In chemical engineering applications, the speed of control action may be limited. For example, a large valve may require several seconds to change position. Rather than model the dynamics of control actuators, it is common practice to include limitations on actuator speed through penalties or constraints on $\Delta u_{k+j|k}$. Aside from the issue of actuator limitations, inclusion of $\Delta u_{k+j|k}$ offers performance advantages in that control action can be limited through restrictions or penalties on $\Delta u_{k+j|k}$ without introducing permanent offset in the output. This issue is discussed further in Section 5.7.

The objective function stage-cost L in Equation 5.3 may be chosen to meet a variety of process objectives including maximization of profits, minimization of operating costs or energy usage, or may be a measure of deviation of the process output from a reference trajectory. In subsequent discussion, we will emphasize the case of regulation to a setpoint using a quadratic error criterion. In this case, the objective function of Equation 5.3 may be written

$$J = \|y_{k+N|k} - y_{\text{ref}}\|_Q^2 + \sum_{j=0}^{N-1} \Big(\|y_{k+j|k} - y_{\text{ref}}\|_Q^2 + \|u_{k+j|k} - u_{\text{ref}}\|_R^2 + \|\Delta u_{k+j|k}\|_S^2\Big) \tag{5.4}$$

In Equation 5.4, Q, R and S indicate weighting matrices in vector norms. When penalties or constraints on $\Delta u_{k+j|k}$ are used, we admit the special notation $u_{k-1|k}$ to indicate the most recently implemented control value, which is *not* a decision variable.

If the process model is linear, MPC with the objective of Equation 5.4 reduces to the linear-quadratic optimal control problem. LQ optimal control theory can be viewed as a (very important) special case of MPC. Many of the key theoretical results from LQ optimal control have analogs or extensions to the nonlinear case. For example, it is possible to guarantee nominal stability of an LQ optimal control in the closed loop based on properties of solutions of the Riccati difference equation [5]. In nonlinear MPC, this concept may be extended to provide stability results for nonlinear MPC based on iterations using the theory of dynamic programming [48].

Another key feature of MPC is the ability to handle constraints directly, including constraints on the control inputs, outputs or internal states. The constraint handling property has been one of the most significant contributors to its success in industrial applications. Some typical constraint formulations are discussed in Section 5.4.

5.2 Some Examples

Before proceeding with more mathematical developments, we wish to introduce two example problems that will be used to illustrate some of the principles of nonlinear MPC. Our examples are both taken from the chemical engineering literature and represent chemical reactors commonly used in industry.

5.2.1 Isothermal CSTR

Here we consider control of a continuous stirred-tank reactor (CSTR) using dilution rate as the manipulated variable. The volume of the reactor is constant. The dynamics and control of this reactor have been studied by several researchers, including Van de Vusse [75], Kantor [31], Kravaris and Daoutidis [38], and Sistu and Bequette [69]. The reactor model is given by

$$\dot{x}_1 = -k_1 x_1 - k_3 x_1^2 + (x_F - x_1)u \tag{5.5}$$

$$\dot{x}_2 = k_1 x_1 - k_2 x_2 - x_2 u \tag{5.6}$$

which models the reactions

$$\mathrm{A} \xrightarrow{k_1} \mathrm{B} \qquad \mathrm{B} \xrightarrow{k_2} \mathrm{C} \qquad 2\mathrm{A} \xrightarrow{k_3} \mathrm{D}$$

The state variables x_1 and x_2 represent the concentrations of species A and B, respectively, x_F represents the concentration of A in the feed and u represents the dilution (feed) rate. The parameters $\{k_1, k_2, k_3, x_F\}$ have nominal values $\left\{50\ \mathrm{hr}^{-1}, 100\ \mathrm{hr}^{-1}, 10\ \mathrm{liter/(mol\ hr)}, 10\ \mathrm{mol/liter}\right\}$. The primary control objective is to maintain the output variable $y = x_2$ at a setpoint of 1.0.

Assuming a perfect model, we can solve Equations 5.5 and 5.6 at steady state to find values of $\{x_1, x_2, u\}$ that give the desired output. Because of the quadratic term in Equation 5.5, there are two solutions: $\{2.5, 1.0, 25\}$ and $\{6.6667, 1.0, 233.33\}$. According to Sistu and Bequette [69], lower feed rates are preferable to achieve higher conversion; therefore as a secondary control objective, we seek to maintain a low dilution rate.

Assuming that the steady-state model solution provides setpoints for $\{x_1, x_2, u\}$, we can form an objective function

$$J = \int_t^{t+T} \gamma_1 \left[x_1(\tau) - 2.5\right]^2 + \gamma_2 \left[x_2(\tau) - 1.0\right]^2 + \gamma_3 \left[u(\tau) - 25\right]^2 d\tau \tag{5.7}$$

in which $\{\gamma_1, \gamma_2, \gamma_3\}$ represent weight factors that may be used for scaling or weighting the relative importance of each contribution.

Equations 5.5, 5.6 and the objective function of Equation 5.7 are defined for continuous time systems. In order to solve with a computer, the differential equations must be discretized and the controller must be expressed in a finite parameterization. We discretized the dynamic equations and the objective function using orthogonal collocation on finite elements. Following the example of Sistu and Bequette [69], each finite element was 0.002 hours (7.2 seconds) long. To solve the model equations, we used three interior collocation points and a point on the element boundaries. (More details on the orthogonal collocation technique are provided in Section 5.4.) In the nonlinear program that solved the MPC problem, we implemented constraints on control and state variables to ensure that they remained non-negative.

Using $\gamma_1 = \gamma_2 = 1000$ and $\gamma_3 = 1$, MPC results for the nominal system are shown in Figure 5.1 for horizon length 5 (five finite elements of 7.2 seconds each). The initial condition was $\{2.9997, 1.11697\}$, which was the same initial condition used by Sistu and Bequette [69]. The points in the diagrams show the sampling times. Between sampling times the curves were computed by integrating the nominal model. Both states and the control converge quickly to the desired target values. Similar results were obtained for simulations with horizon lengths from 1 to 7.

In most processes, all of the internal model states are not measured. If we use only the output variable x_2 in the objective ($\gamma_1 = \gamma_3 = 0$), we find that the MPC controller is able to bring the output to the setpoint, but the controller drives the system to the steady-state values at the higher (undesirable) dilution rate. These results are shown in Figure 5.2. Similar results were also obtained for horizon lengths from 1 to 7.

Since using only the output variable failed to satisfy the secondary control objective of maintaining low dilution rate, a logical attempt to satisfy the secondary objective might be to place an upper limit on the control input. Although not provided in the problem description supplied by Sistu and Bequette [69], we can reasonably assume that the nominal steady-state feed rate (25) lies near the middle of the operating range and we therefore apply a maximum feed rate of 50. The results are provided in Figure 5.3, which shows that the steady-state value of the output x_2 has offset. The same effect is seen with controllers with horizon lengths 1 through 7. The length of the horizon has an effect on the steady-state values observed in the simulations. The steady-state value of the output versus horizon length is shown in Figure 5.4. We can see that longer horizons produce less offset. With "well-behaved" systems, steady-state offset will disappear in the limit as the horizon increases to infinity.

Since none of the closed-loop control values in Figure 5.3 are at the upper limit, it is not obvious why the input constraint causes steady-state offset. To see why this occurs, we must examine the open-loop control profile computed at each time step. For each of the horizon lengths tested in simulation, we found that the input constraint is active for *some* portion of the prediction horizon. For $N = 1$ to 4, the input constraint is active for the initial open-loop control; therefore, the steady-state value of x_2 corresponds to $u = 50$. For $N = 5$, the initial open-loop control is not constrained, but

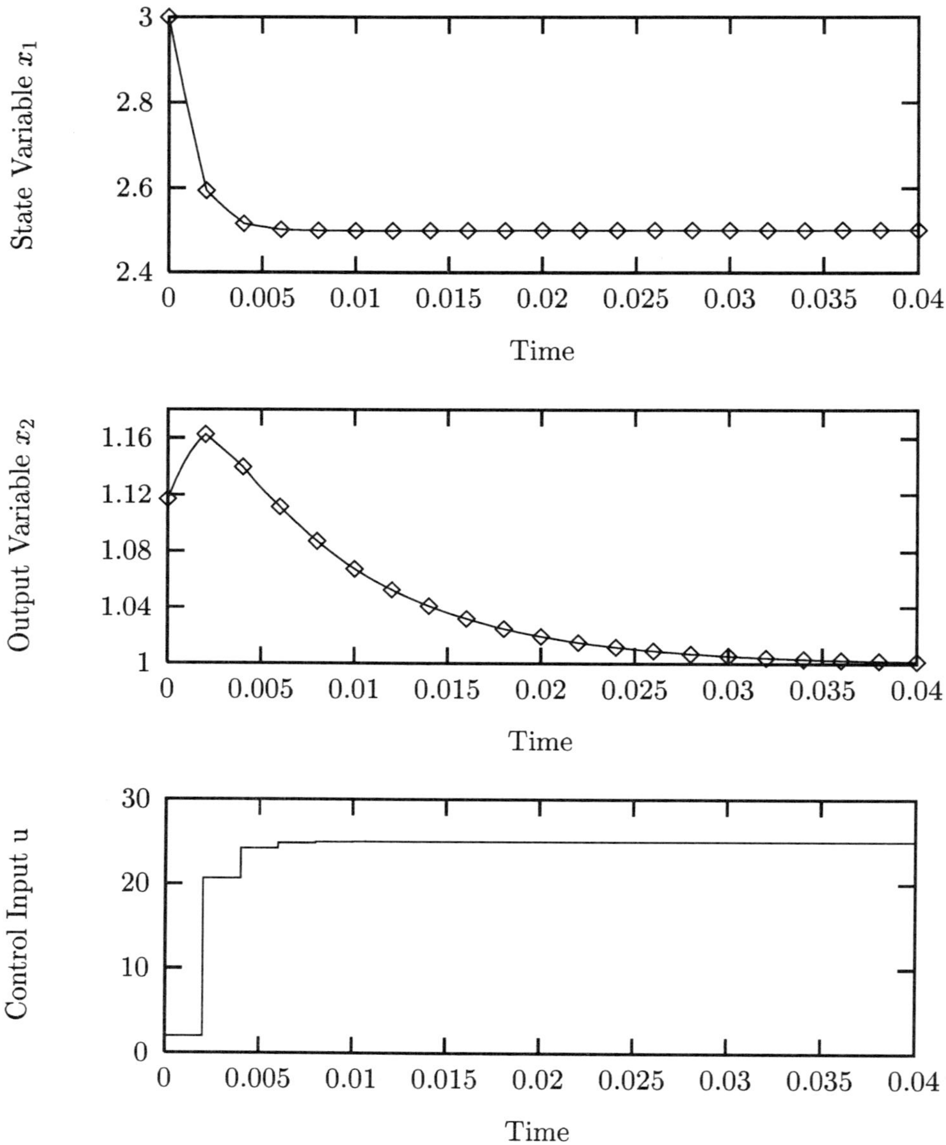

Figure 5.1: Isothermal CSTR: Nominal system with horizon length 5.

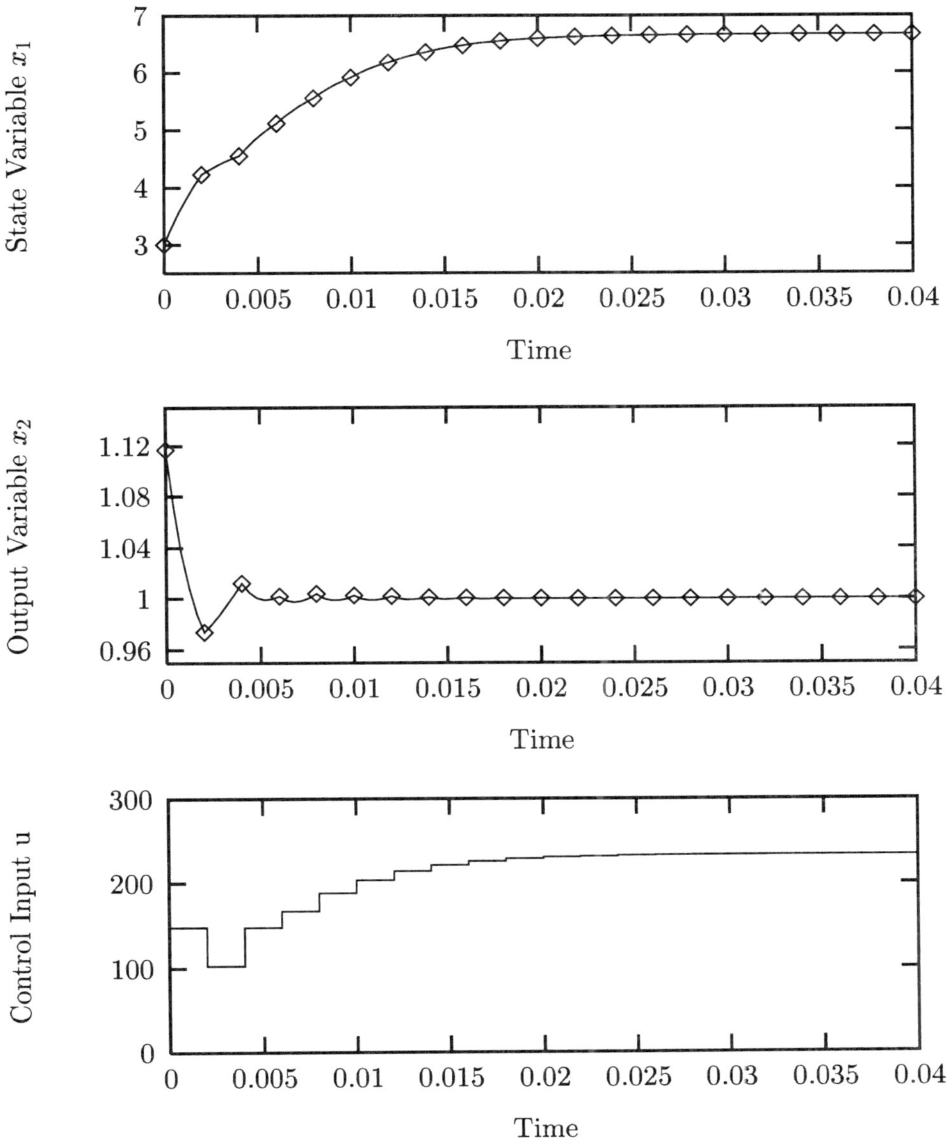

Figure 5.2: Isothermal CSTR: Nominal system with horizon length 5 using only the output.

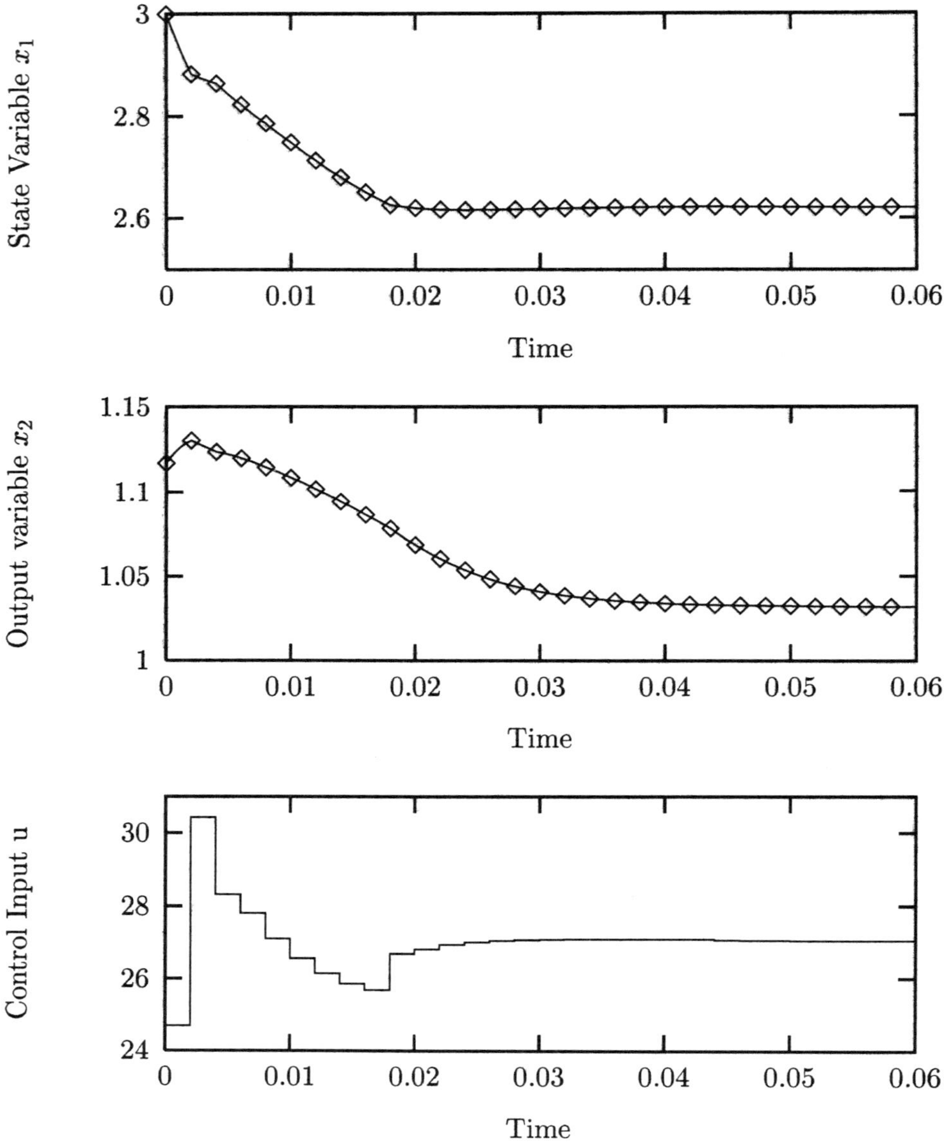

Figure 5.3: Isothermal CSTR: Nominal system with horizon length 5 using only the output with input constraints.

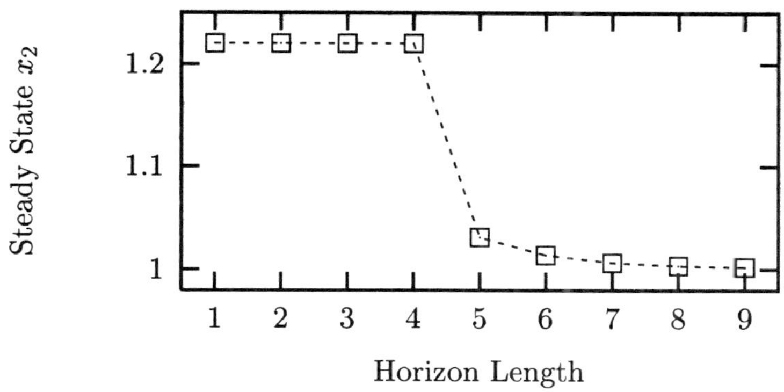

Figure 5.4: Isothermal CSTR: Steady-state output versus horizon length.

subsequent values in the prediction horizon are constrained. With the initial state used in the examples, the initial open-loop control sequence for $N = 5$ is $\{24.7, 50, 50, 50, 50\}$. For horizon lengths greater than 5, the open-loop control profiles start at the *lower* constraint $u_{\min} = 0$ and reach the upper constraint subsequently within the control horizon. For example, with $N = 6$, the open-loop control profile is $\{0, 46.4, 50, 50, 50, 50\}$. To minimize the objective function while satisfying the input constraints, the optimal open-loop control sequence begins at the lower bound and increases during the prediction horizon until it reaches the upper bound.

It should be emphasized that the observed offset is *not* due to model error or unmeasured disturbances. The controller and the simulation both use the nominal model and no noise or disturbances were included. The observed offset is the result of using a controller formulation that is asymptotically stable only in the limit as $N \to \infty$. The control values and corresponding states reflect the best *finite horizon* solution to each constrained optimization problem, but do not necessarily result in zero offset. In the limit as the horizon length approaches infinity, we would expect the closed-loop output to approach its setpoint with zero offset, since steady-state offset would result in an unbounded objective function. This effect can be seen in Figure 5.4.

Since imposing control bounds failed to provide a satisfactory controller, we look for other strategies to improve the performance of the nominal system in closed loop. Figure 5.2 shows that, without the control penalty, the control initially jumps to almost 148 on its way to the final undesirable steady-state value at the undesirable high dilution rate. To prevent excessive jumps in the control, we can include a penalty on $\Delta u_{k+j|k}$. Figure 5.5 shows the effect of including a 1 percent penalty on $\Delta u_{k+j|k}$.

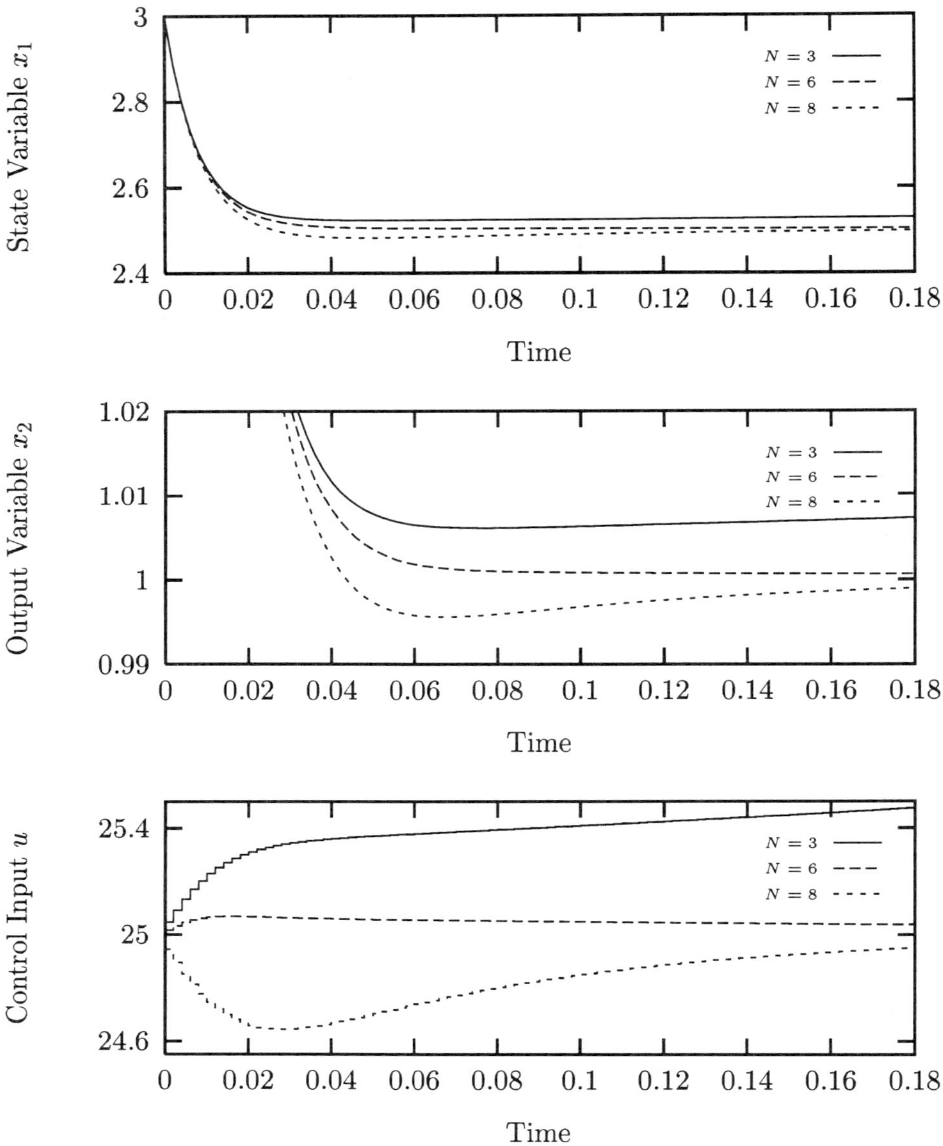

Figure 5.5: Isothermal CSTR: System response with Δu penalty.

For this case, qualitative differences appear in the responses for different horizon lengths. Figure 5.5 shows the results for horizon lengths of 3, 6 and 8. Notice that the speed of response has been greatly slowed due to the penalty on Δu, compared to Figures 5.1 or 5.2. The time scale on the figures has increased by a factor of three from the previous figure. (The use of symbols to identify the sampling instants obscures the results and has been dropped.) We find that for horizons less than 6, the desired steady-state operating point $y = 1.0$ is not asymptotically stable. The case $N = 3$, illustrated in Figure 5.5, shows a continuous increase in the control following the initial transient. The simulation ended before a steady-state value was reached. All horizons shorter that 6 showed similar unstable behavior.

In the other cases illustrated, $N = 6$ and $N = 8$, the controller brings the system to the desired steady state. With $N = 6$, there is no overshoot, while $N = 8$ shows overshoot, both in the control and in the output. The vertical axis of the graph of the output variable, x_2, has been expanded to more clearly show the different behaviors near the setpoint.

When the output variable is x_2 only and the flow rate is low, the linearized reactor model contains a right half-plane zero and therefore exhibits nonminimum phase characteristics. For linear systems, it is known that model predictive control can induce unstable closed-loop behavior for non-minimum phase systems, even if the open-loop system is stable [23]. Some of the linear systems results can be used as a guide to explain the observed behavior of the nominal system under MPC that was examined in this section.

In Figure 5.1, two output were assumed, i.e., $y = [x_1, x_2]^T$. In this case, the system cannot be described as non-minimum phase. Figure 5.2 shows the results of using $y = x_2$, illustrating the non-minimum phase effect. Although the desired setpoint was reached, the control continued to increase until it reached the second steady state at the higher setpoint. This is unstable behavior. A similar non-minimum phase linear system would not have reached an upper steady state, but would have demonstrated control action that continued to increase without limit.

To prevent MPC from returning an unstable controller for non-minimum phase linear systems, it is necessary to penalize control action or decrease the control horizon relative to the prediction horizon [23, 25]. We chose the first course and the results are provided in Figure 5.5.

5.2.2 Fluidized Bed Reactor

In this section, we demonstrate model predictive control through simulation of the fluidized-bed reactor described by Elnashaie, *et al.* [22]. The model describes the catalytic exothermic consecutive reactions

$$\mathrm{A} \xrightarrow{k_1} \mathrm{B} \xrightarrow{k_2} \mathrm{C}$$

in a freely bubbling fluidized-bed reactor. After a series of simplifying assumptions, the dimensionless reactor model is given by the following:

$$\frac{1}{\mathrm{Le}_A}\frac{dx_A}{dt} = \bar{\beta}\left(x_{A,F} - x_A\right) - r_1 \tag{5.8}$$

$$\frac{1}{\mathrm{Le}_B}\frac{dx_B}{dt} = \bar{\beta}\left(x_{B,F} - x_B\right) + r_1 - r_2 \tag{5.9}$$

$$\frac{dT}{dt} = \bar{\beta}\left(T_F - T + u\right) + \beta_1 r_1 + \beta_2 r_2 \tag{5.10}$$

$$r_1 = \alpha_1 \exp\left(-\frac{\gamma_1}{T}\right) x_A \tag{5.11}$$

$$r_2 = \alpha_2 \exp\left(-\frac{\gamma_2}{T}\right) x_B \tag{5.12}$$

In the model, Le_A and Le_B represent Lewis numbers, $\bar{\beta}$ the residence time, α_1, α_2 pre-exponential factors for rate constants, γ_1, γ_2 activation energies, and β_1, β_2 heats of reaction. Concentrations of A and B and the temperature are given by x_A, x_B and T, respectively; $x_{A,F}$, $x_{B,F}$ and T_F represent their values in the inlet stream. For details concerning derivation of the reactor model and the numerical values of the constants, the reader is referred to the original publications [21, 22].

In our version of the model, the control enters the equations in a slightly different way than in the original work of Elnashaie *et al.* [22]. Their paper was concerned with chaotic behavior of this reactor using proportional control. Their original model can be recovered by setting $u = K\left(T_{\text{set}} - T\right)$.

The measured output of the system is the temperature of the reactor effluent T. The control objective is to maintain the effluent temperature at the setpoint T_{set} that corresponds to the maximum yield of species B. Using the nominal model, we computed this optimal (dimensionless) effluent temperature of 0.9213, which will be used as the setpoint for our simulation studies. At this temperature, the nominal steady state for the reactor is $x_A =$

0.2355, $x_B = 0.6448$, which is an unstable steady state. The eigenvalues of the linearized system matrix at this point are $\{-0.2706, -0.0507, 0.6323\}$.

No sampling interval was specified by Elnashaie *et al.*, since their study was not concerned with digital control. We selected a sampling interval of 1.09 dimensionless time units, slightly shorter than the doubling time of the unstable mode. Using the objective function

$$J = 100 \sum_{k=0}^{N-1} \int_{k\,\Delta T}^{(k+1)\,\Delta T} \left[T(\tau) - 0.9213\right]^2 d\tau + \beta\left(\Delta u_k\right)^2$$

with prediction and control horizon equal to 4, simulation results for the nominal system are provided in Figure 5.6. As in the previous example, we employed the orthogonal collocation technique in a simultaneous model solution/optimization approach. Both cases $\beta = 0$ and $\beta = 0.1$ are shown. Without any penalty on control action, Figure 5.6 shows a control range of approximately -2 to 3.5 with a maximum $|\Delta u|$ of approximately 5.5. Based on the values of the numerical results reported by Elnashaie *et al.* [22], we can infer that this control is probably more aggressive than the system can support. Using a 0.1 percent penalty on Δu ($\beta = 0.1$) moderates the control action and yields a slower response by the internal states and the output variable.

Another common approach that dampens excessive control action is to select a control horizon M that is shorter than the prediction horizon N. After the M-th interval, the control is held constant through the end of the prediction horizon. Figure 5.7 shows the results of this approach for $N = 4$ without control penalty ($\beta = 0$), using control horizons $M = 1$ and $M = 2$. Using $M = 1$ greatly dampens the control input and slows the response of the system. Note the increase in scale in Figure 5.7 necessary to illustrate the transient of x_A and x_B. Increasing the prediction horizon from $M = 1$ to $M = 2$ increases both the aggressiveness of the control action and the speed of the system response. With $M = 2$ we observed system behavior very similar to that shown in Figure 5.6 using the 0.1 percent penalty on Δu.

5.3 Models for MPC

The heart of model predictive control is the process model itself. Models can be classified by various features. Since MPC requires the solution of a model to predict future process outputs, the form of the model selected has

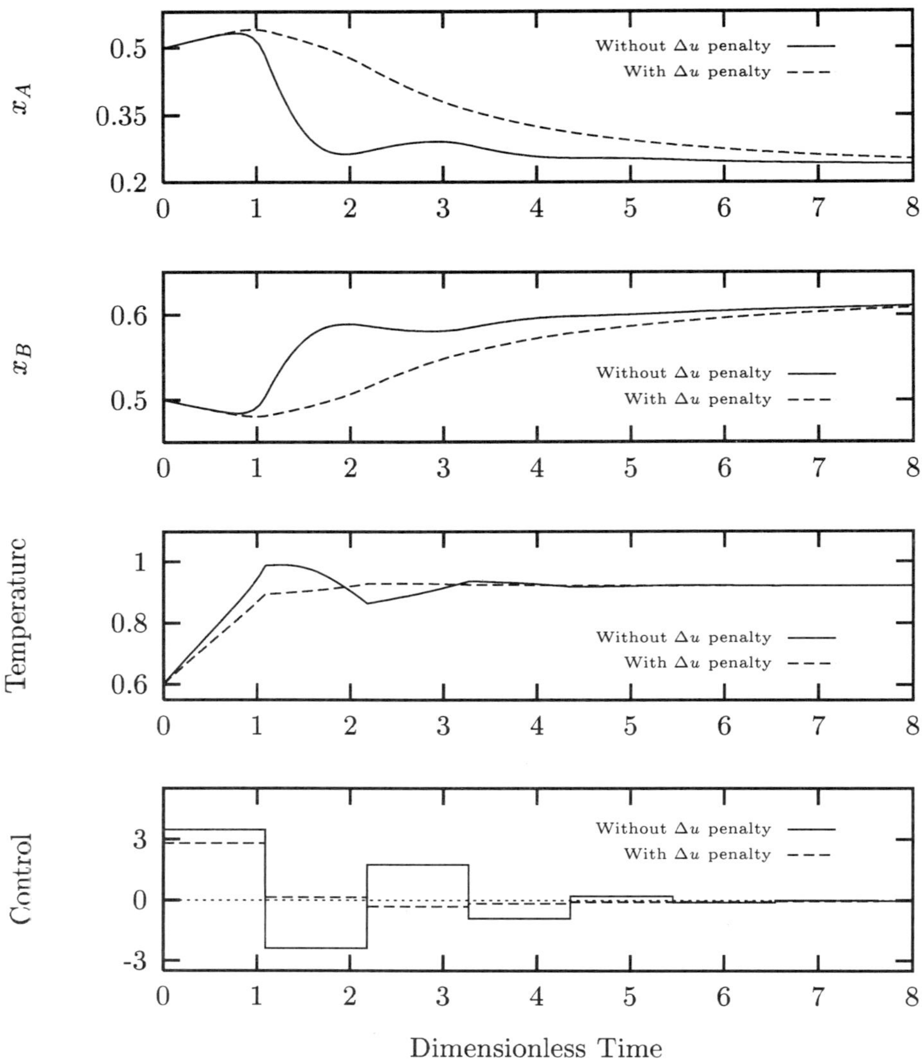

Figure 5.6: Fluidized-bed reactor: Effect of penalty on Δu.

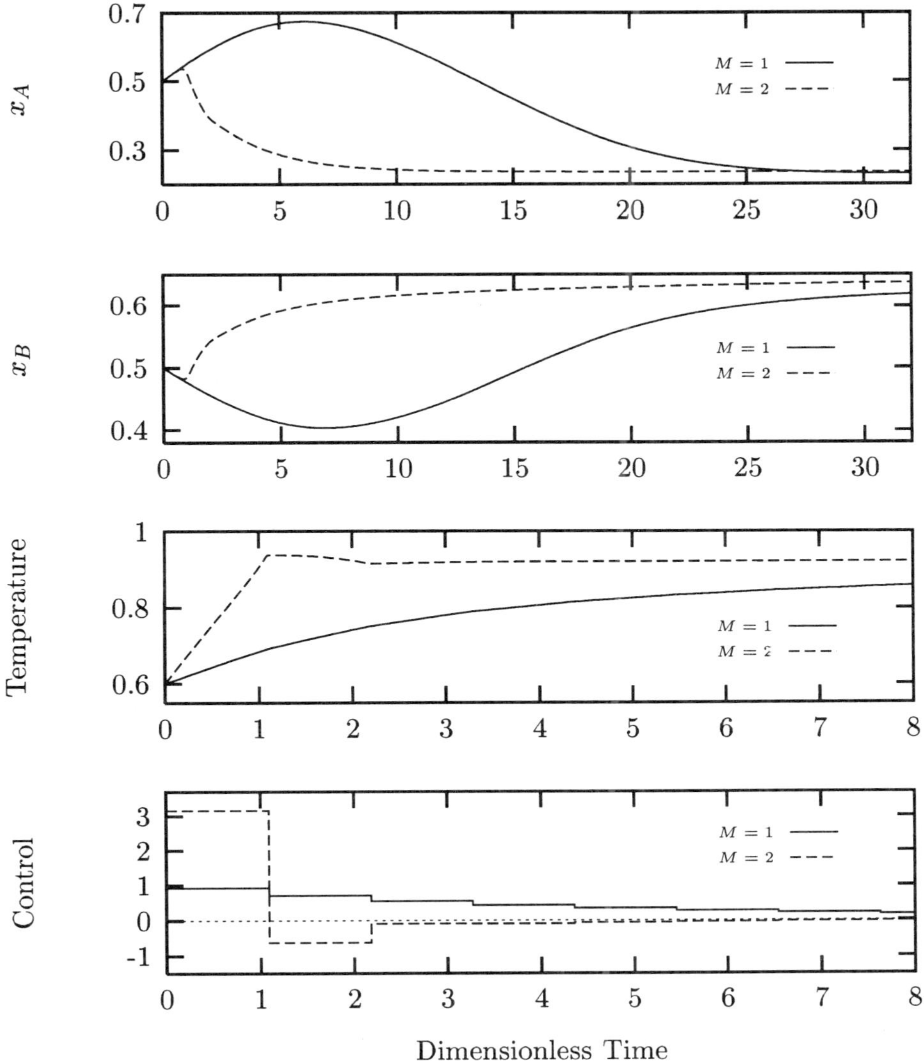

Figure 5.7: Fluidized-bed reactor: Effect of control horizon.

a large impact on our ability to implement MPC. Some specific categories of models are discussed below:

- Linear or nonlinear: The response of linear dynamic systems obeys the principle of superposition, i.e., the response of the system to a linear combination of inputs is a linear combination of their responses to each input separately. Many (perhaps most) systems of engineering interest approximate this behavior for small inputs, which accounts for the universal study and application of linear control theory. In almost any control application, linear design techniques are usually the first to be attempted and are completely satisfactory for many engineering applications, especially those involving regulation about a steady-state operating point. Linear models are used extensively in the industrial practice of MPC.

  Nonlinear models have no specific characteristics except that they don't include the linear case. This makes it difficult to generalize, since nonlinear models can be devised to have almost any characteristic or demonstrate any strange behavior. For many engineering applications, it is usually appropriate to assume that the models are continuous with respect to model parameters and that differential or difference equation models satisfy growth conditions that permit solutions to be computed.

- Continuous-time or discrete-time: Most of the physical laws that are used by engineers to develop models are presented as differential equations with time as the independent variable. A typical representation is of the form $\dot{x} = f(x, u)$. Before the widespread availability of digital computers, differential equation models were the central tool of researchers and control engineers for the study of dynamic systems. If simulations were needed, analog computers could be used. Prediction was rarely used as an element of control engineering, since complex models could not be solved using paper-and-pencil methods.

  With the advent of digital computers, the study of difference equations, which had been relegated to a minor role previously, assumed a new significance. We can write a nonlinear difference equation as $x_{k+1} = f(x_k, u_k)$. Since MPC is universally implemented via digital computer, difference equations are the model of choice. Differential equation models (as in Section 5.2) must be discretized for computer solution. In this chapter, our examples are based on continuous-time models that are discretized using orthogonal collocation.

- Distributed parameter or lumped parameter: A distributed parameter model involves partial differential equations, instead of ordinary differential equations. An example of a distributed parameter model would be that for a plug flow reactor, in which the changes in chemical concentrations in the reactor are subject to both spatial variation and variations in time. Although we do not specifically consider distributed parameter models in this chapter, the basic MPC concept is completely applicable to systems described by distributed parameter models, with a corresponding increase in computations necessary to solve the model.

- Deterministic or stochastic: All physical processes are subject to unpredictable disturbances. Disturbances can affect MPC control design and operation in at least two distinct ways:

  - In the process of model identification, a model is often selected from some class based on experimental results. The selection process implicitly or explicitly uses assumptions about the disturbances to select and evaluate the best model of the class. These assumptions have a direct impact on the model selected.
  - After the model identification phase is complete, the assumptions about the disturbances are sometimes discarded, and control design may be based on this "nominal" model. If the model allows us to predict the statistics of process variables based on assumptions about random effects on the models, we say that it is a stochastic model. We can represent discrete-time stochastic models using w and v to represent state and output disturbances as follows:

$$x_{k+1} = f(x_k, u_k, w_k) \tag{5.13}$$

$$y_k = g(x_k, v_k) \tag{5.14}$$

  In general, better control can be achieved by using explicit stochastic models [16, 17, 51] in the prediction phase of MPC, but with a much higher computational burden. The traditional approach to stochastic MPC is through dynamic programming [2, 3], which suffers from combinatorial growth in the number of optimization variables with increasing horizon length (Bellman's "curse of dimensionality"). Better ways to incorporate stochastic models in MPC are active areas of research.

- Input-output or state-space: As indicated by the name, input-output models provide a relation between the process input and the output, without reference to variables internal to the process. An input-output model of a distillation column, for example, might relate the temperature of a side-product stream to reboiler duty, without consideration of individual tray temperatures or compositions; whereas, a state-space model might include equations relating all internal compositions and flow rates to the reboiler duty, with the side product temperature being provided as a function of the composition. Because most nonlinear state-space models are based on heat, mass and momentum balances, each state has a physical interpretation. States may also be generated as mathematically convenient intermediate variables of an input-output process model. In the examples discussed in this chapter, the states have definite physical significance; however, the principles of MPC are applicable in either case.

  As a special class of input-output model, we should mention that artificial neural network models are becoming important in many engineering applications, including model predictive control, with a number of researchers providing important new results [39, 40, 45, 59, 61, 20]. This chapter does not specifically discuss neural networks in MPC, although the above cited works make clear that artificial neural networks may be successfully used in MPC. Chapter 7 by McAvoy in this book contains further discussion of the use of artificial neural networks in control.

- Frequency domain or time domain: Frequency domain models in continuous time are based on the Laplace transform of continuous linear systems and are not used for model predictive control, except for the linear unconstrained problem. For frequency domain descriptions in discrete time, it is not possible to be so definitive, since the shift operation (q or q^{-1}) is often interchanged with the z transform operator. Depending on the context, these models may be described as input-output or discrete-time frequency domain models. Since few of the concepts of frequency domain analysis are applicable for nonlinear systems, frequency domain concepts play little or no role in nonlinear MPC.

- First-principles or "black box": Models that are derived from heat, mass and momentum balances are frequently called "first-principles" or "fundamental" models, in contrast to other modeling schemes that fit a data set to an arbitrary form. Both approaches have been used successfully in MPC applications. An example of the "black box" approach is that used in model identification using step or impulse response models, in which model coefficients are fitted to process data using statistical methods without regard to underlying physical principles. The first principles approach is reflected in the models used in Section 5.2. Basic principles of chemistry were applied to arrive at a model of the chemical reactions involved and mass balances were used to create an overall reactor model.

The clearest tradeoff between the first-principles and black-box models lies in our confidence in predictions made by the model. A first-principles model presumably can be used to predict over a wide range of conditions, even without prior operating experience, provided that the basic assumptions of the model remain valid. On the other hand, a model based on (for example) artificial neural networks has almost no predictive value outside the range of operating conditions where data has been collected.

The best features of both approaches can perhaps be incorporated into a hybrid approach in which fundamental modeling is used for those portions of a process where the physical phenomena are well understood, and an input-output model used for the remainder of the process model.

Each of these model characteristics has an impact on the implementation of MPC. Since MPC is an on-line control method, the speed of the computer algorithm used is essential. Linear models are very well suited to MPC because they may be solved quickly and the optimization problem may be posed as linear or quadratic programming problems, for which robust and reliable software is available. Using continuous models in MPC also impacts the speed of solution because it usually requires significantly more variables to represent a discretized model. Distributed parameter models require more time to solve because of the additional difficulty in solving partial differential equation models.

From the class of input-output models, step or impulse response models have been favored for industrial application. In view of recent results concerning the use of linear state-space models [55], there appears to be little advantage in continuing with this practice. An equivalent state-space model is no less easy to identify, contains fewer unknown coefficients, can be evaluated more quickly by computer and can represent a wider range of processes.

For the nonlinear case, the decision to use first-principles or input-output models is less clear. For small systems with well-understood physical phenomena, fundamental modeling is preferable, because of the ability of the model to predict beyond the range of existing operating data. On the other hand, model identification is easier for larger systems using black-box models.

In conventional industrial MPC, installation of an MPC controller usually begins with identification of a linear input-output model in discrete time. Little or no effort is made to determine a first-principles model based on mass, energy or momentum balances. The objective of the modeling process is to determine a model that can be numerically evaluated quickly and that adequately describes the process dynamics in a neighborhood of some desired steady-state operating point. MPC based on these models has been extraordinarily successful in industry, but it is only applicable for stable processes that operate continuously near the nominal steady-state values. Although recent research results have extended the use of linear models for unstable processes [55], the range of operating conditions remains limited to those near the nominal steady state.

MPC using nonlinear models is especially suited for batch or semi-batch operations in which process conditions can vary significantly over the course of a batch, or for continuous processes that are expected to experience widely varying operating conditions.

In the past, the expense of computation often prevented even the consideration of the use of nonlinear models. As the power of computers available for on-line computations continues to increase, it has become feasible to consider using more complex models of various kinds. Today, the principle limitations in applying nonlinear models for on-line process control are model identification and robustness of control software and algorithms. A discussion of current research issues is contained in Section 5.9.

5.4 Computational Issues

The MPC control depends on finding solutions to a nonlinear programming problem at each sample time. This section discusses solutions techniques that are readily implementable using current optimization software. Since the complexity of the nonlinear program (NLP) is directly related to the model form, we consider several of the most common forms.

To solve the model predictive control problem, it is necessary to both solve an optimization problem and solve the system model. These two procedures may be implemented either sequentially or simultaneously. For sequential solutions, the model is solved at each iteration of the optimization routine. The controls are the decision variables, which are supplied to a routine that computes the model solution. The model solution is then used to compute the objective function, and the computed value is returned to the optimization code. Since gradient information is used by most modern optimization routines, it can be obtained through numerical differentiation (not recommended, see below), involving repeated calls to the routine that solves the model, or the sensitivity equations can be evaluated along with the model equations using, for example, the software routine DDASAC [7]. Our experience using numerical derivatives obtained through finite differences in the sequential approach has been strongly negative. To obtain gradients using finite difference typically involves differencing the output of an integration routine with adaptive step size. Although tightly bounded by the integration algorithm, the integration error is unpredictable. Differencing such an inexact quantity greatly degrades the quality of the finite difference derivatives. For this reason, using finite difference derivatives in the sequential approach cannot be recommended. See Gill *et al.* [27] for a further discussion of this issue.

In contrast, simultaneous model solution and optimization includes both the model states and controls as decision variables and the model equations are appended to the optimization problem as equality constraints. This can greatly increase the size of the optimization problem, leading to a tradeoff between the two approaches. For small problems with few states and a short prediction horizon, the sequential method using sensitivity equations for derivatives is probably the better approach. For larger problems, we have found that the simultaneous approach generally is more robust, i.e., the optimization software is less likely to fail. For problems with state or output variable constraints in continuous time, the sequential approach is more complicated. An approach to incorporating state variable constraints in continuous time was suggested by Sargent and Sullivan [66] that requires an additional state variable to be defined for each state or output variable constraint.

The remainder of this section discusses implementation issues for MPC using discrete-time and continuous-time models and the simultaneous optimization/model solution approach.

5.4.1 Discrete-Time Models

In this section, we consider the solution of MPC problems posed solely in discrete time. Since the computational issues discussed here are primarily concerned with solving the open-loop, constrained control problem, we will temporarily abandon the double time index. The model is given by Equations 5.1 and 5.2 and the objective function is given by

$$J = \|g(x_N) - y_{\text{ref}}\|_Q^2 + \sum_{j=0}^{N-1} \left(\|g(x_j) - y_{\text{ref}}\|_Q^2 + \|u_j - u_{\text{ref}}\|_R^2 + \|\Delta u_j\|_S^2\right) \tag{5.15}$$

For the unconstrained problem, or problems with only control constraints, it is possible to substitute the model equations directly into Equation 5.15 and to use only the controls as optimization variables (sequential approach). As noted previously, we have observed *significantly* better performance from optimization codes if each state and control is considered an independent optimization variable, with the model equations appended as nonlinear equality constraints:

$$\begin{aligned} x_1 - f(x_0, u_0) &= 0 \\ x_2 - f(x_1, u_1) &= 0 \\ \vdots \quad &\vdots \quad \vdots \\ x_N - f(x_{N-1}, u_{N-1}) &= 0 \end{aligned}$$

Considering states and controls as independent decision variables, the size of the vector of decision variables is $n(N+1) + Nm$. Most optimization software requires the vector to be passed as a single argument. We have found it most convenient to partition the entire vector of decision variables as $\left[x_0^T \; x_1^T \cdots x_N^T \; u_0^T \; u_1^T \; \cdots u_{N-1}\right]$.

MPC is characterized in part by its ability to incorporate constraints. In chemical process applications, constraints are most commonly expressed as simple bounds on the problem variables:

$$\begin{aligned} y_{\min} \leq y_j \leq y_{\max} \\ u_{\min} \leq u_j \leq u_{\max} \\ \Delta u_{\min} \leq \Delta u_j \leq \Delta u_{\max} \end{aligned} \tag{5.16}$$

The control constraints of Equation 5.16 can be expressed as a set of linear inequality constraint

$$\begin{bmatrix} C \\ -C \end{bmatrix} \begin{bmatrix} u_0 \\ u_1 \\ \vdots \\ u_{N-1} \end{bmatrix} \leq \begin{bmatrix} c_{\text{ub}} \\ c_{\text{lb}} \end{bmatrix}$$

in which C, c_{ub} and c_{lb} are given by

$$C = \begin{bmatrix} I & & & & \\ -I & I & & & \\ & -I & I & & \\ & & \ddots & \ddots & \\ & & & -I & I \end{bmatrix}$$

$$c_{\text{ub}} = \begin{bmatrix} u_{-1} + \Delta u_{\max} \\ \Delta u_{\max} \\ \vdots \\ \Delta u_{\max} \end{bmatrix} \qquad c_{\text{lb}} = \begin{bmatrix} -u_{-1} - \Delta u_{\min} \\ -\Delta u_{\min} \\ \vdots \\ -\Delta u_{\min} \end{bmatrix}$$

The form of the above constraint illustrates an example of another commonly used constraint, the general linear inequality constraint of the form

$$\begin{aligned} Dx_k &\leq d \\ Hu_k &\leq h \end{aligned}$$

These constraints are applied to each predicted state or control in the horizon. The complete set of constraints may then be expressed using the Kronecker matrix product

$$(I_N \otimes D) \begin{bmatrix} x_1 \\ x_2 \\ \vdots \\ x_N \end{bmatrix} \leq \mathbf{1}_N \otimes d \qquad (I_N \otimes H) \begin{bmatrix} u_0 \\ u_1 \\ \vdots \\ u_{N-1} \end{bmatrix} \leq \mathbf{1}_N \otimes h$$

in which I_N is the N-dimensional identity matrix and $\mathbf{1}_N$ is an N-dimensional vector whose elements are equal to 1.

Sometimes it provides better closed-loop performance to use a control horizon M that is less than the prediction horizon N, as demonstrated in the simulation results for the fluidized-bed reactor. For $M < N$, we add additional linear equality constraints $u_{N-1-M} = u_{N-M} = \cdots = u_{N-1}$.

The linear bounds discussed above are sufficiently general to include simple bounds on the decision variables. Even so, it is usually preferable to avoid including them in the general formulation for linear constraints, for two reasons: Since most optimization software permits simple bounds to be included in the argument list of the calling routines, it simplifies the programming task to use the available program interface. More importantly, some implementations include special considerations for simple bounds [27] that can reduce the computational effort necessary to solve the nonlinear program.

Most optimization software contain routines to use finite differences to compute gradients. We strongly recommend against use of finite difference gradients except perhaps for the smallest and simplest problems. Gradients (and Jacobians) should be provided if possible to optimization routines to assure good results. For the objective function of Equation 5.15, the gradient may be expressed by

$$\nabla J = \begin{bmatrix} 2G_0^T Q \left[g(x_0) - y_{\text{ref}}\right] \\ 2G_1^T Q \left[g(x_1) - y_{\text{ref}}\right] \\ \vdots \\ 2G_N^T Q \left[g(x_N) - y_{\text{ref}}\right] \\ 2R\,(u_0 - u_{\text{ref}}) + 2S(2u_0 - u_{-1} - u_1) \\ 2R\,(u_1 - u_{\text{ref}}) + 2S(2u_1 - u_0 - u_2) \\ \vdots \\ 2R\,(u_{N-1} - u_{\text{ref}}) + 2S\,(u_{N-1} - u_{N-2}) \end{bmatrix}$$

in which G_k is the Jacobian matrix of $g(x)$ evaluated at (x_k, u_k).

The Jacobian of the equality constraints representing the system equations is sparse and has the following general form:

$$
\begin{bmatrix}
I & & & & & & & & \\
-F_{x,0} & I & & & & -F_{u,0} & & & \\
& -F_{x,1} & I & & & & -F_{u,1} & & \\
& & -F_{x,2} & I & & & & -F_{u,2} & \\
& & & -F_{x,3} & I & & & & -F_{u,3}
\end{bmatrix}
$$

The Jacobian of the inequality constraints is also sparse with the following structure:

$$
\begin{bmatrix}
D & & & & & & & \\
& D & & & & & & \\
& & D & & & & & \\
& & & D & & & & \\
& & & & H & & & \\
& & & & & H & & \\
& & & & & & H & \\
& & & & & & & H \\
& & & & I & & & \\
& & & & -I & I & & \\
& & & & & -I & I & \\
& & & & & & -I & I \\
& & & & -I & & & \\
& & & & I & -I & & \\
& & & & & I & -I & \\
& & & & & & I & -I
\end{bmatrix}
$$

The above examples represent an MPC problem using a prediction horizon of 5, with linear constraints on state, control variable and control increments at each time step.

5.4.2 Continuous-Time Models

For continuous-time systems, we consider here nonlinear models represented as differential equations of the general form

$$\dot{x} = f(x, u, t) \tag{5.17}$$

A controller is sought which minimizes some performance index

$$J = \phi\left[x\left(t_f\right), t_f\right] + \int_{t_0}^{t_f} L\left[x(t), u(t), t\right] dt \tag{5.18}$$

We can find necessary conditions for a minimum using variational techniques, but in general these do not form a practical basis for implementation. We seek an approximate solution using piecewise constant controls with intervals that correspond to some appropriate sampling frequency of the process. Optimal controls for the system of Section 5.2 were obtained using the method of orthogonal collocation on finite elements, a technique based on numerical integration using Gaussian quadrature. We present here a brief tutorial description of the method. See the following for a more detailed theoretical discussion: [4, 15, 12, 77, 78].

Consider the solution of minimizing the objective function of Equation 5.18 over a single time element. In the isothermal CSTR of Section 5.2, the time interval was 0.002 hours or 7.2 seconds. At the beginning of the time interval the state (or state estimate) is known and we seek a single control, constant over the interval, that minimizes the objective function.

The orthogonal collocation technique assumes that the solution to Equation 5.17 can be approximated by an interpolating polynomial, expressed as the sum of Lagrange polynomials [14], with nodes located at the roots of an orthogonal family of polynomials. By this choice of nodes, integration using the orthogonal collocation technique is exact for polynomials of degree $2n-1$, although only a polynomial of degree n is used for the interpolation. See Davis and Rabinowitz [15] for a complete explanation of this remarkable result.

In Section 5.2, all computations were performed using either two or three internal collocation points and one on the boundaries of each finite element. Three internal collocation points are illustrated in Figure 5.8. If the element is scaled so that $t_1 = 0$ and $t_5 = 1$ then the node points are given by $\{t_1, t_2, t_3, t_4, t_5\} = \{0, 0.1127, 0.5, 0.8873, 1.0\}$, which are the roots of the third order Legendre polynomial, augmented by the element endpoints. If $t_1 \neq 0$ or $t_5 \neq 1$, then a linear change of variable is necessary to transform the time interval to $[0, 1]$. When the dynamic model is time-invariant, the only correction necessary is to scale the right-hand side of Equation 5.17 by $t_5 - t_1$.

With these choices of interpolating polynomials and node points, derivatives and integrals can be represented by linear combinations of the state values at the node points. Let the jth state at node i be represented by x_{ij}. We admit the notation $x_{i,*}^T$ to indicate the entire state at node i. With constant control, an approximate solution to the differential equation can be obtained by solving the following nonlinear equation:

$$A_0 X = F(X, u) \tag{5.19}$$

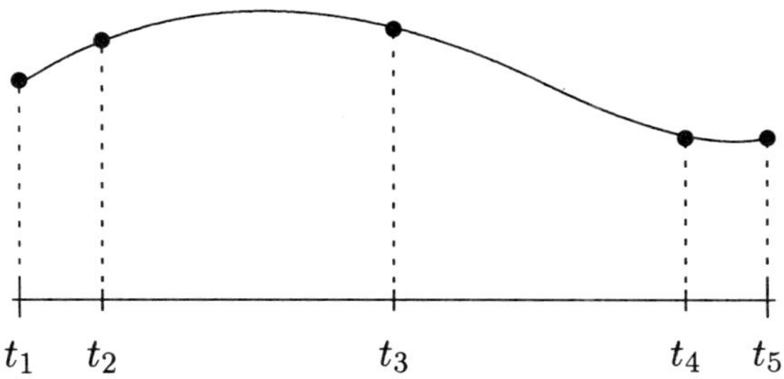

Figure 5.8: Collocation element.

in which

$$X = \begin{bmatrix} x_{1,1} & x_{1,2} & \cdots & x_{1,n} \\ x_{2,1} & x_{2,2} & \cdots & x_{2,n} \\ x_{3,1} & x_{3,2} & \cdots & x_{3,n} \\ x_{4,1} & x_{4,2} & \cdots & x_{4,n} \\ x_{5,1} & x_{5,2} & \cdots & x_{5,n} \end{bmatrix} \qquad F(X,u) = \begin{bmatrix} x_{\text{init}}^T \\ f^T(x_{2,*}^T, u, t_2) \\ f^T(x_{3,*}^T, u, t_3) \\ f^T(x_{4,*}^t, u, t_4) \\ f^T\left(x_{5,*}^T, u, t_5\right) \end{bmatrix}$$

The matrix A_0 contains the first derivative collocation weights:

$$A_0 = \begin{bmatrix} 1 & 0 & 0 & 0 & 0 \\ -5.32 & 3.87 & 2.07 & -1.29 & 0.68 \\ 1.5 & -3.23 & 0 & 3.23 & -1.5 \\ -0.68 & 1.29 & -2.07 & -3.87 & 5.32 \\ 1 & -1.88 & 2.67 & -14.79 & 13 \end{bmatrix}$$

(The subscript 0 is included to emphasize that the first line of the collocation weight matrix is added to account for the initial condition.) The objective function of Equation 5.18 is approximated by

$$J \approx \phi\left(x_{5*}^T, t_5\right) + \sum_{k=1}^{5} c_k\, L\left(x_{k*}^T, u, t_k\right) \tag{5.20}$$

in which the coefficients c_k are the quadrature weights $\left\{0, \frac{5}{18}, \frac{4}{9}, \frac{5}{18}, 0\right\}$ corresponding to the third order Legendre polynomial. For other choices of orthogonal polynomials and node points, the collocation weight matrices and quadrature weights may be computed using the FORTRAN programs of Villadsen and Michelsen [77] or using root-finding techniques using the orthogonal polynomials [15]. In the examples of this chapter, collocation and quadrature weights were computed using the `octave` [18] interface to the Villadsen and Michelson routines.

With these choices, the control is obtained as the solution to the following nonlinear program

$$\min_{u,X} J$$

subject to:

$$A_0 X = F(X, u) \tag{5.21}$$

which represents a discretized form of the MPC problem with horizon length 1.

For horizon lengths greater than 1, we can extend the orthogonal collocation method to a set of finite elements in time, with the solution of the problem approximated by polynomials defined on each element. This situation is illustrated in Figure 5.9. Note that we have chosen an indexing scheme that consecutively numbers nodes across finite element boundaries. It would perhaps be more natural to adopt an additional index to indicate to which finite element a state or control belongs; however, this would result in triple indices for states, controls and outputs, which are not supported in most programming languages or any known optimization software implementations. The structure shown in Figure 5.9 corresponds to the collocation scheme used for the isothermal CSTR example of Section 5.2. The fixed-bed reactor example used one fewer interior collocation point. The matrix A_0 in

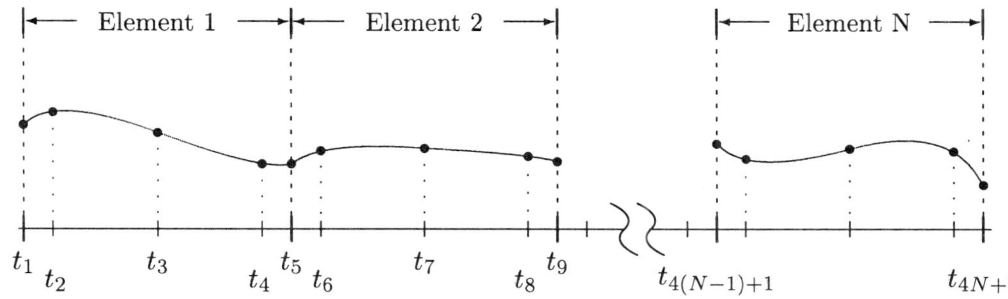

Figure 5.9: Collocation on finite elements.

Equation 5.21 become a partitioned matrix that is almost block diagonal:

$$\tilde{A} = \begin{bmatrix} A_0 & & & & & \\ & A & & & & \\ & & A & & & \\ & & & \ddots & & \\ & & & & A & \\ & & & & & A \end{bmatrix} \tag{5.22}$$

The overlap is due to the presence of the points on the boundaries of two adjacent finite elements. The state vector includes contributions from states in every multiple finite element and so becomes

$$\tilde{X} = \begin{bmatrix} x_{1,1} & x_{1,2} & \cdots & x_{1,n} \\ x_{2,1} & x_{2,2} & \cdots & x_{2,n} \\ x_{3,1} & x_{3,2} & \cdots & x_{3,n} \\ x_{4,1} & x_{4,2} & \cdots & x_{4,n} \\ x_{5,1} & x_{5,2} & \cdots & x_{5,n} \\ \vdots & \vdots & & \vdots \\ x_{4N-2,1} & x_{4N-2,2} & \cdots & x_{4N-2,n} \\ x_{4N-1,1} & x_{4N-1,2} & \cdots & x_{4N-1,n} \end{bmatrix}$$

The $\tilde{F}$ vector is constructed as in the one-element case, except that the value of the control must change across element boundaries:

$$\tilde{F}(\tilde{X}, u_1, u_2, \ldots, u_{N-1}) = \begin{bmatrix} x_{\text{init}}^T \\ f^T(x_{2*}^T, u_0, t_2) \\ f^T(x_{3*}^T, u_0, t_3) \\ f^T(x_{4*}^T, u_0, t_4) \\ f^T\left(x_{5*}^T, u_0, t_5\right) \\ f^T(x_{6*}^T, u_1, t_6) \\ f^T(x_{7*}^T, u_1, t_7) \\ f^T(x_{8*}^T, u_1, t_8) \\ f^T\left(x_{9*}^T, u_1, t_9\right) \\ f^T(x_{10*}^T, u_2, t_{10}) \\ \vdots \\ f^T(x_{(4N-2)*}^T, u_{N-1}, t_{4N-2}) \\ f^T(x_{(4N-1)*}^T, u_{N-1}, t_{4N-1}) \end{bmatrix}$$

Taking $\tilde{c} = \mathbf{1}_N \otimes c$ (the Kronecker product of $\mathbf{1}_N$ and c) the objective function of Equation 5.18 is approximated by

$$J \approx \phi\left(x_{(4N-1)*}^T, t_{4N-1}\right) + \sum_{j=0}^{N-1} \sum_{k=4j+1}^{4j+5} \tilde{c}_k \, L\left(x_{k*}^T, u_j, t_k\right)$$

Taking $U = \left[u_1^T, u_2^T, \ldots, u_{N-1}^T\right]^T$, we can write the orthogonal collocation approximation as

$$\tilde{A}\tilde{X} = \tilde{F}\left(\tilde{X}, U\right) \tag{5.23}$$

Since most optimization codes pass the optimization variables as a single matrix, the matrix equation of Equation 5.23 must be appropriately indexed. In our examples, we were more interested in coding convenience that in optimal performance, so we reconstructed the state matrix $\tilde{X}$ as shown above within the objective function function. Naturally, choices of indexing schemes are at the discretion of the programmer. For on-line application, we would seek the fastest evaluation method possible and would probably include such items as sparse matrix multiplication and optimization of sparse systems.

Since optimization codes require a single vector of decision variables, the task of re-indexing can become tedious. The Jacobian matrix of the equality constraints is quite sparse, becoming more sparse with larger problems, and the overlapped, quasi-block diagonal structure of $\tilde{A}$ is repeated in the Jacobian. This is illustrated more clearly in Figure 5.10, which shows the non-zero elements for the fluidized-bed reactor example with prediction horizon 3, using 5 internal collocation points in each finite element.

5.5 Feedback and State Observers

Like the LQ-optimal controller, the state-space formulation of nonlinear MPC discussed in this chapter uses state feedback. The current state (or state estimate) is used as an initial condition at each iteration to predict the future behavior and select an optimal control sequence. Model predictive control does not prescribe a specific form for the state observer. This section discusses several possibilities and their effects on closed-loop system behavior, including steady-state tracking in the presence of modeling errors and disturbance rejection.

If the full state is measured at each sampling point, an obvious possibility for feedback would be simply to reset the states to the current observations.

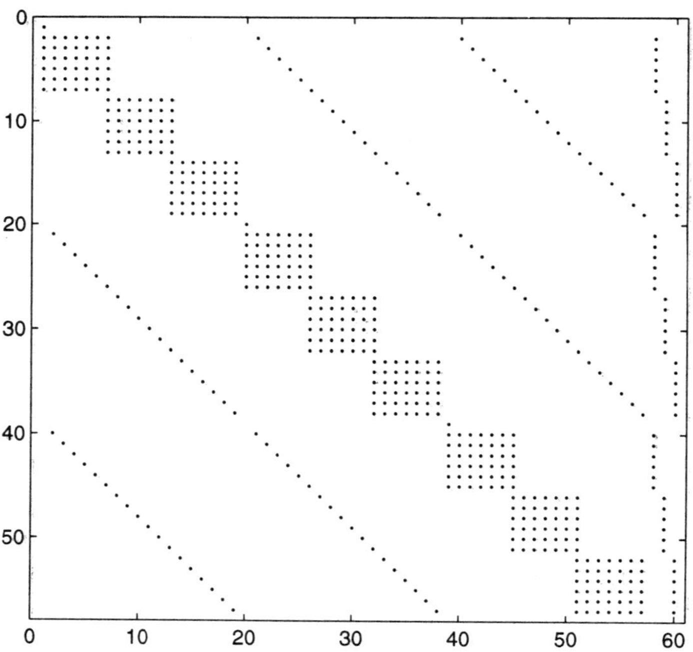

Figure 5.10: Sparse Matrix Structure of Constraint Jacobian.

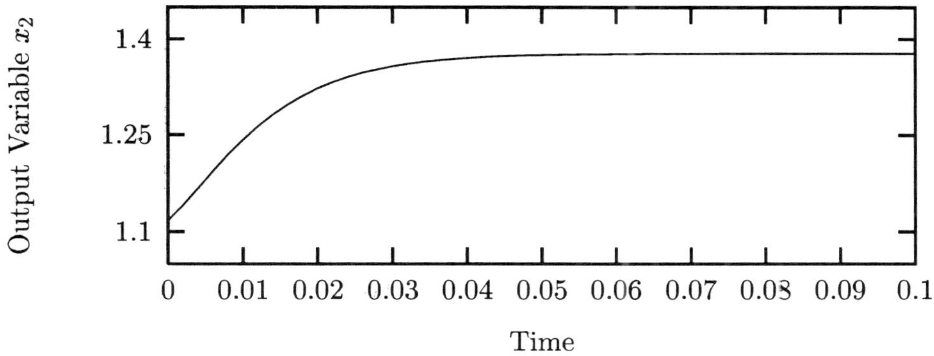

Figure 5.11: Isothermal CSTR: Feedback through resetting of the state.

The updated states then become the initial conditions for the model predictions. For the case without model error, this method works well as shown in Figure 5.1. In the presence of modeling error, simply resetting the state usually produces steady-state offset. We simulated this controller using the model of Section 5.2 and a +50 percent error in the concentration of species A in the the feedstream, x_F. The resulting system output is shown in Figure 5.11. The setpoint was 1.0 and the objective was the same is in Figure 5.5 with horizon length 5. We noted the presence of steady-state offset for horizon lengths 6 through 9 as well. This controller produces offset for *any* finite horizon. Since resetting the state provides such poor results, it is generally not used in practice. We next describe two other approaches with better properties: conventional MPC feedback and MPC with steady-state target optimization.

5.5.1 Conventional MPC Feedback

By far the most common feedback method in MPC is to compare the measured output of the process to the model prediction at time k to generate a disturbance estimate $\hat{d}_k = y_k - \tilde{y}_k$, in which y_k and $\tilde{y}_k$ represent the process measurement and model prediction, respectively. In forming the MPC objective function, the disturbance term is then added to the output prediction over the entire prediction horizon to produce the modified objective

$$J = \sum_{j=0}^{N-1} ||y_{\text{ref}} - \left(y_{k+j|k} - \hat{d}_k\right)||_Q^2 + ||\Delta u_{k+j|k}||_S^2$$

This procedure assumes that differences observed between the process output and the model prediction are due to additive step disturbances in the output that persist throughout the prediction horizon. Although simplistic, this error model offers several related practical advantages:

- It accurately models setpoint changes which often enter feedback loops as step disturbances.
- It approximates slowly varying disturbances. Since errors in the model can appear as slowly varying output disturbances, it provide robustness to model error.
- It provides zero offset for step changes in setpoint.

Coupled with a linear step response model, feedback through the estimation of a step disturbance has been extensively applied in industrial applications. The stability and robustness have been analyzed for linear, unconstrained systems by García and Morari [23]. More details for linear systems with an emphasis on robust design using linear models may be found in the the monograph by Morari and Zafiriou [53].

Figure 5.12 illustrates the effect of feedback using the disturbance term with the isothermal CSTR model. The simulation used the same model mismatch, and the same objective function weights as in Figure 5.11. The results for horizon lengths 6 through 9 are included in Figure 5.12. The MPC controller was not stabilizing for the nominal system with horizons less than 6. Instability was observed in the simulations using these horizon lengths for both cases with and without model mismatch.

Feedback through differencing model prediction and measurement does not require a state-space description. *Linear* MPC methods used in industry do not use a state-space model and incorporate the disturbance directly into the MPC objective function. The feedback mechanism is sometimes obscured by this treatment. A state-space description, even if not used in the controller, allows a more natural interpretation of the conventional MPC feedback procedure. Using an augmented state-space model, conventional MPC feedback can be shown to be a particular form of a state observer for the resulting system:

$$x_{k+1} = f(x_k, u_k) \tag{5.24}$$

$$d_{k+1} = d_k \tag{5.25}$$

$$y_k = g(x_k) + d_k \tag{5.26}$$

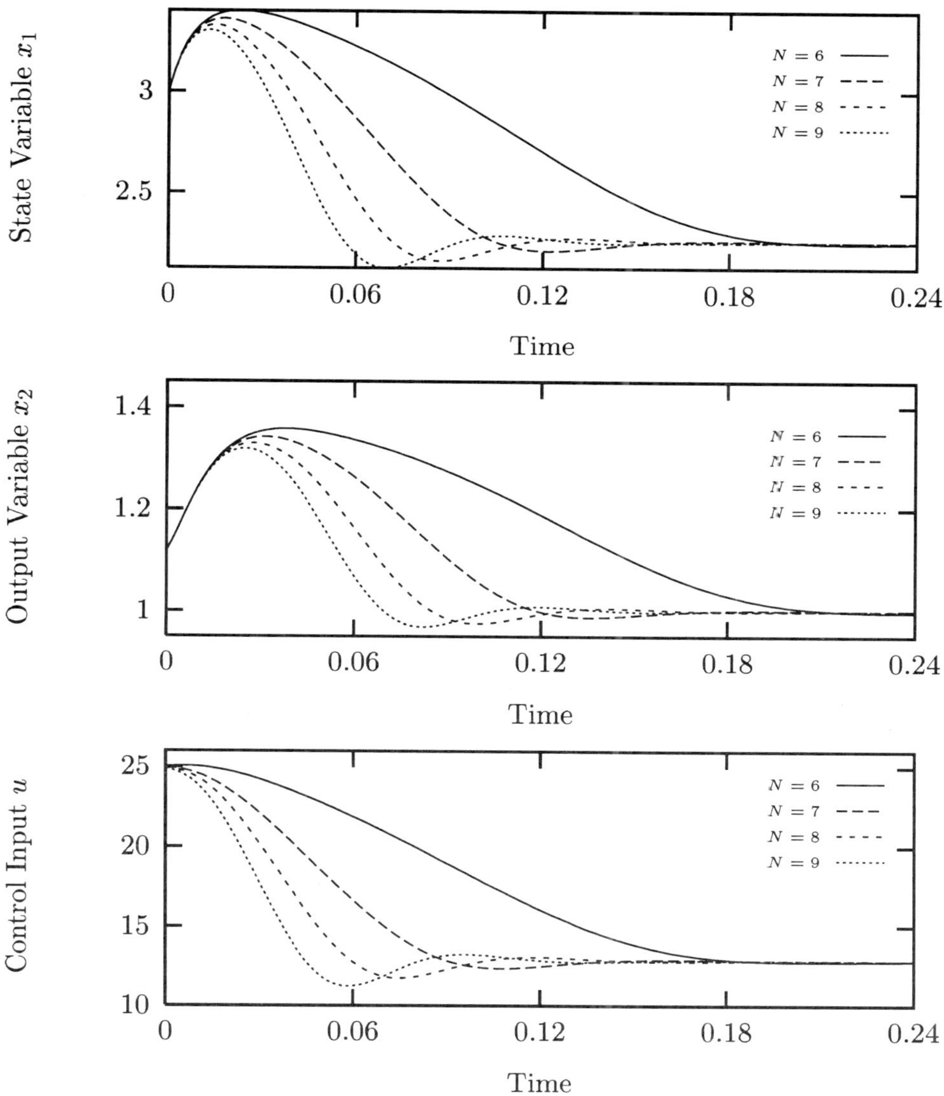

Figure 5.12: Isothermal CSTR: Conventional MPC feedback with model mismatch.

Taking the augmented state to be $\left[x_k^T \;\; d_k^T\right]^T$, we can use a variety of techniques, including the extended Kalman filter or a moving horizon observer [55, 57, 56, 65], to estimate both the state and the disturbance. A general state observer for the augmented system of Equation 5.24 through 5.26 has the form

$$\begin{aligned}\hat{x}_{k+1} &= f(\hat{x}_k, u_k) + h_x\left(\hat{x}_k, \hat{d}_k, y_k, u_k\right)\\ \hat{d}_{k+1} &= \hat{d}_k + h_d\left(\hat{x}_k, \hat{d}_k, y_k, u_k\right)\end{aligned}$$

in which the $\hat{x}_k$ and $\hat{d}_k$ represent the observer estimates. To obtain the usual MPC observer, choose

$$\begin{aligned}h_x\left(\hat{x}_k, \hat{d}_k, y_k, u_k\right) &= 0\\ h_d\left(\hat{x}_k, \hat{d}_k, y_k, u_k\right) &= y_k - g(\hat{x}_k) - \hat{d}_k\end{aligned}$$

This leaves an open-loop state observer for the original states x_k and a deadbeat observer for the disturbance d_k. In the absence of information to construct a disturbance model, this approach is a good choice in view of its favorable features. With additional information and engineering judgment, it is possible to obtain better state estimates using an observer that updates $\hat{x}_k$ as well as $\hat{d}_k$. Whether the extra effort leads to better control depends upon the specific observer design and knowledge about the disturbances. Several researchers have presented results for linear systems showing better control using a full state estimator [26, 46, 55, 64] in the presence of noisy output measurements and model mismatch.

Since the conventional MPC feedback uses a deadbeat observer of the output disturbance, it is sensitive to random fluctuations in the output. This is illustrated in Figure 5.13, which reproduces the situation of Figure 5.12 except that a zero-mean Gaussian disturbance with variance 0.09 has been added to the output measurement. Figure 5.13 show the system response for horizon lengths 7 and 9. The initial portion of the figures shows the expected faster response of the longer horizon lengths. The faster response comes at the expense of poorer disturbance rejection, i.e. noise in the input.

5.5.2 MPC with Steady-State Target Optimization

We next investigate an alternative feedback method that does not directly incorporate the disturbance term in the MPC objective and can penalize

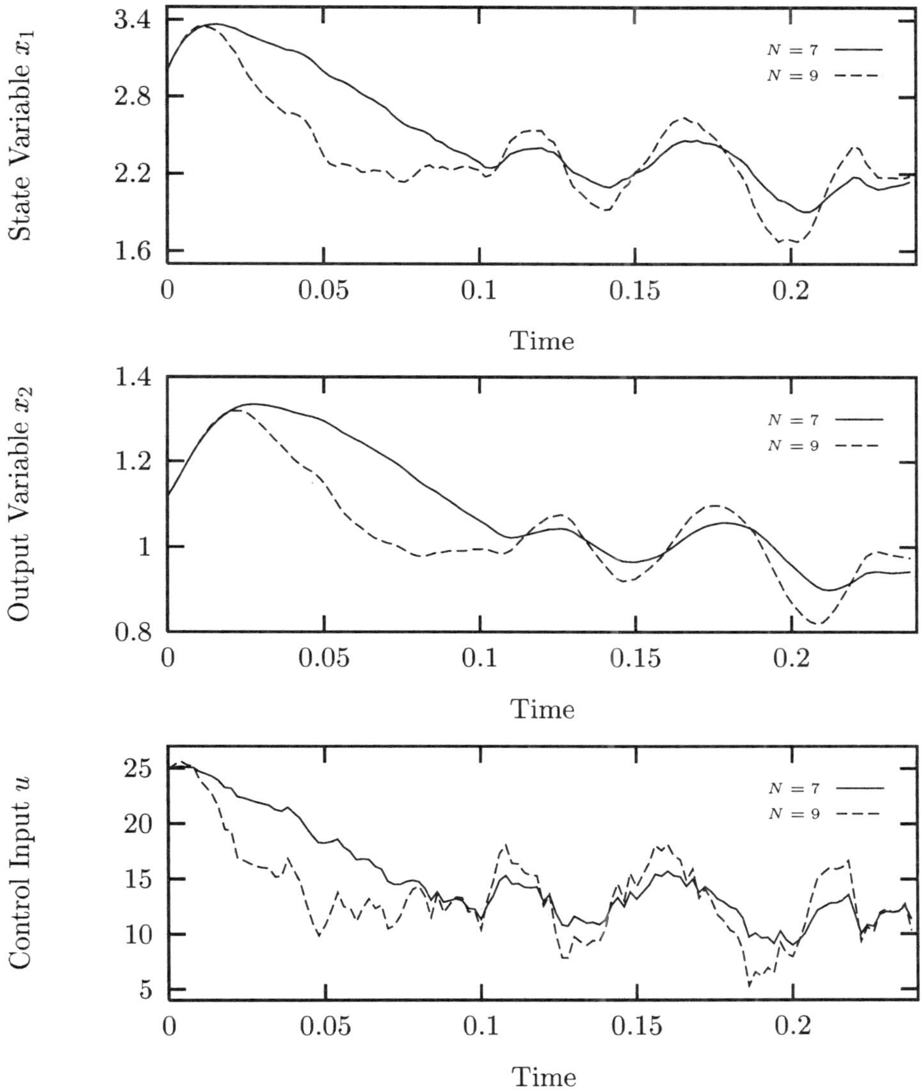

Figure 5.13: Isothermal CSTR: Response of system to Gaussian noise using conventional MPC feedback.

both u and Δu. In this approach, we use a preprocessing step in which new steady-state targets for the state and control, x_t and u_t, are computed based on the disturbance estimate, $\hat{d}_k$, and the output setpoint, y_{ref}. The new targets for the state and control are then incorporated directly into the objective as a penalty on $(x - x_t)$ and $(u - u_t)$. This method is discussed by Muske and Rawlings [55] in the context of linear systems. The method presented here is a direct nonlinear analog.

To apply this method we solve the augmented system of Equations 5.24 through 5.26 at steady state to obtain target values for the internal states, controls or output variable. These are used as setpoints in the MPC objective function, exactly as we did in the first simulation result of Figure 5.1. Figure 5.1 (no model error, no disturbances) shows excellent speed of response without offset. We can duplicate some aspects of that excellent performance using the alternate feedback method.

Using the conventional MPC observer (open-loop state observer with deadbeat disturbance estimator), we can choose steady-state target values by solving the following nonlinear program at each iteration:

$$\min_{u_t, x_t} \|u_{\text{ref}} - u_t\|_R^2 \tag{5.27}$$

subject to:

$$\begin{aligned} x_t &= f(x_t, u_t) \\ y_{\text{ref}} &= g(x_t) + \hat{d}_k \\ x_t &\in \mathcal{X} \\ u_t &\in \mathcal{U} \end{aligned}$$

In the above nonlinear program u_{ref} represents the input setpoint, for problems in which there are extra degrees of freedom in the input vector, i.e. more inputs than output setpoints. The input setpoint is selected with the understanding the the controller will enforce it as a setpoint only in a minimum-norm sense. The solution $\{u_t, x_t\}$ provides target values for the states and controls that account for the current estimate for the output disturbance $\hat{d}_k$. At each iteration, a new estimate of $\hat{d}_k$ is obtained from the state estimator and the target values are readjusted accordingly.

If the above nonlinear programming problem does not have a feasible solution, then the output setpoint cannot be reached exactly; an alternative

NLP may be used to compute the control:

$$\min_{u_t, x_t} \|y_{\text{ref}} - (g(x_t) + \hat{d}_k)\|_Q^2 \tag{5.28}$$

subject to:

$$\begin{aligned} x_t &= f(x_t, u_t) \\ x_t &\in \mathcal{X} \\ u_t &\in \mathcal{U} \end{aligned}$$

In this case, we cannot simultaneously find values of $\{x_t, u_t\}$ to satisfy the steady-state system equations at the desired setpoint. This can happen, for example, if an especially large disturbance enters the system, or if the desired setpoint minus the disturbance does not correspond to a steady state of the model equations. We always assume that the model has at least one real steady-state solution that satisfies the input and output constraints, so Equation 5.28 always has a feasible point.

This technique can give excellent results. Figure 5.14 shows the response of the isothermal CSTR with model mismatch using a 1 percent weight on $(u - u_t)^2$ in relation to the weight on the output. The value of u_t was adjusted at each sampling time using the nonlinear program of Equation 5.27 with $u_{\text{ref}} = 25$, the nominal steady-state value. Unlike the previous examples that used a penalty on Δu, the response is rapid. It satisfies both the primary and secondary control objectives: It brings the system to the setpoint $y = 1$ without offset and it does so at a low feed rate. These results indicate that this is the clear method of choice for control of the reactor of Section 5.2. The curves shown in Figure 5.14 were computed with $N = 5$. Slightly slower response was observed with shorter horizon lengths, but the differences were very small and would not appear as distinct curves if plotted in Figure 5.14. Even with horizon length 1, the response was very similar, and the time necessary to compute the MPC control was much less. Recall that this system was unstable for short horizons using conventional MPC feedback.

Up to this point, all of the isothermal CSTR examples shown have started from the same initial conditions. It is necessary to check the behavior of the controller over a range of initial conditions before deciding on a specific controller design. Figure 5.15 shows trajectories of the system from various initial conditions. The system was controlled using feedback through steady-state target optimization with horizon length 3. The initial conditions were selected to lie along the perimeter of the rectangular region of

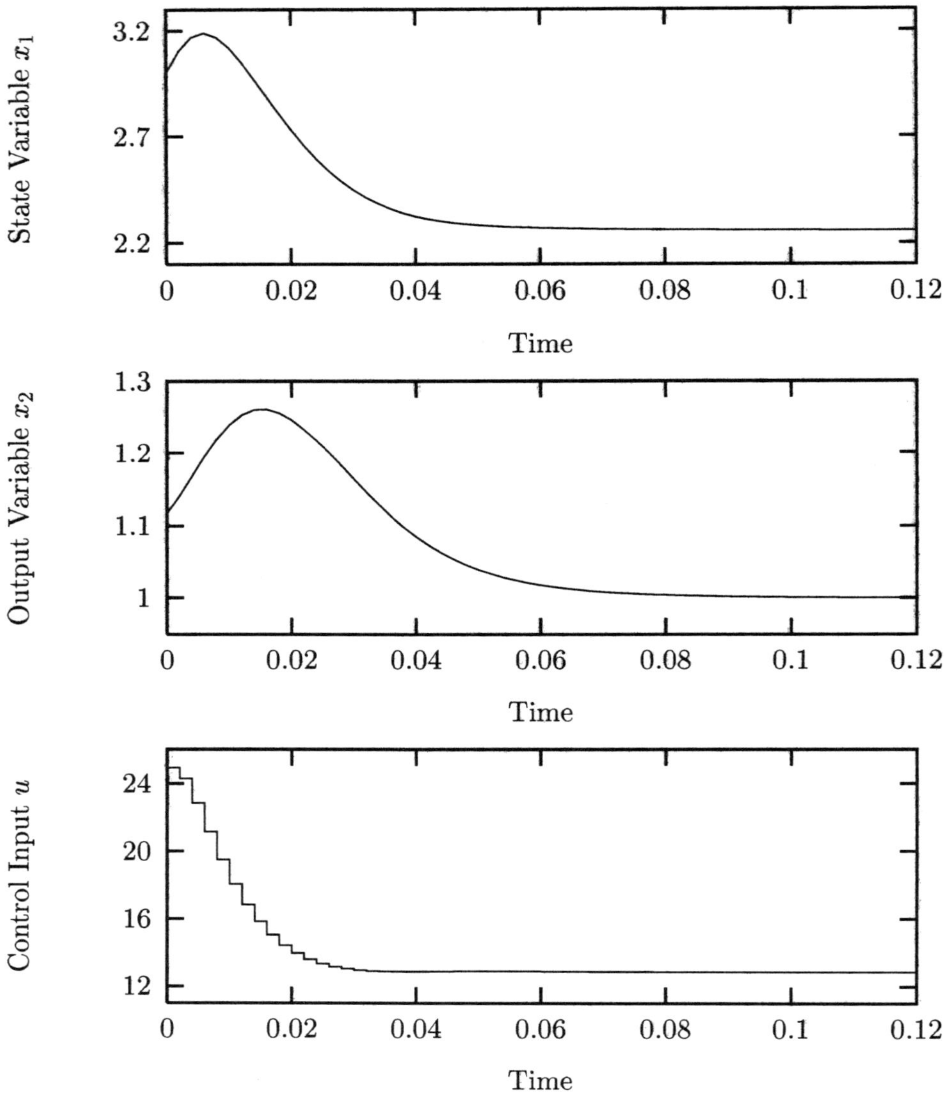

Figure 5.14: Isothermal CSTR: System response using target optimization.

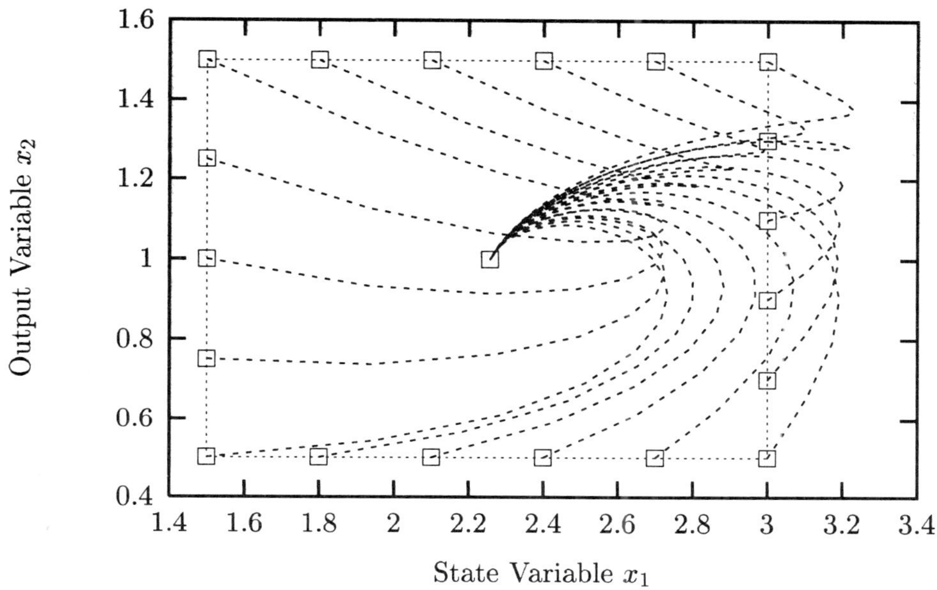

Figure 5.15: Isothermal CSTR: System response from various initial conditions.

the state-space shown in the diagram. The initial conditions and the final value of the controlled system are indicated with boxes. Since the controller is implemented at discrete points in time, the system trajectories can cross, unlike the situation with continuous-time feedback systems in two dimensions. This figure shows that the region of attraction for the MPC controller in the presence of model error is not negligible and that the controller would be expected to perform well from a variety of initial conditions.

Using both the conventional MPC feedback and feedback through target optimization, we compared the ability of the MPC controller to reject zero-mean Gaussian disturbances in the output. As in Figure 5.13, we used discrete, additive, zero-mean Gaussian noise in the output of the process with variance 0.09. To allow direct comparison of various methods, each simulation used the same noise sequence, which we obtained by re-seeding the random number generator with the same seed value.

Figure 5.16 shows the response of the two feedback methods. The target optimization feedback method with horizon length 3 is compared to the conventional DMC controller using horizon length 7. The target optimization feedback is insensitive to the horizon length and provides virtually the same response for horizon lengths from 1 to 5. During the course of the simulations, about 15 percent of the initial nonlinear programming problems of Equation 5.27 had no feasible point (which *never* occurred without the random disturbances). In these cases, the NLP of Equation 5.28 was used to determine the reference values. Figure 5.16 shows that feedback through target optimization with no weighting on Δu is more sensitive to disturbances than the conventional MPC feedback method with the weighting on Δu. This effect clearly shows up in the input. To attenuate the effect of the zero-mean disturbances, one can choose to weight Δu in the controller optimization, but there are other choices as well. For example, we can incorporate a moving average filter to estimate the disturbance term: $\hat{d}_k = 0.9\,\hat{d}_{k-1} + 0.1\,(y_k - g\,(\hat{x}_k))$. This reduces the occurrence of infeasibility in the initial NLP to about 5 percent. The effect of this filter is illustrated in Figure 5.17. As in the previous figure, the conventional MPC feedback controller with horizon length 7 is included for reference. From Figure 5.17, we can see that feedback through target optimization combined with the filter for the disturbance term has greatly improved the disturbance rejection ability of the controller, while retaining the ability of the feedback method to provide robustness and fast response.

Figure 5.17 shows that an MPC controller with a short horizon using target optimization feedback can perform as well as or better than conventional MPC feedback with longer horizons in terms of robustness, speed of response and disturbance rejection; however, the feedback through the target optimization solution is *much* faster to implement. Using horizon length 9 and conventional MPC feedback, the simulations took about 20 hours to simulate 120 time steps on an IBM-compatible 80486/33 MHz computer. At each sampling time, this method required the solution of a nonlinear programming problem involving 99 decision variables. In contrast, using feedback through target optimization, the initial NLP needed to compute the targets involves 3 decision variables and requires an insignificant amount of time to compute. The control computation itself involved as few as 6 decision variables (for $N = 1$) to 18 variable (for $N = 3$) and could easily have been computed in real time using a sampling interval of 7.2 seconds.

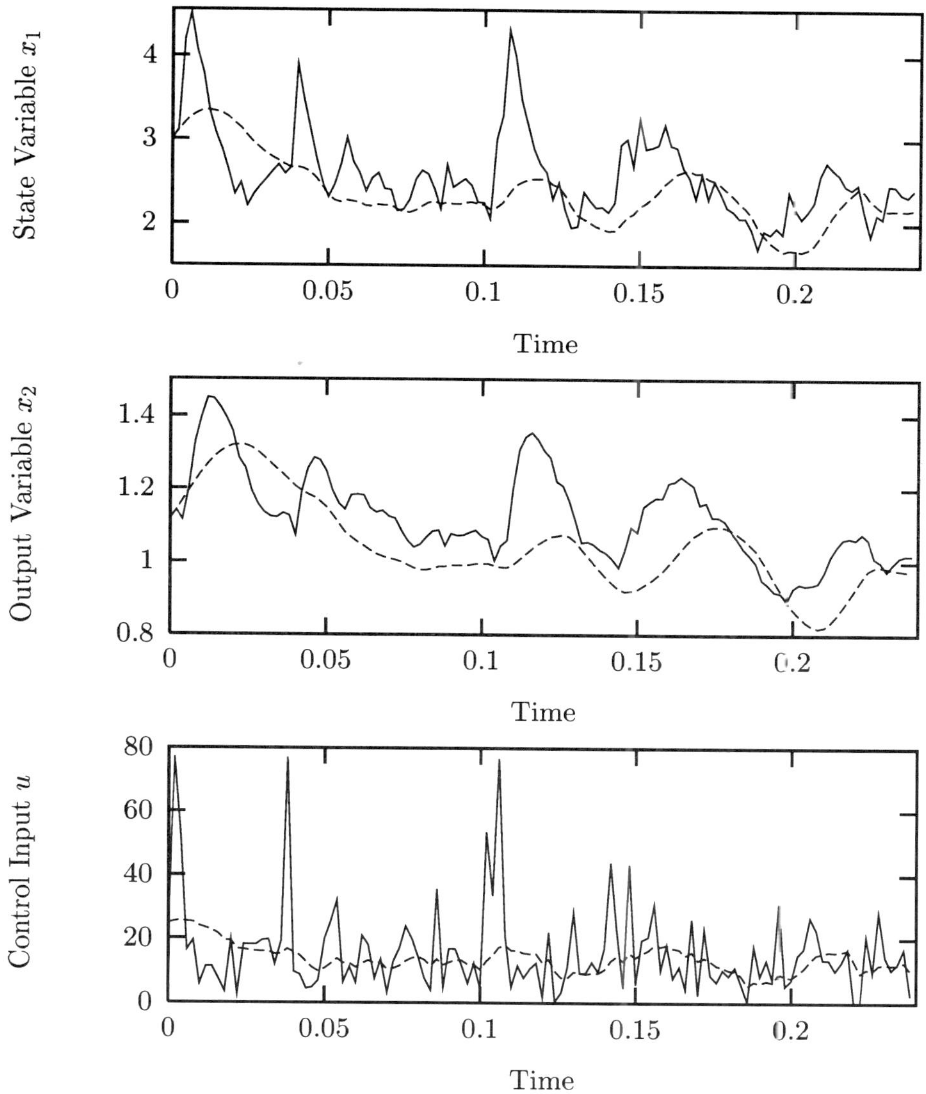

Figure 5.16: Isothermal CSTR: Effect of output disturbances using feedback through target optimization. Solid line: feedback through target optimization. Dashed line: conventional MPC feedback.

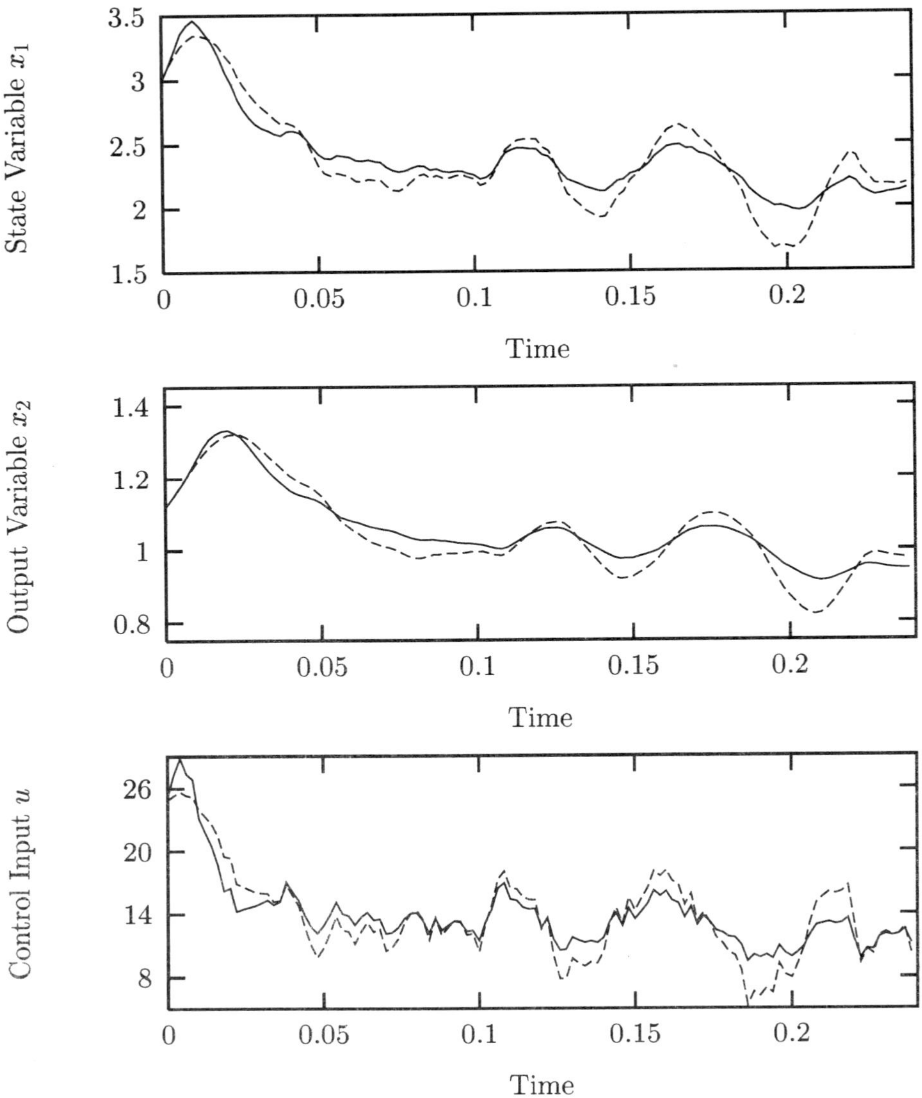

Figure 5.17: Isothermal CSTR: Effect of output disturbances using feedback through target optimization. Solid line: feedback through target optimization. Dashed line: conventional MPC feedback.

5.6 Selection of Tuning Parameters

As illustrated by some of the previous examples, the choices of control horizon, weight matrices and feedback methods have a profound effect on MPC nominal stability and robustness. Ideally, we would prefer to know *a priori* what range of tuning parameters would provide nominal stability and robustness. We would then be able to select the tuning parameters to meet specific process objectives, without being concerned that a poor choice could drive the system to instability. Implementable sufficient conditions that guarantee nominal stability [55] and robustness [1] of MPC are known for linear systems. Unfortunately, the known sufficient conditions for nonlinear systems, such as those described in Section 5.7 (infinite horizon, zero final-state constraint), are usually much too strong to offer practical implementation guidance, and we must resort to a set of heuristics based on the extrapolation of linear systems results, numerical simulations and experiments.

The most significant tuning parameters that must be selected are prediction horizon, control horizon, sampling interval, and penalty weight matrices. In addition, the control engineer must decide whether to penalize Δu or whether to use a filter in the feedback loop. As part of the numerical implementation, the discretization scheme and software must be selected. Some guidelines for selection of tuning parameters are provided below. In the following discussion, we assume that an explicit model is available and that the model has been acceptably verified.

- Sampling Interval: For stable, minimum phase systems, stability does not depend on the sampling interval; however, to ensure good closed-loop performance, the sampling interval should be small enough to capture adequately the dynamics of the process, yet large enough to permit the on-line computations necessary for implementation. Large sampling interval can result in ringing (excessive oscillations) between sample points. An example of this phenomenon is provided by García and Morari [23] using a linear system.

  For unstable systems, robustness depends critically on the sampling interval chosen. Recall that the choice of sampling interval for the fluidized-bed reactor problem of Section 5.2 was based on the doubling time of the unstable mode of the nominal system. The nominal system showed stable behavior under MPC, yet the MPC controller is unstable with model mismatch. We applied a +20 percent error in the pre-exponential factor of the first reaction. Simulation results are presented in Figure 5.18. Using the original sampling interval of 1.09, we see that the reactor experiences a temperature runaway near the end of the third sampling period. This constitutes a catastrophic failure. (Although our integrator bravely attempted to integrate the model past the runaway, we have little confidence in either the model or the integrator beyond this point. Results shown after the runaway must be viewed skeptically.)

After receiving these simulation results, we cut the sampling interval in half to 0.545 and observed similar unstable behavior, with the onset of the runaway occurring slightly later. Halving the sampling interval again to 0.2725 provided the second curve indicated in Figure 5.18, which demonstrates good control without any tendency toward runaway.

For unstable systems, there exists an inverse relationship between the model error and the maximum allowable size of the sampling interval. In the example of Figure 5.18, with the true pre-exponential factor α_1 20 percent higher than the model, we observe a temperature runaway between sampling periods. Since feedback is only incorporated at the sampling points, the sampling interval must be chosen sufficiently small to detect any unexpected increase in temperature before it becomes a runaway situation. This is a difficult design issue that has not been fully explored for nonlinear systems. For stable linear SISO systems, Zafiriou and Morari [79] have provided some criteria for sampling interval selection that later appeared in their book [53]. The interested reader is referred to these works as well as to standard texts on digital control and digital signal processing for further discussion of sampling intervals for linear systems.

- Prediction Horizon: For linear systems, several researchers have provided selection criteria for the prediction horizon that assure stability [23, 62, 67]. The question has not been resolved for nonlinear systems, but we can use the simulation results of Section 5.2 to illustrate some general principles.

  With conventional MPC feedback, Figure 5.12 shows that longer horizons tend to produce more aggressive control action, more overshoot, and faster response. From Figure 5.13, we see also that longer prediction horizons are more sensitive to disturbances, although this effect can be partially mitigated by including a filter in the feedback loop.

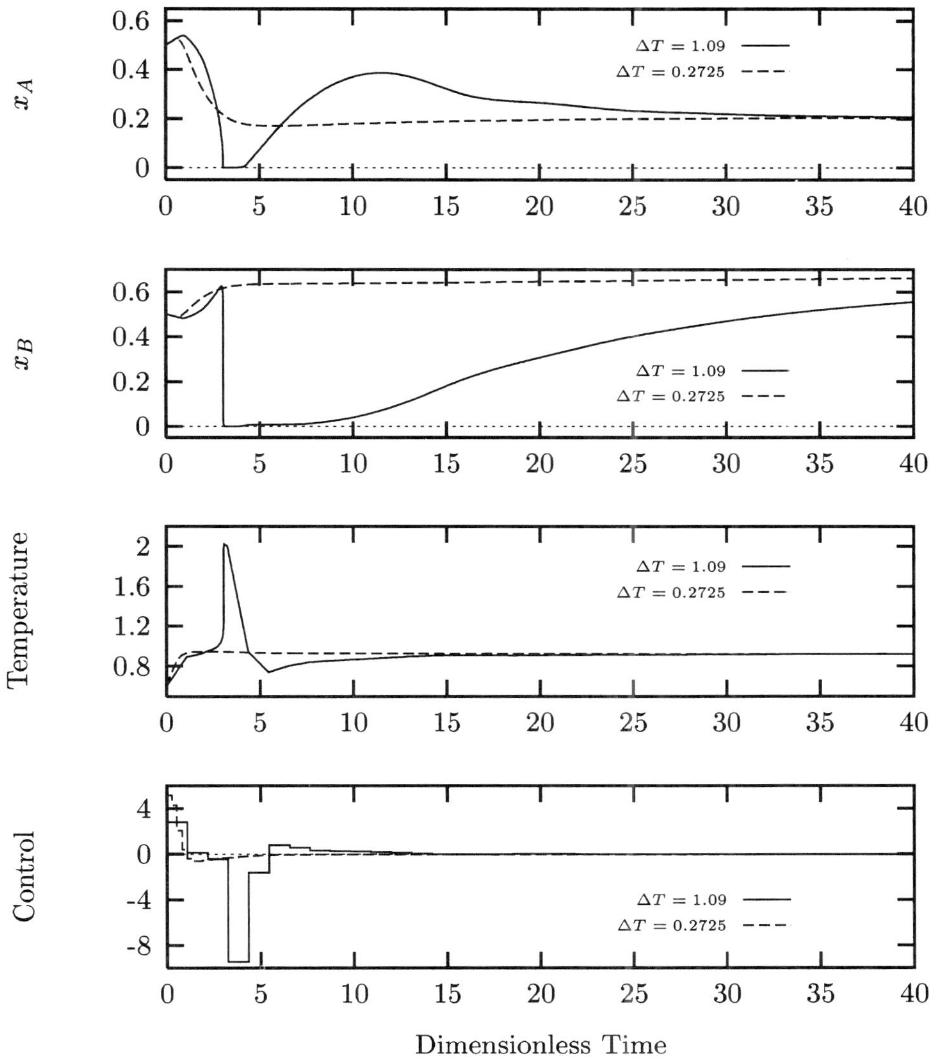

Figure 5.18: Fluidized-bed reactor: Effect of sampling interval.

Nominal stability is strongly affected by the horizon length, as illustrated by Figure 5.5, in which we observed a critical horizon length. Prediction horizons shorter than the critical value produced an unstable closed-loop system.

Based on linear systems results [23] and our simulations, the effect of increasing prediction horizon diminishes as N becomes large, and the advantages of longer horizons are outweighed by the increase in computations required.

With feedback through the nonlinear programming technique, we observed that prediction horizon had almost no effect. Whether this applies more generally is unknown and is topic of current research.

- Control Horizon: Linear systems results [23] indicate that shortening the control horizon relative to the prediction horizon tends to produce less aggressive controllers, slower system response and less sensitivity to disturbances. The effect is very similar to that of increasing the penalty on control action in the MPC objective function. We investigated this effect using the fluidized-bed reactor model of Section 5.2. For the nominal system, Figure 5.7 showed the effects of using prediction horizon $N = 4$ and control horizons $M = 1$ and $M = 2$. Comparing Figure 5.7 with Figure 5.6, we observe that the time scale of the system response is much longer for $M = 1$ or $M = 2$ compared to $M = 4$ with less aggressive control action. Unfortunately, in the case with model mismatch, this is exactly what is *not* needed, as indicated in Figure 5.19. Slowing the control action permits the reactor temperature to run away for $M = 1$. For $M = 2$, the control action is quick enough to prevent the runaway.

- Penalty Weights: For choice of weighting matrices, Bryson and Ho [6] suggest applying weights that are inversely related to the maximum acceptable range of the variables being penalized. We have generally found that this technique rarely provides good results because it over-penalizes control action. To provide good performance by the optimization algorithm, we recommend applying a penalty to setpoint deviations that places the objective in the interval 1-100 for the range of expected conditions, and then applying small penalties (less than 10 percent of the output penalty) on control or control increments, to achieve good closed-loop performance.

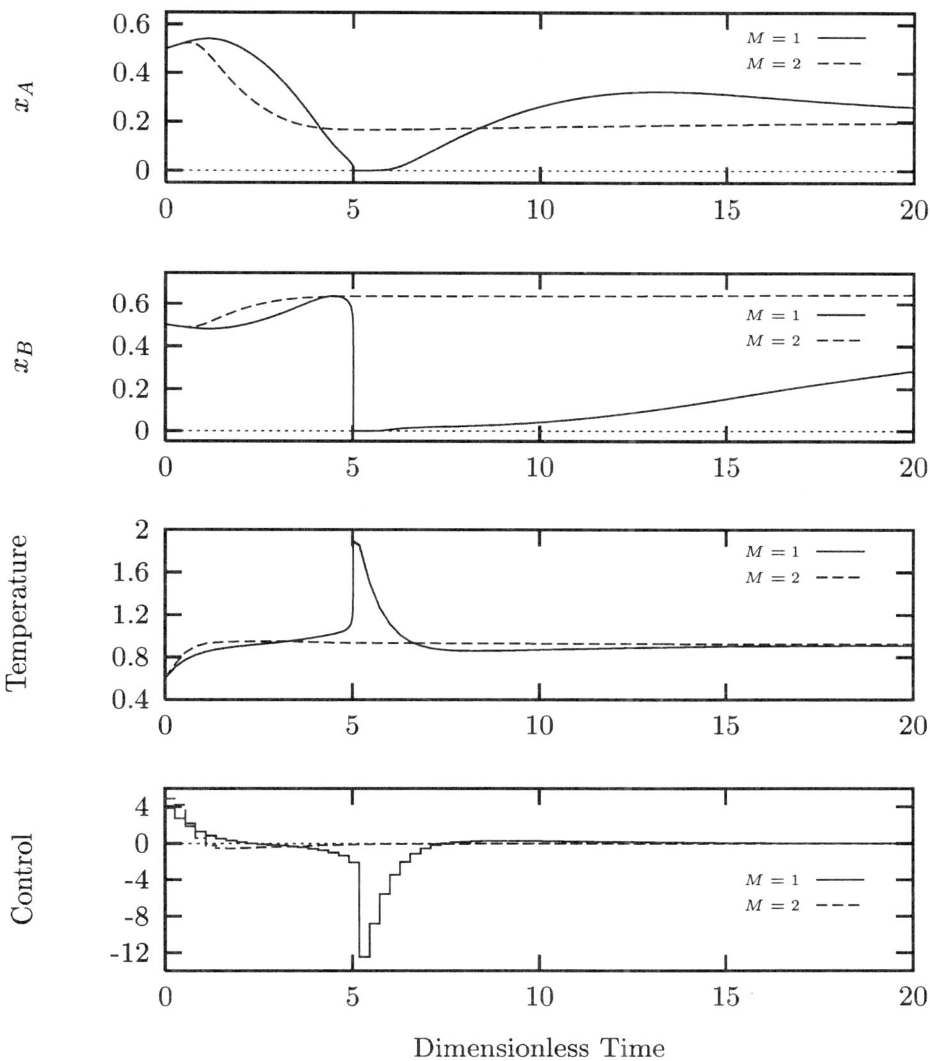

Figure 5.19: Fluidized-bed reactor: Effect of control horizon.

In the simulations results presented in this chapter, we typically applied a penalty weight of 100 on the integral square deviation of the output from the setpoint. Relative to the penalty on setpoint deviations, in our simulation studies we applied a 0.1 to 1 percent penalty to the control or control increments. Naturally, these weight must be adjusted if the terms which contribute to the objective have widely differing magnitudes arising from dimensional differences. See Gill, *et al.* [27] for a discussion of scaling issues in optimization.

- Collocation approximation: For the simulations using the isothermal CSTR model, we used 3 internal collocation points in the interior of each element, as well as points on the element boundaries. For the fixed-bed reactor, we generally used two interior collocation points to reduce the computation time. In one case, we repeated the fixed-bed reactor simulation in the presence of model mismatch using both 1 and 4 interior collocation points. The higher order approximation gave virtually identical results as the case with 2 interior collocation points. The lower order approximation resulted in a reactor runaway.

  Clearly, there are relationships between the size of the sampling interval, model error, the degree of the interpolating polynomial, and the performance and stability of MPC using orthogonal collocation. These relationships have not been quantified, so we can provide only a set of heuristic guidance based on our own experience. We suggest at least two interior collocation points be used, and no more than about 5. Unless the model predicts rapidly varying solutions within one finite element, a smaller number is desirable to reduce computation time.

- Filter in feedback loop: In Section 5.5, we showed that filtering the feedback signal when using steady-state target optimization provided good disturbance rejection and fast system response. The choice and effect of the disturbance filter is strongly system dependent. In the example of Section 5.5, we (naively) chose a very simple moving average filter with relatively slow response and observed excellent results. Applying the same procedure to the unstable fixed-bed reactor problem, the filter can produce catastrophic failure, depending on the speed of the linear filter chosen.

  As in Section 5.5, we used the filter $\hat{d}_k = \alpha_f \hat{d}_{k-1} + (1-\alpha_f)\,(y_k - g\,(\hat{x}_k))$ in which $\alpha_f = 0.9$. The same filter applied to the fixed-bed reactor problem produced a runway in simulations that included a zero-mean

Gaussian disturbance in the output. Figure 5.20 shows the effect of the disturbance, both with and without the filter. The variance of the Gaussian disturbance is 0.01. As before, we do not place any confidence in the post-runaway simulation results; however, the computed temperature following the runaway illustrates that the algorithm with the filter slows and smooths the computed control input. We previously observed that increasing the control penalty or decreasing the control horizon relative to the prediction horizon also slowed control action and also resulted in reactor runaway.

We attempted to avoid the runaway situation by using a faster filter. Using $\alpha_f = 0.5$ produced the simulation results shown in Figure 5.21. We have succeeded in avoiding the runaway, the filter is effective in reducing noise sensitivity in the control signal, and slightly less effective in reducing variation in the temperature, which is the controlled output. If we consider data in the final portion of the time interval of Figure 5.21, the controller is essentially in a regulator mode to maintain the setpoint in the presence of disturbances. A rough measure of the effectiveness of the filter is the ratio of standard deviation with the filter to the standard deviation without the filter. After subtracting linear trends, the filter with $\alpha_f = 0.5$ produced standard deviation ratios of 0.538, 0.720 and 0.577 for the control, temperature and feedback ($\hat{d}_k$) signals, respectively. (For comparison, the same filter used to attenuate a Gaussian noise signal alone produces a ratio of 0.578.)

These results indicate that a filter in the feedback loop can be effective in reducing the controller sensitivity to output disturbances; however, like other methods that slow system response, the filter can induce unstable closed-loop behavior with unstable processes in the presence of uncertainty. Simulation studies are essential to detect such behavior before implementation.

5.7 Stability and Performance

5.7.1 Nominal Stability

The first property of a control system that should be satisfied is nominal stability, i.e., stability for systems free from modeling errors or disturbances. For linear systems without constraints, stability can be verified by checking the eigenvalues of the closed-loop system. For nonlinear systems, there

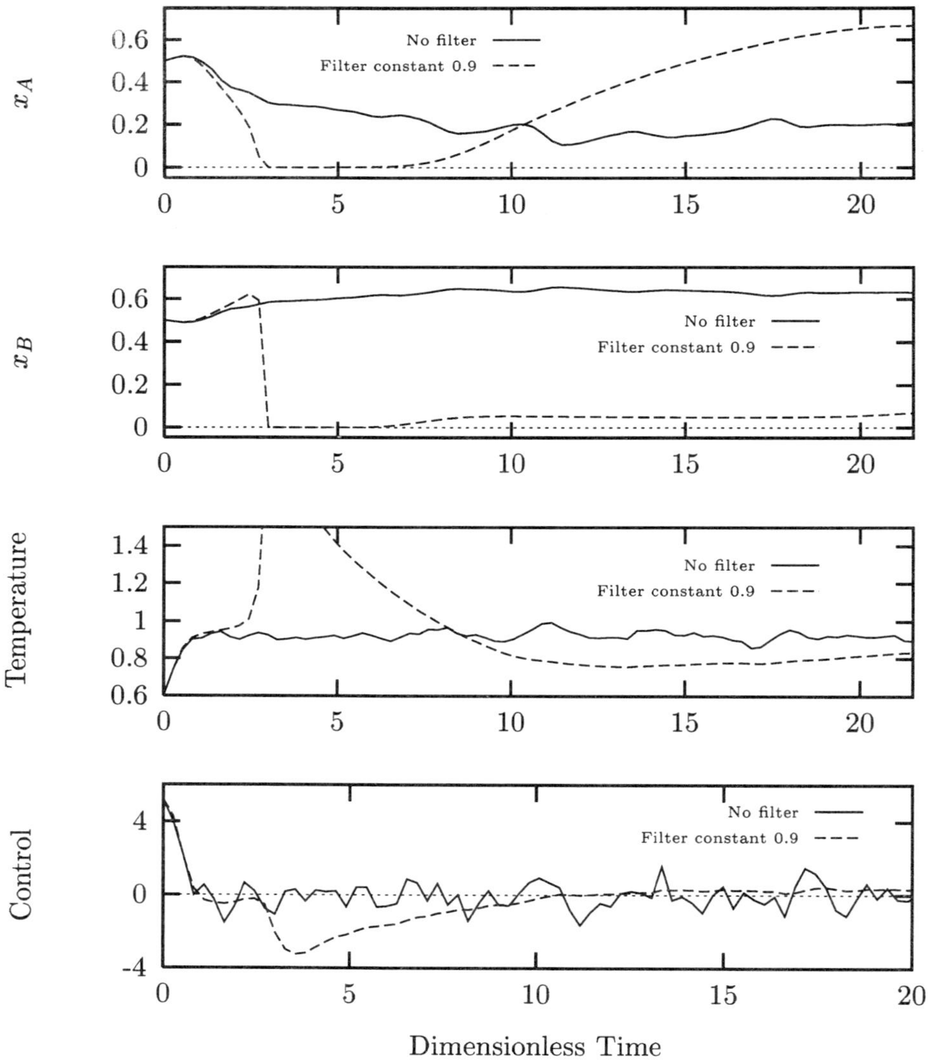

Figure 5.20: Fluidized-bed reactor: Disturbances and filtering effects.

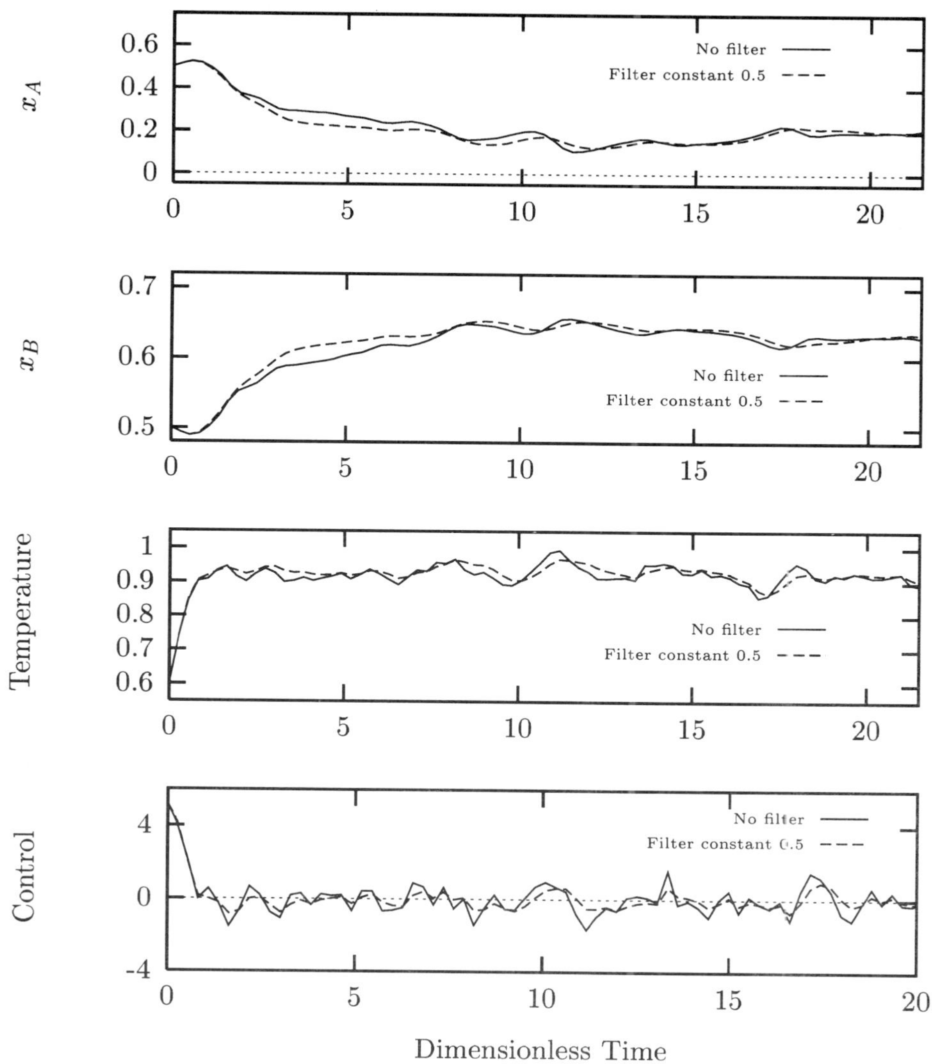

Figure 5.21: Fluidized-bed reactor: Disturbances and filtering effects.

is no simple equivalent criterion. Since closed-loop solutions to MPC control problems for nonlinear systems are almost never available, the Second Method of Lyapunov has been extensively used to analyze the stability of MPC. With sufficiently strong conditions on the nonlinear program that provides the MPC controller, we can view MPC as a method for generating Lyapunov functions for nonlinear systems.

Since Lyapunov analysis assumes such a prominent role, we repeat the basic theorem below. The following is adapted from Vidyasagar [76]:

Theorem 5 (Lyapunov Stability Theorem) *Let x_k be a sequence in $\Re^n$ indexed by $k \in \{0, 1, 2 \ldots\}$. The dynamic process x_k is asymptotically stable if there exist a function $V : \Re^n \to \Re_+$ and three functions α, β and γ of class $\mathcal{K}$ such that*

$$\alpha(\|x_k\|) \leq V(x_k) \leq \beta(\|x_k\|) \tag{5.29}$$

$$\Delta V(x_k) \triangleq V(x_{k+1}) - V(x_k) \leq -\gamma(\|x_k\|) \tag{5.30}$$

Class $\mathcal{K}$ functions are a specific class of positive definite functions. Using the definition of Vidyasagar [76], a class $\mathcal{K}$ function $\alpha : \Re_+ \to \Re_+$ is continuous and strictly increasing and satisfies $\alpha(0) = 0$. If α, β and γ have the properties of class $\mathcal{K}$ functions only in a neighborhood the origin, then the Lyapunov stability theorem provides only a local result.

Model predictive control provides an objective function that satisfies the sufficient conditions of Theorem 5 if we include the final state constraint $x_{k+N|k} = 0$. For simplicity, we will use a quadratic stage cost $\|x_{k+j|k}\|_Q^2 + \|u_{k+j|k}\|_R^2$ with Q and R positive definite. Without loss of generality, we can assume that $f(0,0) = 0$ through a simple change of coordinates. Using the MPC objective function

$$J^*(x_k) = \min_{u_{k+j|k}} \sum_{j=0}^{N-1} \left(\|x_{k+j|k}\|_Q^2 + \|u_{k+j|k}\|_R^2\right) + \|x_{k+N|k}\|_Q^2$$

we see that $J^*(x_k) \geq \|x_k\|_Q^2$ and the quadratic function

$$\alpha(\|x_k\|) = \sigma_{\min}(Q)\, \|x_k\|^2$$

(in which $\sigma_{\min}(Q)$ is the smallest singular value of Q), satisfies the lower bound of Equation 5.29. Note that x_k and u_k are equivalent to $x_{k|k}$ and $u_{k|k}$, respectively.

Equation 5.30 requires a bound for the difference between values of the Lyapunov Function for subsequent states. Using the MPC objective function as our candidate Lyapunov function, Equation 5.30 is also satisfied by including the final state constraint $x_{k+N|k} = 0$. If we represent a sequence of controls that corresponds to the optimal $J^*(x_k)$ by

$$\left\{u^*_{k|k}, u^*_{k+1|k}, \ldots u^*_{k+N-1|k}\right\}$$

and let J' be defined by

$$\begin{aligned} J' &= J^*(x_k) - \left(\|x_k\|^2_Q + \|u^*_k\|^2_R\right) & (5.31) \\ &= \sum_{j=1}^{N-1} \left(\|x_{k+j|k}\|^2_Q + \|u^*_{k+j|k}\|^2_R\right) \end{aligned}$$

The value of J' is simply $J^*(x_k)$ without the contribution of the first stage. In this case there is also no contribution from the final state to the summation since it is constrained to be zero.

After the control $u^*_{k|k}$ is implemented, the state moves to $x_{k+1|k} = f\left(x_k, u^*_{k|k}\right)$. Since we are considering the nominal system, the state moves exactly as predicted. At the new time $k+1$, the optimal control moves computed at time k still constitute a feasible sequence to which we can append a zero control without affecting the value of J'. This feasible sequence $\left\{u^*_{k+1|k}, \ldots u^*_{k+N-1|k}, 0\right\}$ may or may not be optimal at time $k+1$, but it provides an upper bound for the optimal objective $J^*(x_{k+1})$ so that we have $J' \geq J^*(x_{k+1})$. Combining this inequality with Equation 5.31, we obtain

$$\begin{aligned} J^*(x_k) - J^*(x_{k+1}) &\geq \|x_k\|^2_Q & (5.32) \\ &\geq \sigma_{\min}(Q)\,\|x_k\|^2 & (5.33) \end{aligned}$$

which satisfies the inequality of Equation 5.30 of Theorem 5 after a change in sign. Notice that it was not necessary to consider the effects of constraints. For the nominal system, the existence of a solution to the nonlinear program at some initial time is sufficient for the existence of subsequent feasible solutions. Figure 5.22 provides an intuitive basis for the stability arguments above. After the control at time k is selected and implemented, a new problem is defined over the subsequent "window" as shown in the figure. The new objective includes state and control contributions on the interval $[k+1, k+N+1]$. The value of the optimal objective at $k+1$ must be less

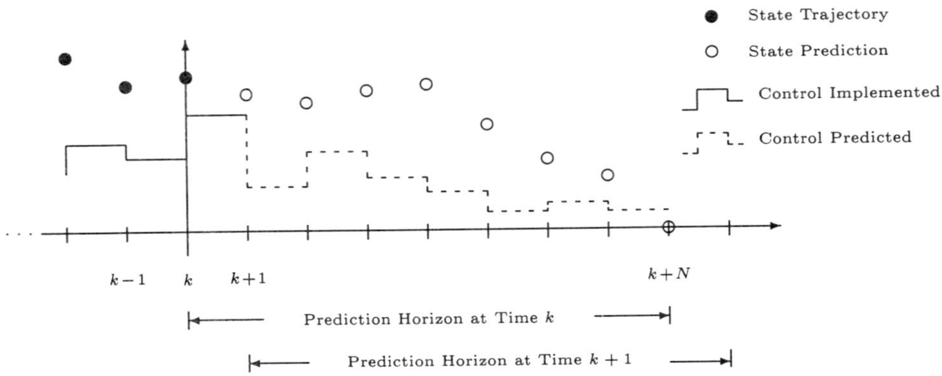

Figure 5.22: Model predictive control concept.

than that at k, because of the final state constraint. The decrease is lower bounded by $\|x_k\|_Q^2$, since the contribution of the state at k is no longer included.

To apply the Lyapunov Stability Theorem to MPC, it remains to be shown that there exist an upper bound $\beta(\|x\|)$ on the MPC objective. The upper bound is closely related to the null-controllability of the system. In general, *a priori* evaluation of null controllability is difficult for nonlinear systems; the case of constrained nonlinear systems is even more difficult. Suppose, however, that we knew that the origin could be reached using feasible controls from any state in some specific subset $\mathcal{S}$ of $\Re^n$ and the state trajectory under this sequence of controls remained in the feasible set. Then the MPC objective function could then be evaluated using the feasible control sequence and would provide an upper bound on the optimal MPC objective. The argument proceeds as follows: Let x_k be some state in $\mathcal{S}$ such that the control sequence $\left\{u_{k|k}, u_{k+1|k}, \dots u_{k+N-1|k}\right\}$ produces a final state $x_{k+N|k} = 0$. We can construct a (non-optimal) MPC objective

$$J(x_k) = \sum_{j=0}^{N-1} \|x_{k+j|k}\|_Q^2 + \|u_{k+j|k}\|_R^2 \tag{5.34}$$

Again, there is no contribution of the final state because of the zero end-state. If $J(x_k)$ is continuous, then it is a class $\mathcal{K}$ function, since it is lower bounded by $\|x_k\|_Q^2$ which is a class $\mathcal{K}$ function. The arguments above may be formalized in the following theorem:

Theorem 6 (Stability of MPC) *For the nominal system*

$$x_{k+1} = f(x_k, u_k)$$
$$f(0,0) = 0$$

if the MPC objective function is continuous, then the origin is an asymptotically stable fixed point under MPC with a region of attraction that consists of the subset of $\Re^n$ *that is constrained null controllable.*

A proof of the theorem follows from the previous discussion.

It is possible to generalize the previous discussion by substituting a general class $\mathcal{K}$ stage cost for the quadratic functions used above. In this case, asymptotic stability is retained. It is also possible to arrive at a similar stability result using an infinite horizon in which the final state constraint may be viewed as a finite parameterization of the more general infinite horizon problem. Details are provided by Meadows *et al.* [49]. Similar results were also reported by Keerthi and Gilbert, who considered both the infinite horizon and the final state constraint [34]. Stronger results are available. If the stage cost is quadratic, as discussed above, then MPC produces exponential stability. The result follows from the fact that the bounding class $\mathcal{K}$ functions α, β and γ are quadratic [68]. See also Vidyasagar [76] for a continuous time proof.

We have not directly addressed the question of constrained null controllability. As a practical matter, null controllability is usually verified in the process of solving the nonlinear program for the MPC controller. If no feasible point can be found, then for practical purposes, the system is not null-controllable. A numerical algorithm is not equivalent to a mathematical proof, but as a practical matter, if optimization software cannot find a feasible point for some initial state of interest, the system is a poor candidate for control with MPC. This is an aesthetically unsatisfactory approach to the verification of nonlinear controllability; however, it must be noted that the same problem exists with respect to linear MPC controllers subject to constraints, and has not prevented their widespread implementation.

The topic of nonlinear constrained controllability is an active research area which has not yet produced a complete theory. For further discussion of nonlinear controllability concepts see the following: [28, 29, 58, 70, 71, 76].

5.7.2 Steady-State Performance

In addition to nominal stability, an evaluation of a control method should consider robustness to model error. In this section we state some steady-state results that hold for MPC controllers applied to **arbitrary plants** rather than the nominal plant. The price we pay for the generality of considering arbitrary plants is that the results are in the form of **alternative conditions**, i.e. **either** the closed loop does not approach a steady state **or** the plant goes to a steady state with certain guaranteed properties, such as zero offset. The value of these results is that they characterize the steady state that the closed loop exhibits for any plant, in terms of the nominal model only. The steady-state properties can therefore be examined without knowledge of the plant.

We are concerned in this section with the target optimization feedback form of MPC. Some of the methods we use subsequently can also be applied to develop analogous results for the conventional MPC feedback scheme, but we do not investigate that here. Our opinion is that the target optimization provides a more general framework to handle a larger variety of process models than can be addressed with the conventional MPC feedback, so we focus our attention there. First we summarize the three parts of the controller discussed in Section 5.5.2

State estimator. The state estimator takes the current control and plant measurement and updates the state and disturbance estimates via

$$\begin{aligned} \hat{x}_{k+1} &= f(\hat{x}_k, u_k) \\ \hat{d}_{k+1} &= y_k - g(\hat{x}_k) \end{aligned}$$

Steady-state target optimization. The steady-state target optimization takes $\hat{d}_k$ and y_{ref} and produces as output x_t, u_t through solution of

$$\min_{u_t, x_t} \|u_{\text{ref}} - u_t\|_R^2 \tag{5.35}$$

subject to:

$$\begin{aligned} x_t &= f(x_t, u_t) \\ y_{\text{ref}} &= g(x_t) + \hat{d}_k \\ x_t &\in \mathcal{X} \\ u_t &\in \mathcal{U} \end{aligned}$$

Notice that we consider explicitly any controller input and state (output) constraints in the target optimization as well. If this optimization problem does not have a feasible point, the following nonlinear program is used to compute the targets

$$\min_{u_t, x_t} \|y_{\text{ref}} - (g(x_t) + \hat{d}_k)\|_Q^2 \tag{5.36}$$

subject to:

$$\begin{aligned} x_t &= f(x_t, u_t) \\ x_t &\in \mathcal{X} \\ u_t &\in \mathcal{U} \end{aligned}$$

We always assume that this optimization problem has a feasible point, i.e. the model has at least one steady-state solution satisfying the constraints.

MPC control law. Finally, the MPC controller takes x_t, u_t and $\hat{x}_k$ and produces the control input u_k from the solution of

$$\min_{u_{k+j|k}} \sum_{j=0}^{N-1} \|x_{k+j|k} - x_t\|_Q^2 + \|u_{k+j|k} - u_t\|_R^2 \tag{5.37}$$

subject to:

$$\begin{aligned} x_{k|k} &= \hat{x}_k \\ x_{k+N|k} &= x_t \\ x_{k+j|k} &\in \mathcal{X} \\ u_{k+j|k} &\in \mathcal{U} \end{aligned}$$

The first optimal input is then applied to the plant, $u_k = u^*_{k|k}$.

The first result states that if the closed-loop system goes to steady state, then the steady-state plant output and input satisfy the target optimization problem.

Theorem 7 (Steady-state properties) *If*

1. *The steady-state model, $x = f(x, u)$, has a solution, $x \in \mathcal{X}$, $u \in \mathcal{U}$.*

2. *The MPC control law is nominally asymptotically stabilizing.*

3. The closed-loop goes to a steady state, $y_k \to y_s$, $u_k \to u_s$, $x_t \to x_{ts}$, $u_t \to u_{ts}$.

Then either $\|u_{\text{ref}} - u_s\|_R$ *is minimized and* $y_s = y_{\text{ref}}$ *if Equation 5.35 has a feasible solution at steady state or* $\|y_{\text{ref}} - y_s\|_Q$ *is minimized.*

Proof: At steady state (Assumption 3), the estimator and target optimization yield

$$\begin{aligned}
\hat{x}_s &= f(\hat{x}_s, u_s) \\
\hat{d}_s &= y_s - g(\hat{x}_s) \\
x_{ts} &= f(x_{ts}, u_{ts}) \\
x_{ts} &\in \mathcal{X} \\
u_{ts} &\in \mathcal{U}
\end{aligned}$$

The target optimization is well defined because of Assumption 1. We next show that because the controller is nominally asymptotically stabilizing (Assumption 2), $\hat{x}_s = x_{ts}$, i.e. the state estimate converges to the steady-state target. Consider the control problem with steady-state input and state targets (Assumption 3). The state estimate evolves according to

$$\hat{x}_{k+1} = f(\hat{x}_k, u_s)$$

Because the controller is nominally asymptotically stabilizing and the state estimator evolves according to the nominal model, the stage cost in Equation 5.37 goes to zero, which implies $\hat{x}_s = x_{ts}$ and $u_s = u_{ts}$. From the disturbance estimate equation

$$y_s = g(x_{ts}) + \hat{d}_s$$

Therefore, if at the steady state Equation 5.35 is being solved then the steady input minimizes $\|u_{\text{ref}} - u_s\|_R$ and $y_s = y_{\text{ref}}$. If this target problem has no feasible solution at steady state, then Equation 5.36 applies and the steady plant output minimizes $\|y_{\text{ref}} - y_s\|_Q$.

Because of the constraints, plant-model mismatch, and plants with more setpoints than inputs, it may not be possible to satisfy the conditions necessary for zero steady-state offset. We can add further restrictions to provide conditions that do ensure zero steady-state offset.

Corollary 1 (Zero steady-state offset) *If the conditions of Theorem 7 are satisfied and there exists x_t, u_t satisfying*

$$\begin{aligned} x_t &= f(x_t, u_t) \\ y_{\text{ref}} &= g(x_t) + \hat{d}_k \\ x_t &\in \mathcal{X} \\ u_t &\in \mathcal{U} \end{aligned}$$

for every y_{ref} and $\hat{d}_k$, then there is zero steady-state offset, $y_s = y_{\text{ref}}$.

Proof: The added conditions of the corollary ensure that the target optimization in Equation 5.35 has a feasible point. Therefore, zero steady-state offset follows from Theorem 7.

5.8 Discussion

MPC has its roots in the development of optimal control theory in the late 1950's, in which the "state-space" approach was considered a modern alternative to the frequency-domain techniques that had dominated control theory and practice since the 1930's. Kalman's landmark paper [30] consolidated and extended much of the then current work in the field, and provided rigorous definitions for the now-common concepts of controllability and observability. Although Kalman reported work for linear systems, many of the key concepts may be extended to nonlinear systems.

It was recognized early that optimality in the linear-quadratic problem is not sufficient for stability. Kalman [30] provided sufficient conditions for stability for this case. Several other workers considered formulations that could provide a stabilizing LQ optimal controller. One of these was Kleinman [36], whose "easy way" to stabilize a linear system was equivalent to a finite-horizon LQ optimal control that is subject to the final state constraint $x(T) = 0$. Kleinman followed up his continuous-time result with a discrete-time analog [37] in which he provided a criterion for selection of an initial condition for the matrix Riccati Difference Equation (RDE) that arises from the LQ optimal control problem in discrete time.

Thomas and coworkers[72, 73, 74] recognized that the zero final state constraint could be incorporated into the LQ optimal regulator by inverting the Riccati Differential (or Difference) Equation and integrating (or iterating) the result from an initial condition of zero. Conceptually, this is equivalent to an infinite final state weighing matrix in a finite-horizon LQ optimal control problem. Kwon and Pearson provided a firm theoretical basis for using the zero final-state constraint to provide stability for time-varying and invariant linear systems, including results for continuous time [42] and discrete time systems [43]. To implement MPC in continuous time requires that a solution of the optimization be available continuously. For linear systems, this means that the solution of matrix Riccati differential equation must be available. Kwon *et. al* [41] investigated the time-varying case and showed that a stabilizing solution to the matrix Riccati difference equation could be updated based on the system equations, without the need for a new solution at each time. Although not presented, the authors pointed out that the results are also applicable to discrete-time systems.

By the late 1970's, a substantial body of research results had been developed that provided a solid theoretical foundation for implementation of model predictive control. At about the same time, the price instability of feedstocks and energy in the process industries demanded higher efficiency from existing plants. Rather than control processes in midrange of operations where traditional linear controllers would suffice, it became necessary to operate closer to limits using models that could be generated quickly with a minimum of modeling effort. In 1978, Richalet et. al [63] presented IDCOM and in 1979, Cutler and Ramaker [13] presented Dynamic Matrix Control (DMC), both of which responded to these new demands. These approaches used either step or impulse response models in discrete time whose coefficients can be determined from plant step test data. Since processes are assumed to be time-invariant, the plant dynamics can be described by a matrix of constant coefficients, hence the acronym DMC. The DMC controller used the linear system represented by the dynamic matrix to predict future process output based on current and future control action. The control was selected that minimized a quadratic objective based on control and process variables predicted by the model. Constraints on the control variable were explicitly incorporated by the optimization algorithm in subsequent formulations [24].

In the chemical engineering literature, model predictive control came to be identified with the methods of DMC, including linear step response models, finite prediction horizons, and the use of the DMC disturbance model. DMC assumes that the difference between current processes measurements and the model prediction is due to an unmodeled disturbance that maintains the same value throughout the prediction horizon; to obtain a prediction, one must use that model equations and add to the predicted variables the value of the current disturbance estimate. The effect is extremely useful, since it eliminates steady-state offset for step changes in setpoint, as discussed in Section 5.5.

The basic DMC algorithm has undoubtedly been a phenomenal success by any measure; however, its origins as a somewhat *ad hoc* method necessarily left some questions unanswered. For example, with the introduction of output variable constraints, some stable processes can become closed-loop unstable. The multivariable nature of DMC allows for a large number of tuning variables in the objective function. The length of the prediction horizon is a tuning parameter. Often, the control horizon is different from the prediction horizon–yet another tuning parameter is available.

The influential paper by García and Morari [23] strengthened the trend toward identification of DMC and its descendants with MPC in general. For the unconstrained case, García and Morari provided some stability results and suggested tuning parameters for the DMC algorithm.

The step response model used in DMC admits only stable processes. Seeking to extend the applicability of the step response models and to incorporate the large body of existing results for state space models, Li and coworkers [46] and Lee *et al.* [44] produced MPC algorithms that converted the step response models into a state-space form. The new formulations were based on the step response models and incorporated the step response coefficients. Although Eaton and Rawlings [19] and Lee *et al.* extended the basic method to include integrating processes, formulations that depend on step response coefficients are unsuitable for general unstable processes.

In parallel, Clarke and coworkers opted for a transfer function model of the process and formulated the approach known as generalized predictive control (GPC) [9, 10, 11]. Part of the aim of this development was to be able to handle unstable plants.

Rawlings and Muske [54, 55, 62] departed from the step response models common in chemical engineering practice to use the standard linear state space model to formulate a stabilizing constrained MPC method. Their control formulation guaranteed nominal stability using any valid tuning parameters.

Inequality constraints in any MPC method lead to a non-linear feedback control law. A natural extension of existing results for linear systems involves MPC with explicitly nonlinear models. In the chemical engineering literature, Patwardhan and coworkers [60] and Eaton and Rawlings [19] propose a simultaneous optimization-model solution approach that provides a continuous-time MPC controller. Although no stability proofs were provided, numerical results indicated stability using chemical reactor example problems. Patwardhan, *et al.*, also demonstrate robustness to parameter error.

Chen and Shaw [8] provide an early discussion of model predictive control with the end point constraint for control of nonlinear systems. A rigorous and comprehensive approach to the nonlinear problem in discrete time was presented by Keerthi and Gilbert in a series of papers in the mid-1980's [32, 33, 34, 35]. They consider a general nonlinear, time-varying dynamic system with a general cost function defined on either a finite or infinite horizon. Finite horizon problems include the zero final state constraint. Sufficient conditions are provided that ensure asymptotic stability of the nonlinear system using an MPC control law. These include a modified controllability condition on the nonlinear system. Keerthi and Gilbert also provided sufficient conditions for convergence of finite horizon results to the infinite horizon case. Meadows and coworkers [49] present similar results for the time-invariant case and show that any feedback linearizable system can be stabilized using MPC. In the same work, they also show that there exist discrete-time systems that require discontinuous feedback laws for stability and that nonlinear MPC can generate such control laws.

The work of Keerthi and Gilbert and Meadows *et al.* provides sufficient conditions for stability of nonlinear MPC. The key difficulty in implementation arises from the final state constraint; it is difficult or even impossible to find the set of states for which the final state constraint is feasible *a priori*. As we argued previously, we can shift the burden of verifying feasibility to the optimization routines, but even so, the range of feasible states may be unacceptably small.

With these limitations in mind, Mayne and Michalska [47, 52] offer an approach that retains the stability property of MPC using the final state constraint, while expanding the set of initial feasible states. Rather than adopt the strict final state equality constraint $x_N = 0$, Mayne and Michalska propose using a "dual-mode" controller whose final state is constrained to lie within a region of the state space $W \subset \Re^n$ that contains the origin. Outside the region, a nonlinear MPC controller is used that includes the final state constraint $x_N \in W$. Within W, a stabilizing controller is used whose design is based on the system's linearization at the origin. The region W is defined as the largest elliptical region in the state space in which the linearized controller produces a decreasing (and therefore asymptotically stable) Lyapunov function. It can be determined off-line through the solution of a single initial nonlinear programming problem. Mayne and

Michalska presented detailed analyses of stability and robustness, with special emphasis on providing an implementable algorithm. A disadvantage of the Mayne-Michalska algorithm is that it requires a system model that has a stabilizable linearization at the origin, but this condition is usually satisfied in systems of engineering interest.

5.9 Future Directions

Nonlinear MPC offers a number of interesting challenges. Among these we can list a few:

- Robustness: We have seen that MPC used with the deadbeat disturbance estimator is robust to model structural and parameter uncertainty. To date, most results in this area indicate good performance in the presence of model uncertainty; however, a rigorous analysis that can provide sufficient conditions for robustness has not been completed.

- Estimator/Controller Interactions: Unlike the linear systems result, for general nonlinear systems there is no separation principle. Controller performance is intimately linked to estimator design. In the CSTR example, without state feedback it was difficult to achieve the desired steady state even though the setpoint was satisfied. Meadows *et al.* [50] and Scokaert and Rawlings [68] showed that with constrained linear systems, if the closed-loop dynamic system satisfies a Lipschitz condition, a weak form of separability can be demonstrated. More work is needed for the general nonlinear case.

- Nominal Stability: There continues to exist a gap between observed results and theory in nonlinear MPC. Experimental results, in terms of both simulations and applications, continue to indicate good performance of MPC on a wide class of problems, but tuning the algorithm to obtain stability is usually necessary. A theory to guide the tuning process is not yet available. Current theoretical results require infinite horizons or the final state constraint $x_N = 0$ to assure stability, conditions that can be much too strong (the final state constraint) or unachievable in practice (infinite horizon).

  As an alternative to these two approaches, it is possible to guarantee stability by using a final state penalty function rather than a constraint. Consider the case in which an infinite horizon is used. If this problem could be solved, it would provide a stabilizing controller. If we had an analytical expression for the optimal infinite horizon objective function $J_\infty^*(x)$, we could devise an equivalent finite horizon function that used $J_\infty^*(x)$ as a final state penalty function:

$$\Psi^*(x_k) = \min_{u_{k+j|k}} \sum_{j=1}^{N-1} \left(\|x_{k+j|k}\|_Q^2 + \|u_{k+j|k}\|_R^2 \right) + J_\infty^*(x_N)$$

Provided that the stage cost used to compute J_∞^* is the same as in Ψ^*, the principle of optimality guarantees that $\Psi^*(x) = J_\infty^*(x)$; therefore, MPC using the objective function $J_\infty^*(x_k)$ is stabilizing.

It is probably difficult or impossible to provide a closed-form expression for J_∞^*; however, if we are able to compute a bound on $J_\infty^*(x) \leq \tilde{J}^*(x)$ such that $\tilde{J}^*(0) = 0$, then the MPC controller computed using $\tilde{J}^*$ is also stabilizing, subject to some technical conditions. Some preliminary results in this area have already been presented [48] and an article containing more details concerning this approach is in preparation.

Acknowledgments

The authors wish to thank Dr. Pierre O. M. Scokaert for helpful discussions. The financial support of NSF, through grant CTS–9311420, and the industrial members of the Texas–Wisconsin Modeling and Control Consortium (TWMCC) is gratefully acknowledged. The first author was financially supported during the writing of this chapter from a fellowship from the Brazilian government, acting through the Conselho Nacional de Desenvolvimento Científico e Tecnológico (National Council of Scientific and Technological Development). This support is gratefully acknowledged.

5.10 Notation

A	matrix of first derivative collocation weights without initial condition
$\tilde{A}$	matrix of first derivative collocation weight in orthogonal collocation approximation with multiple finite elements
A_0	matrix of first derivative collocation weight, modified to include the initial condition equation

C	matrix used to form linear inequality constraints on Δu
c_k	vector of quadrature weights for integration in orthogonal collocation approximation
$\tilde{c}_k$	vector of quadrature weights for integration in orthogonal collocation approximation
$c_{\text{lb}}, c_{\text{ub}}$	upper and lower bounding vectors for linear inequality constraints on control increments
D, H	matrices used to express linear constraints on state and control variables
d, h	right-hand sides of linear inequality constraints on state and control variables
d_k	disturbance term used in conventional MPC feedback
$\hat{d}_k$	disturbance estimate obtained from state observer
F	right-hand side of matrix equation representing a differential equation discretized by orthogonal collocation approximation with a single finite element
$\tilde{F}$	right-hand side of matrix equation representing a differential equation discretized by orthogonal collocation approximation with multiple finite elements
$F_{x,k}, F_{u,k}$	Jacobian matrix of $f(x, u)$ with respect to x or u, evaluated at (x_k, u_k)
f	state transition function in dynamic system equation
G_k	Jacobian matrix of $g(x)$ evaluated at (x_k, u_k)
g	function relating state vector to output vector
h_x, h_d	correction terms for state update in state observer
J	MPC objective function
J_∞	infinite horizon optimal MPC objective function
J^*	optimal MPC objective function
$\tilde{J}^*_\infty$	upper bound on infinity horizon optimal MPC objective function
j	indexing variable for time within prediction horizon for discrete system

K	constant for proportional control in fluidized-bed reactor model
$\mathcal{K}$	set of functions used in Lyapunov stability theorem
k	independent variable representing time in discrete system
k_1, k_2	reaction rate constants in fluidized-bed reactor model
k_1, k_2, k_3	reaction rate constants in isothermal CSTR model
L	stage cost, penalty incurred at each time in prediction horizon as function of outputs, controls or control increments
$\mathrm{Le}_A, \mathrm{Le}_B$	Lewis number in fluidized-bed reactor model
M	MPC control horizon, when different from N
m	dimension of control vector
N	MPC prediction horizon
n	dimension of state vector
Q, R, S	weighting matrices for vector norms used in MPC objective function
q, q^{-1}	shift operators in discrete-time systems
$\mathcal{S}$	set or null-controllable initial states
T	temperature in fluidized-bed reactor model
T	prediction horizon in continuous-time MPC
T_F	feed temperature in fluidized-bed reactor model
T_{set}	setpoint temperature in fluidized-bed reactor model
ΔT	sampling interval
t	independent variable time in continuous-time systems
t_f	final time in prediction horizon for continuous-time MPC problems
t_0	initial time in prediction horizon for continuous-time MPC problems
u_k	control vector at time k
u_t	input target vector obtained from target optimization
u^*	solution of nonlinear program for feedback through steady-state target optimization

Δu_k	control increments, i.e., $u_k - u_{k-1}$
$\mathcal{U}$	subset of $\Re^m$ used as general control constraint
V	Lyapunov function
W	in the algorithm of Mayne and Michalska, the region near the origin in which a linear control law is stabilizing
w_k	random input variable in discrete-time stochastic model
X	matrix containing elements $x_{i,j}$
$\tilde{X}$	matrix of states $x_{i,j}$ in orthogonal collocation approximation with multiple finite elements
$\mathcal{X}$	subset of $\Re^n$ used as general state constraint
x_1, x_2	state variables in the isothermal CSTR model
x_A, x_B	concentrations in fluidized-bed reactor model
$x_{A,F}, x_{B,F}$	concentrations of species A and B in fluidized-bed reactor model
x_F	feed concentration of species A in isothermal CSTR model
$x_{i,j}$	in problems discretized with orthogonal collocation, the j-th element of the state vector at discretization point i
$x_{i,*}$	in problems discretized with orthogonal collocation, the entire state vector at discretization point i
x_{init}	initial condition for state in orthogonal collocation approximation
$\hat{x}_k$	state estimate obtained from state observer
x_k	state vector at time k
x_t	state target vector obtained from target optimization
y_k	output vector at time k
y_{ref}	reference trajectory in MPC objective
$\tilde{y}_k$	model output, used when necessary to distinguish model output from true process output

Greek Symbols

α_f	filter constant in experiments with fluidized-bed reactor

α_1, α_2	pre-exponential factors in fluidized-bed reactor model
α, β, γ	class $\mathcal{K}$ function used in Lyapunov Stability Theorem
β	control weight in MPC objective function
$\bar{\beta}$	residence time in fluidized-bed reactor model
β_1, β_2	heats of reaction in fluidized-bed reactor model
$\gamma_1, \gamma_2, \gamma_3$	weighting constants in MPC objective function of Equation 5.7
γ_1, γ_2	activation energies in fluidized-bed reactor model
μ	feedback control law resulting from MPC
ϕ	final state penalty function
Ψ^*	finite horizon objective function equivalent to an infinite horizon objective
$\sigma_{\min}$	smallest singular value
τ	independent variable time in MPC objective function of Equation 5.7

Subscripts

A, B	refers to species A and B in fluidized-bed reactor model
F	value in feed stream in example models
k	value at time k
$j\|k$	value at time j, given information up to and including time k
i, j	in combination, refers to j-th element of vector at collocation point i
$i, *$	in combination, refers to entire vector at collocation point i
init	initial condition in orthogonal collocation approximation
lb, ub	upper and lower bounding vector for linear constraints
max, min	used to indicate upper and lower limits in linear inequality constraints
min	used in conjunction with $\sigma_{\min}$ to indicate the smallest singular value
ref	reference values, often indicates setpoint

set	setpoint in fluidized-bed reactor model
x, d	in state observer, refers to terms which account for the state and disturbance estimate, respectively
$+$	only used in conjunction with $\Re_+$ to indicate the subset of real numbers $[0, \infty)$
∞	infinite horizon value

Operators and Special Symbols

$\|\cdot\|$	vector norm
$\|\cdot\|_Q$	Q-weighted vector norm
$\Re^n$	the n-th Cartesian product of the real numbers
$\Re_+$	the semi-infinite subset of the real numbers $[0, \infty\}$
$\otimes$	Kronecker matrix product

References

[1] T. A. Badgwell. Robust stability conditions for a SISO model predictive control algorithm. Submitted for publication in *Automatica*, 1995.

[2] R. E. Bellman. *Dynamic Programming.* Princeton University Press, Princeton, New Jersey, 1957.

[3] R. E. Bellman and S. E. Dreyfus. *Applied Dynamic Programming.* Princeton University Press, Princeton, New Jersey, 1962.

[4] L. T. Biegler. Solution of dynamic optimization problems by successive quadratic programming and orthogonal collocation. *Comput. Chem. Eng.*, 8(3/4):243–248, 1984.

[5] R. R. Bitmead, M. Gevers, and V. Wertz. *Adaptive Optimal Control, The Thinking Man's GPC.* Prentice-Hall, Englewood Cliffs, New Jersey, 1990.

[6] A. E. Bryson and Y. Ho. *Applied Optimal Control.* Hemisphere Publishing, New York, 1975.

[7] M. Caracotsios. *Model Parametric Sensitivity Analysis and Nonlinear Parameter Estimation. Theory and Applications.* PhD thesis, University of Wisconsin, Madison, 1986.

[8] C. C. Chen and L. Shaw. On receding horizon control. *Automatica*, 16(3):349–352, 1982.

[9] D. W. Clarke and C. Mohtadi. Properties of generalized predictive control. *Automatica*, 25(6):859–875, 1989.

[10] D. W. Clarke, C. Mohtadi, and P. S. Tuffs. Generalized predictive control—Part I. The basic algorithm. *Automatica*, 23(2):137–148, 1987.

[11] D. W. Clarke, C. Mohtadi, and P. S. Tuffs. Generalized predictive control—Part II. Extensions and interpretations. *Automatica*, 23(2):149–160, 1987.

[12] J. E. Cuthrell and L. T. Biegler. On the optimization of differential-algebraic systems. *AIChE J.*, 33(8):1257–1270, 1987.

[13] C. R. Cutler and B. L. Ramaker. Dynamic matrix control—a computer control algorithm. In *Proceedings of the Joint Automatic Control Conference*, 1980.

[14] P. J. Davis. *Interpolation and Approximation.* Dover Books on Advanced Mathematics. Dover Publications, Inc., New York, 1975.

[15] P. J. Davis and P. Rabinowitz. *Methods of Numerical Integration.* Academic Press, Orlando, Florida, 2nd edition, 1984.

[16] S. E. Dreyfus. *Dynamic Programming and the Calculus of Variations.* Academic Press, New York, 1965.

[17] S. E. Dreyfus and A. M. Law. *The Art and Theory of Dynamic Programming.* Academic Press, New York, 1977.

[18] J. W. Eaton. Octave—a high-level interactive language for numerical computations. Technical Report CCSR-93-003, The University of Texas Center for Control and Systems Research, Austin, Texas 78712, 1993.

[19] J. W. Eaton and J. B. Rawlings. Model-predictive control of chemical processes. *Chem. Eng. Sci.*, 47(4):705–720, 1992.

[20] J. W. Eaton, J. B. Rawlings, and L. H. Ungar. Stability of neural net based model predictive control. In *Proceedings of the 1994 American Control Conference*, pages 2481–2485, 1994.

[21] S. S. E. H. Elnashaie. Multiplicity of the steady states in fluidized-bed reactors. III. yield of the consecutive reaction, A→B→C. *Chem. Eng. Sci.*, 32:295–301, 1977.

[22] S. S. E. H. Elnashaie, M. E. Abashar, and F. A. Teymour. Chaotic behaviour of fluidized-bed catalytic reactors with consecutive exothermic chemical reactions. *Chem. Eng. Sci.*, 50(1):49–67, 1995.

[23] C. E. García and M. Morari. Internal model control. 1. A unifying review and some new results. *Ind. Eng. Chem. Proc. Des. Dev.*, 21:308–323, 1982.

[24] C. E. García and A. M. Morshedi. Quadratic programming solution of dynamic matrix control (QDMC). *Chem. Eng. Commun.*, 46:73–87, 1986.

[25] C. E. García, D. M. Prett, and M. Morari. Model predictive control: Theory and practice—a survey. *Automatica*, 25(3):335–348, 1989.

[26] G. Gattu and E. Zafiriou. Nonlinear quadratic dynamic matrix control with state estimation. *Ind. Eng. Chem.*, 31(4):1096–1104, 1992.

[27] P. E. Gill, W. Murray, and M. H. Wright. *Practical Optimization.* Academic Press, London, 1981.

[28] R. Hermann and A. J. Krener. Nonlinear controllability and observability. *IEEE Trans. Auto. Cont.*, 22(5):728–740, 1977.

[29] A. Isidori. *Nonlinear Control Systems.* Springer-Verlag, Berlin, 2nd edition, 1989.

[30] R. E. Kalman. Contributions to the theory of optimal control. *Bull. Soc. Math. Mex.*, 5:102–119, 1960.

[31] J. C. Kantor. Stability of state feedback transformations for nonlinear systems–some practical considerations. In *Proceedings of the 1986 American Control Conference*, page 1014, Seattle, Washington, June 1986.

[32] S. S. Keerthi. *Optimal Feedback Control of Discrete-Time Systems with State-Control Constraints and General Cost Functions.* PhD thesis, University of Michigan, 1986.

[33] S. S. Keerthi and E. G. Gilbert. An existence theorem for discrete-time infinite-horizon optimal control problems. *IEEE Trans. Auto. Cont.*, 30(9):907–909, September 1985.

[34] S. S. Keerthi and E. G. Gilbert. Moving-horizon approximations for a general class of optimal nonlinear infinite-horizon discrete-time systems. In *Proceedings of the 20th Annual Conference on Information Sciences and Systems*, pages 301–306, Princeton, NJ, 1986. Princeton University.

[35] S. S. Keerthi and E. G. Gilbert. Optimal infinite-horizon feedback laws for a general class of constrained discrete-time systems: Stability and moving-horizon approximations. *J. Optim. Theory Appl.*, 57(2):265–293, May 1988.

[36] D. L. Kleinman. An easy way to stabilize a linear constant system. *IEEE Trans. Auto. Cont.*, 15(12):692, December 1970.

[37] D. L. Kleinman. Stabilizing a discrete, constant, linear system with application to iterative methods for solving the riccati equation. *IEEE Trans. Auto. Cont.*, 19:252–254, June 1974.

[38] C. Kravaris and P. Daoutidis. Nonlinear state feedback control of second-order nonminimum-phase nonlinear systems. *Comput. Chem. Eng.*, 14(4/5):439–449, 1990.

[39] C.-C. Ku and K. Y. Lee. Diagonal recurrent neural networks for dynamic systems control. *IEEE Trans. Neural Net.*, 6(1):144–156, January 1995.

[40] J. G. Kuschewski, S. Hui, and S. H. Zak. Application of feedforward neural networks to dynamical system identification and control. *IEEE Trans. Cont. Sys. Tech.*, 1(1):37–49, March 1993.

[41] W. H. Kwon, A. M. Bruckstein, and T. Kailath. Stabilizing state-feedback design via the moving horizon method. *Int. J. Control*, 37(3):631–643, 1983.

[42] W. H. Kwon and A. E. Pearson. A modified quadratic cost problem and feedback stabilization of a linear system. *IEEE Trans. Auto. Cont.*, 22(5):838–842, October 1977.

[43] W. H. Kwon and A. E. Pearson. On feedback stabilization of time-varying discrete linear systems. *IEEE Trans. Auto. Cont.*, 23(3):479–481, June 1978.

[44] J. H. Lee, M. Morari, and C. E. García. State-space interpretation of model predictive control. *Automatica*, 30(4):707–717, 1994.

[45] A. U. Levin and K. S. Narendra. Control of nonlinear dynamical systems using neural networks: controllability and stabilization. *IEEE Trans. Neural Net.*, 4(2):192–206, March 1993.

[46] S. Li, K. Y. Lim, and D. G. Fisher. A state space formulation of model predictive control. *AIChE J.*, 35(2):241–249, February 1989.

[47] D. Q. Mayne and H. Michalska. Receding horizon control of nonlinear systems. *IEEE Trans. Auto. Cont.*, 35(7):814–824, July 1990.

[48] E. S. Meadows. *Stability and Continuity of Nonlinear Model Predictive Control.* PhD thesis, The University of Texas at Austin, 1994.

[49] E. S. Meadows, M. A. Henson, J. W. Eaton, and J. B. Rawlings. Receding horizon control and discontinuous state feedback stabilization. To appear in *International Journal of Control.*

[50] E. S. Meadows, K. R. Muske, and J. B. Rawlings. Constrained state estimation and discontinuous feedback in model predictive control. In *Proceedings of the 1993 European Control Conference*, pages 2308–2312. European Automatic Control Council, June 1993.

[51] E. S. Meadows and J. B. Rawlings. Topics in model predictive control. In R. Berber, editor, *NATO Advanced Study Institute on Methods of Model Based Control Proceedings*, Antalya, Turkey, July 1994. Kluwer Academic Publishers.

[52] H. Michalska and D. Q. Mayne. Robust receding horizon control of constrained nonlinear systems. *IEEE Trans. Auto. Cont.*, 38(11):1623–1633, 1993.

[53] M. Morari and E. Zafiriou. *Robust Process Control.* Prentice-Hall, Englewood Cliffs, New Jersey, 1989.

[54] K. R. Muske and J. B. Rawlings. Linear model predictive control of unstable processes. *J. Proc. Cont.*, 3(2):85–96, 1993.

[55] K. R. Muske and J. B. Rawlings. Model predictive control with linear models. *AIChE J.*, 39(2):262–287, 1993.

[56] K. R. Muske and J. B. Rawlings. Nonlinear moving horizon state estimation. In R. Berber, editor, *NATO Advanced Study Institute on Methods of Model Based Control Proceedings*, Antalya, Turkey, July 1994. Kluwer Academic Publishers.

[57] K. R. Muske, J. B. Rawlings, and J. H. Lee. Receding horizon recursive state estimation. In *Proceedings of the 1993 American Control Conference*, pages 900–904, June 1993.

[58] H. Nijmeijer and A. J. van der Schaft. *Nonlinear Dynamical Control Systems.* Springer-Verlag, New York, 1990.

[59] R. E. Nordgrenn and P. H. Merckl. An analytical comparison of a neural network and a model-based adaptive controller. *IEEE Trans. Neural Net.*, 4(4):685–694, July 1993.

[60] A. A. Patwardhan, J. B. Rawlings, and T. F. Edgar. Nonlinear model predictive control. *Chem. Eng. Commun.*, 87:123–141, 1990.

[61] S.-Z. Qin, H.-T. Su, and T. J. McAvoy. Comparison of four neural net learning methods for dynamic system identification. *IEEE Trans. Neural Net.*, 3(1):122–130, January 1992.

[62] J. B. Rawlings and K. R. Muske. Stability of constrained receding horizon control. *IEEE Trans. Auto. Cont.*, 38(10):1512–1516, October 1993.

[63] J. Richalet, A. Rault, J. L. Testud, and J. Papon. Model predictive heuristic control: Applications to industrial processes. *Automatica*, 14:413–428, 1978.

[64] N. L. Ricker. Use of quadratic programming for constrained internal model control. *Ind. Eng. Chem. Proc. Des. Dev.*, 24:925–936, 1985.

[65] D. G. Robertson, J. H. Lee, and J. B. Rawlings. A moving horizon-based approach for least squares state estimation. Submitted for publication in *AIChE J.*, 1994.

[66] R. W. H. Sargent and G. R. Sullivan. The development of an efficient optimal control package. In J. Stoer, editor, *Lecture Notes in Control and Information Sciences 7*, pages 158–168, Berlin, 1978. 8th IFIP Conference on Optimization Techniques, Springer-Verlag.

[67] R. Scattolini and S. Bittanti. On the choice of the horizon in long-range predictive control—some simple criteria. *Automatica*, 26(5):915–917, 1990.

[68] P. O. Scokaert and J. B. Rawlings. Stability of model predictive control under perturbations. In *IFAC Symposium on Nonlinear Control System Design, Tahoe City, California*, 1995.

[69] P. B. Sistu and B. W. Bequette. Model predictive control of processes with input multiplicities. *Chem. Eng. Sci.*, 6(6):921–936, 1995.

[70] E. D. Sontag. An algebraic approach to bounded controllability of nonlinear systems. *Int. J. Control*, 39:181–188, 1984.

[71] H. J. Sussmann and V. Jurdjevic. Controllability of nonlinear systems. *J. Diff. Eq.*, 12:95–115, 1972.

[72] Y. Thomas and A. Barraud. Commande optimale a horizon fuyant. *Revue RAIRO*, J(1):126–140, April 1974.

[73] Y. Thomas, D. Sarlat, and L. Shaw. Receding-horizon approach to the synthesis of nonlinear multivariable regulators. *Electron. Lett.*, 13(11):329–331, May 1977.

[74] Y. A. Thomas. Linear quadratic optimal estimation and control with receding horizon. *Electron. Lett.*, 11:19–21, January 1975.

[75] J. G. Van de Vusse. Plug-flow type reactor versus tank reactor. *Chem. Eng. Sci.*, 19:964, 1964.

[76] M. Vidyasagar. *Nonlinear Systems Analysis*. Prentice-Hall, Inc., Englewood Cliffs, New Jersey, 2nd edition, 1993.

[77] J. Villadsen and M. L. Michelsen. *Solution of Differential Equation Models by Polynomial Approximation.* Prentice-Hall, Englewood Cliffs New Jersey, 1978.

[78] J. V. Villadsen and W. E. Stewart. Solution of boundary-value problems by orthogonal collocation. *Chem. Eng. Sci.*, 22:1483–1501, 1967.

[79] E. Zafiriou and M. Morari. Design of robust digital controllers and sampling-time selection for SISO systems. *Int. J. Control*, 44(3):711–735, 1986.

Chapter 6

Nonlinear State Estimation

Kenneth R. Muske
Computational Science Methods Group
Los Alamos National Laboratory
Los Alamos, NM 87545

Thomas F. Edgar
Department of Chemical Engineering
The University of Texas at Austin
Austin, TX 78712-1062

6.1 Introduction

The implementation of the nonlinear model predictive control techniques discussed in the previous chapter requires knowledge of the current state of the nonlinear system in order to compute the solution to the open-loop optimal control problem formulated at each control interval. Feedback in these controllers comes from the update of the current state. Since the full state of the nonlinear system is not directly measurable in most applications, some method of reconstructing the current state of the system from the measured outputs must be employed. In this chapter, several methods of state estimation for nonlinear systems are presented.

The estimation of the state of a linear system is performed using techniques based on well-established optimal linear estimation theory. Due to the mathematical complexity introduced by a nonlinear model, nonlinear optimal state estimation is much less established in practice. However, a number of sub-optimal techniques that are computationally tractable have been proposed for solving the nonlinear problem. Each of these techniques is based on some simplification of or approximation to the underlying nonlinear stochastic system. Consequently, they have their associated advantages and limitations. It is these techniques, which can reasonably be implemented in real-time applications, that are discussed in this chapter.

We begin with a review of linear state estimation to set the stage for the nonlinear approaches. Linearized filters for the nonlinear system are the next topic and include extended Kalman filtering and statistical approximation methods. Nonlinear observers are then presented followed by a nonlinear moving horizon state estimator. We close with a brief discussion of combined state and parameter estimation. The goal of this chapter is to summarize the methods currently available for nonlinear state estimation from an applications-oriented viewpoint. Since these techniques will be implemented using a process control computer, we emphasize techniques in which the measured outputs are available at discrete sampling times.

Before proceeding with our discussion of state estimation, let us define the state, inputs, parameters, and outputs of a nonlinear dynamic system. The *state* is the minimum amount of information necessary at the current time to uniquely determine the dynamic behavior of the system at all future times given the inputs and parameters. An *input* is an independent variable of the system analogous to the forcing function of a dynamic differential equation. Chemical process examples include the feed rate and composition to a process unit and the inlet cooling medium temperature to a heat exchanger. *Parameters* consist of the physical properties used in the description of the system. For chemical processes, parameters include thermodynamic properties such as density and heat capacity, physical dimensions such as volume and area, and rate terms such as heat transfer coefficients and reaction rate constants. In the discussion that follows, we assume that the inputs and parameters of the system are known. An *output* of the system is a variable that can be measured directly. Outputs may be either a state variable or some function of the state. State estimation is the process of determining the state from the output measurements given a dynamic model of the system. We illustrate this terminology with the following example.

Example 6.1

Consider an exothermic reaction carried out in a perfectly mixed batch reactor with a constant temperature cooling coil. The reaction is second-order and irreversible in which chemical species A dimerizes to form species B.

$$\mathrm{A} + \mathrm{A} \longrightarrow \mathrm{B}$$

The rate of reaction of species A is expressed as follows.

$$r = -kC_{\mathrm{A}}{}^2, \qquad k = k_0 \exp(-E_a/RT)$$

The rate coefficient, k, is described by an Arrhenius relation in which k_0 is the frequency factor, E_a is the activation energy, T is the reactor temperature, and R is the gas constant. Heat is generated due to the heat of reaction and removed via heat transfer through the cooling coil in which ΔH is the heat of reaction, V is the volume of the reactor, and U, A, and T_c are the overall heat transfer coefficient, heat transfer area, and temperature of the cooling coil.

$$h = -\Delta H V k C_{\mathrm{A}}{}^2 + UA(T_c - T)$$

This expression assumes that the heat capacity of the cooling fluid is large and the resistance through the coil wall is negligible such that the coil temperature remains constant at the cooling fluid temperature.

A dynamic model of the batch reactor can be obtained from mass and energy balances in which ρ and C are the density and heat capacity of the fluid in the reactor, respectively.

$$\frac{dC_{\mathrm{A}}}{dt} = -k_0 \exp(-E_a/RT) C_{\mathrm{A}}{}^2 \tag{6.1}$$

$$\frac{dT}{dt} = -\frac{\Delta H}{\rho C} k_0 \exp(-E_a/RT) C_{\mathrm{A}}{}^2 + \frac{UA}{V\rho C}(T_c - T) \tag{6.2}$$

This model assumes that the heat of reaction, density, and heat capacity are independent of temperature.

For this batch reactor model, the concentration of species A, C_{A}, and the reactor fluid temperature, T, are the state variables, the cooling coil temperature, T_c, is the input, and the remaining items are parameters. The dynamic model consists of the two differential equations in Eqs. 6.1 and 6.2 that determine the behavior of the state variables. In a typical application, the temperature is measured and used to estimate the concentration.

6.2 Linear Systems

In this section, the basic concepts of state estimation are discussed using linear models. These concepts are extended to nonlinear models in Section 6.4.

We adopt this approach in order to introduce the topic within a simpler framework. In addition, several nonlinear state estimation techniques are based on the application of linear state estimation on a linearization of the nonlinear model. More information on linear system theory and state estimation can be found in standard linear system theory reference texts such as Kwakernaak and Sivan [34] and Kailath [26].

6.2.1 Linear System Models

A continuous-time linear system model can be constructed using the following system of linear differential equations.

$$\begin{aligned} \dot{x}(t) &= Ax(t) + Bu(t) \\ y(t) &= Cx(t) \end{aligned} \tag{6.3}$$

This representation is referred to as a linear *state-space* model in which $y \in \Re^p$ is the output, $u \in \Re^m$ is the input, $x \in \Re^n$ is the state of the system, and $A \in \Re^{n \times n}$, $B \in \Re^{n \times m}$, and $C \in \Re^{p \times n}$ are the system matrices. We restrict the input to bounded, integrable functions in the following discussion.

If the output measurements are available only at equally spaced time intervals or *sampling periods*, referred to as Δt, a discrete linear state-space model can be used to describe the dynamic behavior of the system. In this chapter, we represent discrete-time models with the discrete time index k specified as a subscript and the system matrices denoted by an overbar.

$$\begin{aligned} x_{k+1} &= x\left((k+1)\Delta t\right) = \bar{A}x_k + \bar{B}u_k \\ y_k &= y\left(k\Delta t\right) = \bar{C}x_k \end{aligned} \tag{6.4}$$

The discrete-time system matrices are obtained from the continuous-time matrices in Eq. 6.3 as follows.

$$\bar{A} = e^{A\Delta t}, \qquad \bar{B} = \left(\int_0^{\Delta t} e^{A\tau} d\tau\right) B, \qquad \bar{C} = C$$

The difference equation in Eq. 6.4 is valid when the input remains constant over the sampling period and the output is sampled at the same time the input is injected [34]. Due to the long process response times for most chemical processes, sampling periods much larger than the computational time required by the controller are used. Therefore, the validity of Eq. 6.4 typically can be assumed for chemical process computer control applications.

The linear system models presented in Eqs. 6.3 and 6.4 are referred to as *time-invariant* since the system matrices are independent of time. For continuous-time systems, *time-varying* models, in which the system matrices are functions of time, are represented as follows.

$$\begin{aligned} \dot{x}(t) &= A(t)x(t) + B(t)u(t) \\ y(t) &= C(t)x(t) \end{aligned} \tag{6.5}$$

For the discrete-time system in Eq. 6.4, the time dependence of the system matrices is represented by a subscript indicating the discrete time index.

$$\begin{aligned} x_{k+1} &= \bar{A}_k x_k + \bar{B}_k u_k \\ y_k &= \bar{C}_k x_k \end{aligned} \tag{6.6}$$

6.2.2 Observability of Linear Systems

Before proceeding with a discussion of state estimation, we first consider whether it is possible to estimate the state from the output measurements for a given linear system model. If $C = 0$, for example, state estimation is clearly not possible because the output measurements are always zero and contain no information about the state. A system in which the state can be determined uniquely from the output measurements is referred to as *observable.* Observability for linear systems is a global property of the system matrices that does not depend on the specific state, input, or output values. A characterization of observable linear systems is developed as follows.

Consider the output of the discrete linear system in Eq. 6.4 at successive sample times up to sample time $k = n - 1$. The outputs are related to the initial state of the system, x_0, and the previous inputs, $\{u_0, u_1, \ldots, u_{n-2}\}$.

$$\begin{aligned} y_0 &= \bar{C}x_0 \\ y_1 &= \bar{C}\bar{A}x_0 + \bar{C}\bar{B}u_0 \\ y_2 &= \bar{C}\bar{A}^2x_0 + \bar{C}\bar{A}\bar{B}u_0 + \bar{C}\bar{B}u_1 \\ \vdots & \qquad \vdots \\ y_{n-1} &= \bar{C}\bar{A}^{n-1}x_0 + \bar{C}\bar{A}^{n-2}\bar{B}u_0 + \ldots + \bar{C}\bar{B}u_{n-2} \end{aligned}$$

These relations can be combined into a linear system of equations, $\bar{\mathcal{O}}x_0 = \mathcal{Y}$, in which the $pn \times n$ matrix $\bar{\mathcal{O}}$ is called the *observability matrix.*

$$\bar{\mathcal{O}} = \begin{bmatrix} \bar{C} \\ \bar{C}\bar{A} \\ \vdots \\ \bar{C}\bar{A}^{n-1} \end{bmatrix}, \quad \mathcal{Y} = \begin{bmatrix} y_0 \\ y_1 - \bar{C}\bar{B}u_0 \\ \vdots \\ y_{n-1} - \bar{C}\bar{A}^{n-2}\bar{B}u_0 - \ldots - \bar{C}\bar{B}u_{n-1} \end{bmatrix} \tag{6.7}$$

The initial state x_0 can be determined uniquely if $\bar{\mathcal{O}}$ is full rank. In this case, the linear system in Eq. 6.4 is observable or reconstructable. Since $\bar{\mathcal{O}}$ is constructed from only $\bar{A}$ and $\bar{C}$, $(\bar{C}, \bar{A})$ is referred to as observable. If $\bar{\mathcal{O}}$ is not full rank, any initial state with a component in the null space of $\bar{\mathcal{O}}$ cannot be uniquely determined from the output measurements. In this case, the linear system in Eq. 6.4 is not observable and the null space of $\bar{\mathcal{O}}$ is called the *unobservable subspace* of the linear system. If the unobservable subspace does not contain unstable modes of A, the system is *detectable.*

The linear system model in Eq. 6.4 can be transformed into *observer canonical* or modal form using the nonsingular transformation matrix V. This transformation matrix is partitioned into the observable and unobservable subspaces of the linear system. The observable subspace, V_o, is composed of the basis vectors of the range of $\bar{\mathcal{O}}$. The unobservable subspace, V_u, is composed of the basis vectors of the null space of $\bar{\mathcal{O}}$.

$$z_k = Vx_k, \qquad V = \begin{bmatrix} V_o \\ V_u \end{bmatrix}, \qquad \begin{aligned} z_{k+1} &= V\bar{A}V^{-1}z_k + V\bar{B}u_k \\ y_k &= \bar{C}V^{-1}z_k \end{aligned}$$

The advantage of observer canonical form is that the observable modes, z^o, and the unobservable modes, z^u, are apparent by inspection of the partitioned system.

$$\begin{aligned} \begin{bmatrix} z^o_{k+1} \\ z^u_{k+1} \end{bmatrix} &= \begin{bmatrix} \tilde{A}_o & 0 \\ \tilde{A} & \tilde{A}_u \end{bmatrix} \begin{bmatrix} z^o_k \\ z^u_k \end{bmatrix} + \begin{bmatrix} \tilde{B}_o \\ \tilde{B}_u \end{bmatrix} u_k \\ y_k &= \begin{bmatrix} \tilde{C}_o & 0 \end{bmatrix} \begin{bmatrix} z^o_k \\ z^u_k \end{bmatrix} \end{aligned}$$

Since the canonical form is generated by a similarity transform that results in a lower triangular matrix, the characteristic values of the system are composed of the eigenvalues of $\tilde{A}_o$ and $\tilde{A}_u$. The linear system is detectable if and only if the eigenvalues of $\tilde{A}_u$, which are the characteristic values of the unobservable modes, have moduli strictly less than one. Detectability ensures that the modes of the system that cannot be observed or reconstructed from the output measurements are stable.

The continuous-time linear system in Eq. 6.3 is observable if there exists some δ, $0 < \delta < \infty$, such that the state $x(t-\delta)$ can be uniquely determined from $y(\tau)$ and $u(\tau)$ for all $u(\tau)$ and $t - \delta \leq \tau \leq t$. It can be shown that when the observability matrix in Eq. 6.7 is constructed from the continuous-time system matrices A and C, the same conditions for observability and detectability apply for continuous-time systems. A proof using the derivatives of the output and following an argument similar to that presented here for discrete-time systems is available in [26]. The continuous-time observer canonical form is constructed in the same manner as that for discrete-time systems. In this case, the system is detectable if and only if the eigenvalues of $\tilde{A}_u$ are strictly negative.

For time-varying linear systems, the conditions for observability are more complicated and involve the *state transition matrix*, Φ, and the ***observability** Gramian matrix*, M. The time-varying, continuous-time system in Eq. 6.5 is observable if and only if for all $t > 0$ there exists a $\sigma > 0$ such that the observability Gramian matrix $M(t-\sigma, t)$ is nonsingular. The continuous-time observability Gramian is a symmetric, nonnegative definite matrix computed from the state transition matrix $\Phi(t, t_o)$.

$$M(t-\sigma, t) = \int_{t-\sigma}^{t} \Phi^T(\tau, t-\sigma)C^T(\tau)C(\tau)\Phi(\tau, t-\sigma)d\tau \qquad (6.8)$$

The continuous-time state transition matrix is the solution to the matrix differential equation

$$\frac{d\Phi(t,t_o)}{dt} = A(t)\Phi(t,t_o), \qquad \Phi(t_o,t_o) = I \qquad (6.9)$$

that describes the dynamic behavior of the following unforced or *autonomous* linear system.

$$\dot{x}(t) = A(t)x(t), \qquad x(t) = \Phi(t,t_o)x(t_o)$$

Note that when A is a time-invariant matrix, the state transition matrix can be determined from the matrix exponential for $\bar{A}$ in Eq. 6.4.

The time-varying discrete linear system in Eq. 6.6 is observable if and only if for every k there exists a $j \leq k-1$ such that the discrete observability

Gramian matrix

$$\bar{M}(j,k) = \sum_{i=j+1}^{k} \bar{\Phi}^T(i,j+1)\bar{C}_i^T\bar{C}_i\bar{\Phi}(i,j+1) \tag{6.10}$$

is nonsingular. The observability Gramian matrix is computed using the discrete state transition matrix, $\bar{\Phi}(k,j)$, which is the solution to the following difference equation.

$$\bar{\Phi}(k+1,j) = \bar{A}_k\bar{\Phi}(k,j), \qquad \bar{\Phi}(j,j) = I \tag{6.11}$$

This matrix describes the dynamic behavior of the autonomous discrete linear system.

$$x_{k+1} = \bar{A}_k x_k, \qquad x_k = \bar{\Phi}(k,0)x_0$$

A more detailed discussion of observability for linear systems can be found in Kwakernaak and Sivan [34] and Kailath [26].

6.3 Linear State Estimation

We denote estimated variables with the symbol "^". Since these variables are determined using the output measurements, we will also indicate the time index of the output measurements along with the time index of the estimate. For the discrete-time system in Eq. 6.4, $\hat{x}_{j|k}$ represents the estimate of the state at sample time j given output measurements up to sample time k. For continuous-time systems, $\hat{x}\,(\tau \mid k)$ represents the estimate of the state at time τ given discrete output measurements up to sample time k.

The estimate $\hat{x}_{j|k}$ is a *smoothed* state estimate when $j < k$, a *filtered* state estimate when $j = k$, and a *predicted* state estimate when $j > k$. The filtered estimate is the current estimate based on all of the available output measurements. It is this estimate of the state that is normally used for control applications. Predicted estimates are obtained by using the model in Eq. 6.4 to extrapolate the filtered estimate into the future. These estimates typically are used to compute the objective or cost function in model predictive control formulations. Smoothed estimates are based on both past and future output measurements and provide a more accurate estimate than the filtered estimate. Smoothed estimates usually are computed off-line from previously collected process data and used in process analysis and diagnosis. For continuous-time systems, the estimate $\hat{x}\,(\tau \mid k)$ is a smoothed estimate when $\tau < k\Delta t$, a filtered estimate when $\tau = k\Delta t$, and a predicted estimate when $\tau > k\Delta t$.

A very simple conceptual approach to discrete linear state estimation is to solve Eq. 6.7 for the unknown initial state x_0. The filtered state at sample period k is then easily computed from the known inputs. However, in any practical application there are unmeasured disturbances present such that the solution to Eq. 6.7 may not exist or may result in an unreliable estimate. Therefore, an approach that considers disturbances in the determination of the state estimate is more useful.

6.3.1 Least Squares Estimator

We assume that disturbances to the state and measurement can be represented by additive terms to the state-space model. The discrete linear system in Eq. 6.4 with disturbances is expressed as

$$\begin{aligned} x_{k+1} &= \bar{A}x_k + \bar{B}u_k + w_k \\ y_k &= \bar{C}x_k + v_k \end{aligned} \tag{6.12}$$

in which $w \in \Re^n$ is the process or state disturbance vector and $v \in \Re^p$ is the measurement disturbance vector. The process disturbance vector models unmeasured disturbances to the process that effect the state. The measurement disturbance vector represents the error in the measuring device caused by instrument noise.

One approach that can be considered to estimate the state of the system in Eq. 6.12 is to minimize the estimated process and measurement disturbances in a least squares sense. The following least squares minimization problem is solved to obtain an estimate of the initial state for the discrete linear system with disturbances.

$$\min_{\{\hat{w}_{-1|k},\ldots,\hat{w}_{k-1|k}\}} \hat{w}_{-1|k}^T Q_0^{-1} \hat{w}_{-1|k} + \sum_{j=0}^{k-1} \hat{w}_{j|k}^T Q^{-1} \hat{w}_{j|k} + \sum_{j=0}^{k} \hat{v}_{j|k}^T R^{-1} \hat{v}_{j|k} \tag{6.13}$$

$$\text{Subject to:} \quad \begin{aligned} \hat{x}_{0|k} &= \bar{x}_0 + \hat{w}_{-1|k} \\ \hat{x}_{j+1|k} &= \bar{A}\hat{x}_{j|k} + \bar{B}u_j + \hat{w}_{j|k} \\ \hat{v}_{j|k} &= y_j - \bar{C}\hat{x}_{j|k} \end{aligned}$$

In this approach, $\hat{x}_{0|k}$ is the estimate of the initial state given k output measurements, $\bar{x}_0$ is an *a priori* estimate of the initial state, $\hat{w}_{j|k}$ are the estimated state disturbances in which $\hat{w}_{-1|k}$ is a correction to the *a priori* estimate, and $\hat{v}_{j|k}$ are the estimated measurement disturbances. The weighting matrices, Q_0^{-1}, Q^{-1}, and R^{-1}, specify the relative contribution of each of the terms in the quadratic objective. The inverse is used for notational consistency in the succeeding sections.

The optimal solution to this least squares problem at time k, $\hat{w}^*_{j|k}$, is used to compute the state estimate at time j given the k output measurements, $\hat{x}_{j|k}$, as follows.

$$\hat{x}_{j|k} = A^j \bar{x}_0 + \sum_{i=0}^{j} \bar{A}^{j-i} \hat{w}^*_{i-1|k} + \sum_{i=1}^{j} \bar{A}^{j-i} \bar{B} u_{i-1} \tag{6.14}$$

This expression computes a smoothed state estimate when $j < k$, a filtered state estimate when $j = k$, and a predicted state estimate when $j > k$.

The choice of the coefficients in the weighting matrices is a compromise between minimizing the estimated process disturbances versus minimizing the estimated measurement disturbances. This choice is based on the expected magnitudes of each of these disturbances. If the output measurements are reliable, then R^{-1} is chosen to be large relative to Q^{-1}. This choice tends to reduce the deviations between the measured output and the predicted output by increasing the estimated process disturbances. In this case, process disturbances are estimated to bring the model into agreement with the output measurements. On the other hand, if the output measurements are poor, then R^{-1} is chosen to be small relative to Q^{-1}. This choice prevents uncertain measurements from causing large estimated state disturbances. In this case, the model is assumed to be a more reliable indication of the state than the output measurements.

6.3.2 Recursive Estimation

The least squares approach presented in the previous section is referred to as non-sequential or *batch* estimation. It requires the solution of a least squares problem using all k previous output measurements to obtain a filtered state estimate at each sample time. As time progresses, the number of decision variables in this optimization problem quickly becomes prohibitive, making it a rather cumbersome procedure that cannot be implemented in practice. However, a sequential or *recursive* solution to the batch estimation problem posed in Eq. 6.13 that requires information only from the previous sample time can be constructed.

$$\hat{x}_{k|k} = \hat{x}_{k|k-1} + L_k(y_k - \bar{C}\hat{x}_{k|k-1}) \tag{6.15}$$

$$\hat{x}_{k+1|k} = \bar{A}\hat{x}_{k|k} + \bar{B}u_k \tag{6.16}$$

$$\hat{x}_{0|0} = \bar{x}_0$$

The filtered state estimate is computed from the predicted state estimate and the output measurement at each sample time. A gain matrix, L_k, multiplies the difference between the current measured output, y_k, and the predicted output, $\bar{C}\hat{x}_{k|k-1}$, to form a correction to the predicted state estimate. The gain matrix is a function of the system and weighting matrices used in the batch least squares estimator as shown in the following section.

6.3.3 Kalman Filter

The least squares recursive estimator in Eq. 6.15 can also be derived by minimization of the mean square reconstruction error [2]. The reconstruction error is a measure of the deviation between the actual and estimated state.

$$e_{k|k-1} = x_k - \hat{x}_{k|k-1} \tag{6.17}$$

When w_j and v_j are independent, zero mean, normally distributed random variables with covariances Q and R, respectively, and $\bar{x}_0$ is an independent, normally distributed random variable with covariance Q_0, the recursive estimator in Eq. 6.15 produces the minimum variance estimate of the state. For linear Gaussian systems this is also the most probable or *maximum likelihood* estimate. This recursive estimator is referred to as the discrete *Kalman filter*. A detailed probabilistic derivation of the Kalman filter can be found in Jazwinski [25] and Lewis [36].

In the probabilistic formulation, the covariances specify the expected magnitudes of the disturbances to the output measurement and the state. If the covariance of the measurement noise, R, is small relative to the covariance of the process noise, Q, the measurements are relatively noise free and the deviations between the measured output and the predicted output should be small. If the measurement noise covariance is large relative to the process noise covariance, the measurements are relatively uncertain and the feedback correction to the model prediction should be small.

The probabilistic approach provides a rigorous way to determine the tuning parameters for the estimator provided that the assumptions concerning the stochastic process are valid and the covariances are known or can be determined. The filter gain is computed at each sampling time using the covariance of the state estimate in Eq. 6.16, which we denote as P_k.

$$L_k = P_k\bar{C}^T(\bar{C}P_k\bar{C}^T + R)^{-1} \tag{6.18}$$

This covariance is propagated using the discrete filtering Riccati equation.

$$\begin{aligned} P_{k+1} &= \bar{A}P_k\bar{A}^T + Q - \bar{A}P_k\bar{C}^T(\bar{C}P_k\bar{C}^T + R)^{-1}\bar{C}P_k\bar{A}^T \\ P_0 &= Q_0 \end{aligned} \tag{6.19}$$

The Riccati equation contains a term, $\bar{A}P_k\bar{A}^T + Q$, that increases the covariance of the state estimate at each sampling time due to the presence of state or process noise. The remaining term represents the contribution due to the output measurement and generally decreases the covariance. The recursion in Eq. 6.19 tends to a constant matrix at large k when $(\bar{C}, \bar{A})$ is detectable, Q and R are positive definite, and Q_0 is non-negative definite.

$$P_\infty = \bar{A}(P_\infty - P_\infty\bar{C}^T(\bar{C}P_\infty\bar{C}^T + R)^{-1}\bar{C}P_\infty)\bar{A}^T + Q \tag{6.20}$$

This matrix is called the steady-state discrete filtering Riccati matrix. The steady-state Kalman filter gain is computed by using P_∞ in Eq. 6.18.

For the time-varying discrete linear system in Eq. 6.6, the optimal filter gain is computed from the solution of the discrete filtering Riccati equation at each sampling time using the following expressions.

$$L_k = P_k\bar{C}_k^T(\bar{C}_kP_k\bar{C}_k^T + R)^{-1} \tag{6.21}$$

$$\begin{aligned} P_{k+1} &= \bar{A}_kP_k\bar{A}_k^T + Q - \bar{A}_kP_k\bar{C}_k^T(\bar{C}_kP_k\bar{C}_k^T + R)^{-1}\bar{C}_kP_k\bar{A}_k^T \\ P_0 &= Q_0 \end{aligned} \tag{6.22}$$

There is no general steady-state solution to this Riccati equation or steady-state filter gain for time-varying systems.

A stochastic model for the time-varying, continuous-time linear system in Eq. 6.5 can also be constructed.

$$\begin{aligned} \dot{x}(t) &= A(t)x(t) + B(t)u(t) + w(t) \\ y(t) &= C(t)x(t) + v(t) \end{aligned} \tag{6.23}$$

When $v(t) \in \Re^p$ and $w(t) \in \Re^n$ are independent, zero mean, white noise processes with spectral densities Q and R, respectively, a continuous-time Kalman filter similar to that for discrete systems is obtained [21].

$$\dot{\hat{x}}(t \mid t) = A(t)\hat{x}(t \mid t) + B(t)u(t) + L(t)(y(t) - C(t)\hat{x}(t \mid t)) \tag{6.24}$$

$$L(t) = P(t)C(t)^T R^{-1} \tag{6.25}$$

$$\dot{P}(t) = A(t)P(t) + P(t)A(t)^T + Q - P(t)C(t)^T R^{-1} C(t)P(t) \tag{6.26}$$

$$P(0) = Q_0$$

The result in Eq. 6.24 is the minimum variance filtered estimate in which the filter gain, $L(t)$, is computed via the differential equation in Eq. 6.26 and $P(t)$ is the covariance of the state estimate $\hat{x}(t \mid t)$. There is again a term for the increase in the covariance due to the presence of state noise, $A(t)P(t) + P(t)A(t)^T + Q$, and a remaining term, $P(t)C(t)^T R^{-1} C(t)P(t)$, that generally decreases the covariance due to the contribution of the measurement.

The Kalman filter in Eq. 6.15 for discrete-time systems or Eq. 6.24 for continuous-time systems produces the minimum variance and maximum likelihood state estimate provided that the assumptions concerning the stochastic process are valid. In this case, the Kalman filter is the optimal linear state estimator. It is nominally stable provided R is positive definite, Q and Q_0 are positive semidefinite, and the corresponding linear system is detectable. *Nominal stability* ensures that the filtered state will converge to the true state in the absence of disturbances for any initial reconstruction error given an exact model of the process.

In addition to the standard linear system references previously cited, optimal linear state estimation is discussed in Åström [2] and Bryson and Ho [8]. A discrete-time presentation is provided in Catlin [10]. A discussion of square-root algorithms, which result in more accurate numerical computations, is presented in Park and Kailath [50]. An excellent practical review of linear filtering and smoothing is contained in Gelb [21].

For the linear discrete-time optimal estimator, the gain in Eq. 6.15 must be computed at each sampling time. A common approximation to the optimal estimator that avoids this computation is to use the steady-state solution of the discrete filtering Riccati equation in Eq. 6.20 to determine the steady-state Kalman filter gain matrix. The steady-state gain is then used at each sampling time. Another common approximation is to discontinue the update of the gain matrix at a given sample time and use this gain at each successive sampling time. This technique avoids the computation of the steady-state gain matrix, but does not guarantee nominal stability when the gain calculation is stopped at an arbitrary sample time [15].

6.3.4 Luenberger Observer

Another method of constructing a recursive estimator is to use a constant gain matrix in Eq. 6.15 that achieves some desired performance criteria.

$$\hat{x}_{k|k} = \hat{x}_{k|k-1} + L(y_k - \bar{C}\hat{x}_{k|k-1}) \tag{6.27}$$

The performance criterion is based on the dynamic behavior of the reconstruction error and ignores the stochastic properties of the disturbances w_k and v_k. This recursive estimator is referred to as a deterministic or Luenberger *observer* [40]. For discrete-time systems, the observer is nominally stable if and only if the eigenvalues of the $n \times n$ matrix $\bar{A} - \bar{A}L\bar{C}$ have moduli strictly less than one. This result comes from the dynamic response of the reconstruction error from an initial nonzero value with no disturbances.

$$e_{k+1} = (\bar{A} - \bar{A}L\bar{C})e_k \tag{6.28}$$

The gain matrix can be determined by choosing the eigenvalues of $\bar{A} - \bar{A}L\bar{C}$, which are referred to as the *closed-loop observer poles*. This technique is referred to as *pole placement* and requires that the system be observable. If the gain is chosen such that all of the eigenvalues are zero, the result is a *deadbeat* observer. The choice of the observer poles is a compromise between rapid decay of the reconstruction error, which requires that the poles be placed close to the origin, and sensitivity to measurement noise and modeling error, which increases as the poles are moved toward the origin.

An observer for continuous-time systems can be constructed analogous to that for discrete-time systems. The reconstruction error for the estimate in Eq. 6.24 behaves as follows.

$$\begin{aligned} \dot{e}(t) &= (A - LC)e(t) \\ e(t) &= x(t) - \hat{x}(t \mid t) \end{aligned} \tag{6.29}$$

In this case, stability of the observer requires that the gain be selected such that the eigenvalues of $A - LC$, which are the closed-loop observer poles, are strictly negative. As the poles are moved further into the left half plane, the rate of convergence of the reconstruction error increases. For the continuous-time observer, there is also a compromise between the rapid decay of the reconstruction error and sensitivity to measurement noise and modeling error.

The Luenberger observer is a deterministic estimator that is employed when the stochastic model of the process is unknown or if the optimal filter exhibits poor performance. It can also be employed to reduce the computational requirements of the estimator. For modern chemical process control applications, however, computational time is generally not an issue with linear systems except in large-scale applications. Further discussion of observers for linear systems can be found in Kwakernaak and Sivan [34] and Kailath [26].

6.3.5 Moving Horizon Estimator

A recursive form of the batch state estimation problem in Eq. 6.13 can be constructed using a moving horizon approach. In this approach, the state estimate at time k is determined recursively from the solution of the following least squares problem that uses the predicted estimate at time $k-N-1$, $\hat{x}_{k-N|k-N-1}$, and the most recent $N+1$ output measurements.

$$\min_{\{\hat{w}_{k-N-1|k},\ldots,\hat{w}_{k-1|k}\}} \quad \hat{w}_{k-N-1|k}^T P_{k-N}^{-1} \hat{w}_{k-N-1|k} + \sum_{j=k-N}^{k-1} \hat{w}_{j|k}^T Q^{-1} \hat{w}_{j|k} + \sum_{j=k-N}^{k} \hat{v}_{j|k}^T R^{-1} \hat{v}_{j|k} \tag{6.30}$$

Subject to:

$$\begin{aligned} \hat{x}_{k-N|k} &= \hat{x}_{k-N|k-N-1} + \hat{w}_{k-N-1|k} \\ \hat{x}_{j+1|k} &= \bar{A}\hat{x}_{j|k} + \bar{B}u_j + \hat{w}_{j|k} \\ \hat{v}_{j|k} &= y_j - \bar{C}\hat{x}_{j|k} \end{aligned}$$

The moving horizon of output measurements allows for a finite number of decision variables at each sampling time. This procedure is initialized by computing the first N predicted estimates using the batch estimator in Eq. 6.13. The state estimate at time $n+j$ given k output measurements, in which $n = k - N$, can then be computed from the predicted state estimate at sample time n and the solution of the least squares problem in a manner similar to the batch estimator.

$$\hat{x}_{n+j|k} = A^j \hat{x}_{n|n-1} + \sum_{i=0}^{j} A^{j-i} \hat{w}_{n-1+i|k}^* + \sum_{i=1}^{j} A^{j-i} B u_{n-1+i} \tag{6.31}$$

This expression computes a smoothed state estimate when $j < N$, a filtered state estimate when $j = N$, and a predicted state estimate when $j > N$. It can be shown that the filtered estimate from the moving horizon estimator is the optimal filtered estimate for the discrete stochastic linear system in Eq. 6.12 in which v_k, w_k and $\bar{x}_0$ follow the same assumptions made in the Kalman filter, the weighting matrices Q^{-1} and R^{-1} are the inverses of the covariance matrices for w_k and v_k, respectively, and the weighting matrix P_{k-N}^{-1} is the inverse of the discrete filtering Riccati matrix at sampling time $k-N$ [47]. This matrix is the covariance of the state estimate $\hat{x}_{k-N|k-N-1}$ and is computed using the recursion in Eq. 6.19.

Constraints

Since the moving horizon estimator produces the same estimate as the Kalman filter, there is no incentive to implement this approach due to the increased computational effort required to solve the least squares problem in Eq. 6.30. The motivation for employing the moving horizon formulation is the addition of constraints on the estimated states and state disturbances.

$$h_{\min} \leq \quad H\hat{x}_{j|k} \quad \leq h_{\max}, \qquad j = k-N+1, \ldots, k \tag{6.32}$$

$$w_{\min} \leq \quad \hat{w}_{j|k} \quad \leq w_{\max}, \qquad j = k-N, \ldots, k-1 \tag{6.33}$$

The estimated state constraint in Eq. 6.32 specifies maximum and minimum limits on the estimate of the state. These constraints are applied to prevent physically unrealistic state estimates, such as negative concentrations, that can be due to spurious output measurements. The estimated state disturbance constraint in Eq. 6.33 specifies limits on the estimated state disturbances. These limits can be viewed as altering the probability distribution function of the state disturbances such that the probability of any state disturbance outside of the constraints is zero. This constraint prevents estimated state disturbances that cannot realistically occur in the process.

The constraints are chosen to define a convex region in the $\hat{w}$ space containing the origin. For systems forced by an input, the state constraints are shifted by the nominal trajectory due to the input. A feasible constraint set can always be achieved since $\hat{w}_{k-N-1|k}$ is unconstrained at each sample time. Due to these constraints, this formulation is a nonlinear estimator for the linear system that requires the solution of the quadratic program consisting of Eqs. 6.30 through 6.33. Muske *et al.* [47] construct this constrained moving horizon observer quadratic program. Meadows *et al.* [43] present an example of moving horizon constrained state estimation in a linear model predictive control application.

6.4 Nonlinear Systems

Our presentation of nonlinear state estimation parallels the discussion for linear systems in the previous section. We begin with a definition of observability for nonlinear systems. Nonlinear recursive filters and observers analogous to the linear estimators follow. We then present nonlinear moving horizon state estimation. Combined state and parameter estimation, a topic not addressed in our presentation of linear systems, followed by a summary of the nonlinear state estimation techniques closes our presentation.

Nonlinearity introduces additional complexity to the state estimation problem. This complexity applies to the derivation as well as the implementation of the nonlinear state estimator. Since the optimal estimator generally is not available for the nonlinear problem, nonlinear techniques are based on sub-optimal approaches. These approaches lead to estimators that are quite different from each other. There are also restrictions on the nonlinear systems that each approach can adequately address.

6.4.1 Nonlinear System Models

Continuous-time nonlinear systems are described by the following nonlinear system of differential equations in which $y(t) \in \Re^p$ is the output, $u(t) \in \Re^m$ is the input, $x(t) \in \Re^n$ is the state of the system, $\mathbf{f} : \Re^n \times \Re^m \times \Re^+ \rightarrow \Re^n$ is the system function, and $\mathbf{g} : \Re^n \times \Re^+ \rightarrow \Re^m$ is the measurement function.

$$\begin{aligned} \dot{x}(t) &= \mathbf{f}(x, u, t) \qquad (6.34) \\ y(t) &= \mathbf{g}(x, t) \end{aligned}$$

We assume that the model parameters are known and do not include them in the argument list for the system and measurement functions. We also restrict the input to bounded, integrable functions. Discrete-time nonlinear systems are described by the following difference equation in which the discrete system function is $\bar{\mathbf{f}} : \Re^n \times \Re^m \times \mathcal{Z}^+ \rightarrow \Re^n$ and the discrete measurement function is $\bar{\mathbf{g}} : \Re^n \times \mathcal{Z}^+ \rightarrow \Re^m$.

$$\begin{aligned} x_{k+1} &= \bar{\mathbf{f}}(x_k, u_k, k) \qquad (6.35) \\ y_k &= \bar{\mathbf{g}}(x_k, k) \end{aligned}$$

In general, there exists no functional representation for the discrete sampling of a continuous-time nonlinear system such as Eq. 6.4 for linear systems. In this case, the continuous-time nonlinear model in Eq. 6.34 with a discrete sampling period is used to model the system as a discrete process. The output at each sample time is determined from the solution of the nonlinear system of differential equations.

$$\begin{aligned} \dot{x}(\tau) &= \mathbf{f}(x(\tau), u(\tau), \tau), \qquad x\left(k\Delta t\right) = x_k \\ x_{k+1} &= x\left((k+1)\Delta t\right) \\ y_k &= \mathbf{g}(x_k, k\Delta t) \end{aligned} \tag{6.36}$$

When the input remains constant over the sampling period, $u(\tau)$ can be represented as a discrete input in which k is the discrete sampling time.

$$u(\tau) = u_k, \qquad k\Delta t \leq \tau < (k+1)\Delta t \tag{6.37}$$

Although we assume a uniform sampling period in the following discussion, we are not restricted to this assumption.

6.4.2 Observability of Nonlinear Systems

Observability of a linear system is a global property that can be determined either from the rank of the observability matrix for time-invariant systems or from the rank of the observability Gramian matrix for time-varying systems. In contrast, observability of a nonlinear system is determined locally about a given state or equilibrium point. In this section, we present a brief description of the requirements for local observability of nonlinear systems. The reader is referred to Hermann and Krener [22], Sontag [57], and Krener and Respondek [33] for a more detailed discussion on continuous-time systems. A further discussion on discrete systems is contained in Nijmeijer [49], Keerthi and Gilbert [28], and Lee and Nam [35].

Continuous-time Systems

Let $y(t, t_0, x_0, u(t))$ denote the output trajectory from an initial state x_0, initial time t_0, and input $u(t)$ for the continuous-time nonlinear system in Eq. 6.34. Two states x_0 and z_0 are defined to be *indistinguishable* if $y(t, t_0, x_0, u(t)) = y(t, t_0, z_0, u(t))$ for $t_0 < t < T$, $T < \infty$, and all admissible input trajectories $u(t)$. The nonlinear system is observable at x_0 if the set of states indistinguishable from x_0 contains only x_0 [22]. A nonlinear system is observable if it is observable at x for all $x \in \Re^n$. A weaker form of observability is obtained if we require that it only be necessary to distinguish

x_0 from its neighbors. In this case, a nonlinear system is ***weakly observable*** at x_0 if there exists a neighborhood $\mathcal{S}^n$ of x_0 such that the set of states in $\mathcal{S}^n$ that are indistinguishable from x_0 contains only x_0 [22]. A nonlinear system is weakly observable if it is weakly observable at x for all $x \in \Re^n$. These are global concepts of observability in which it might be necessary to compute a trajectory that is very far from x_0 or in which T is very large. By also restricting the state trajectories to a neighborhood of x_0, ***local observability*** can be defined. A nonlinear system is ***locally weakly observable*** at x_0 if for every neighborhood $\mathcal{X}^n$ of x_0 there exists another neighborhood $\mathcal{S}^n$ contained in $\mathcal{X}^n$ such that the set of states in $\mathcal{S}^n$ that are indistinguishable from x_0 for all input trajectories in which $x(t, t_0, z_0, u(t)) \in \mathcal{X}^n$ and $z_0 \in \mathcal{S}^n$ contains only x_0 [57]. This definition is slightly different from that presented in [22]. A nonlinear system is locally weakly observable if it is locally weakly observable at x for all $x \in \Re^n$.

Local weak observability implies that states close to x_0 are distinguishable from x_0 without large excursions. The advantage to this concept is that a linearized algebraic test similar to that for linear systems can be developed. The nonlinear system in Eq. 6.34 is observable at x_0 if there exists a neighborhood of x_0 and a p-tuple of integers $(k_1, k_2, \ldots, k_p)$, referred to as ***observability indices***, such that

i) $k_1 \geq k_2 \geq \ldots \geq k_p \geq 0, \quad \sum_{i=1}^{p} k_i = n$

ii) The n row vectors of $\{L_f^{j-1}(dg_i) \; : \; i = 1, \ldots, p; \; j = 1, \ldots, k_i\}$ are linearly independent.

in which $L_f^j(dg_i)$ is the jth Lie derivative of the gradient of $\mathbf{g}_i$ by the vector field $\mathbf{f}$ [33]. An observability matrix can be constructed using the row vectors $L_f^{j-1}(dg_i)$ in which full rank of this matrix implies local weak observability.

$$\mathcal{O} \;=\; \left[L_f^{j-1}(dg_i)\right], \qquad i = 1, \ldots, p; \;\; j = 1, \ldots, k_i \tag{6.38}$$

Note that when $\mathbf{f}$ and $\mathbf{g}$ are linear functions, this matrix is equivalent to the observability matrix for continuous-time linear systems. If the continuous-time system in Eq. 6.34 is observable, the continuous-time system with discrete measurements in Eq. 6.36 is also observable for sufficiently small sampling periods [57]. Certain unobservable systems can be decomposed into local observable and unobservable subspaces by a state transformation [62].

A simpler observability test that is commonly applied to nonlinear systems can be constructed by linearizing the system about x_0 using the first order terms of the following Taylor series expansion [52].

$$\begin{aligned}\dot{x} &= \mathbf{f}(x_0, u, t) + \left.\frac{\partial \mathbf{f}}{\partial x}\right|_{x=x_0} (x - x_0) \;+\; \ldots \\ y &= \mathbf{g}(x_0, t) + \left.\frac{\partial \mathbf{g}}{\partial x}\right|_{x=x_0} (x - x_0) \;+\; \ldots\end{aligned} \tag{6.39}$$

Observability is determined using the linear system techniques discussed in Section 6.2.2 with the linearized system matrices $A(t) = \partial \mathbf{f}/\partial x$ and $C(t) = \partial \mathbf{g}/\partial x$. Note that this is an approximation to the nonlinear observability matrix presented in Eq. 6.38. Observability of this linearized system may not imply local weak observability of the original nonlinear system.

Discrete-time Systems

For the discrete nonlinear system in Eq. 6.35, consider any admissible input sequence $\{u_0, u_1, \ldots, u_{k-1}\}$ and an initial state x_0. Let x_j, $j = 1, \ldots, k$, denote the state sequence generated by the model, $x_j = \bar{\mathbf{f}}(x_{j-1}, u_{j-1}, j-1)$, and $y_j(x_0)$ denote the output trajectory corresponding to x_0, $y_j(x_0) = \bar{\mathbf{g}}(x_j, j)$. Two states x_0 and z_0 are indistinguishable if $y_j(x_0) = y_j(z_0)$ for all j and all admissible input sequences. A nonlinear system is observable at x_0 if the set of states indistinguishable from x_0 contains only x_0 [49]. This definition of observability requires an infinite test and, therefore, is not useful in practice. A finite time test can be constructed in which the discrete nonlinear system is *finite time observable* or *strongly observable* at x_0 if for all $z_0 \in \Re^n$ and any admissible input sequence $\{u_0, u_1, \ldots u_{n-1}\}$, $y_j(z_0) = y_j(x_0)$, $j = 1, \ldots, n-1$ implies $z_0 = x_0$. The system is strongly observable if it is strongly observable at x for all $x \in \Re^n$. A local concept of observability for discrete systems can also be defined if we require that it only be necessary to distinguish x_0 from its neighbors. In this case, the discrete nonlinear system is *strongly locally observable* at x_0 if there exists a neighborhood $\mathcal{S}^n$ of x_0 such that for all $z_0 \in \mathcal{S}^n$ and all admissible input sequences, $y_j(z_0) = y_j(x_0)$, $j = 1, \ldots, n-1$ implies $z_0 = x_0$ [49]. The system is strongly locally observable if it is strongly locally observable at x for all $x \in \Re^n$.

A stronger observability condition than finite time observability is *uniform observability*. The discrete nonlinear system in Eq. 6.35 is uniformly observable at x_0 if for all $z_0 \in \Re^n$ and all admissible input sequences, there exists an $N \in [n-1, \infty)$ and a function $\alpha : \Re^+ \rightarrow \Re^+$ such that

$$\sum_{j=0}^{N} \|y_j(x_0) - y_j(z_0)\| \geq \alpha(\|x_0 - z_0\|) \tag{6.40}$$

in which α is a continuous, increasing function with $\alpha(0) = 0$ [46]. The system is uniformly observable if it is uniformly observable at x for all $x \in \Re^n$. This condition is similar to the uniform observability property in [28] and is a weaker discrete-time version of that presented in [44].

A linearized algebraic test for strong local observability of discrete nonlinear systems is developed as follows. The discrete nonlinear system in Eq. 6.35 is observable at x_0 if there exists a neighborhood of x_0 and a p-tuple of integers $k_1 \geq k_2 \geq \ldots \geq k_p \geq 0$, $\sum_{i=1}^{p} k_i = n$ such that the observability matrix $\bar{\mathcal{O}}$ for the discrete nonlinear system is full rank in which $\bar{\mathbf{F}}^i$ is the ith composite of the system function $\bar{\mathbf{f}}$ [35].

$$\bar{\mathcal{O}} = \begin{bmatrix} \bar{\mathcal{O}}_1 \\ \vdots \\ \bar{\mathcal{O}}_p \end{bmatrix}, \qquad \bar{\mathcal{O}}_i = \frac{\partial}{\partial x} \begin{bmatrix} \bar{\mathbf{g}}_i(x_0) \\ \bar{\mathbf{g}}_i(\bar{\mathbf{F}}^1(x_0)) \\ \vdots \\ \bar{\mathbf{g}}_i(\bar{\mathbf{F}}^{k_i-1}(x_0)) \end{bmatrix} \tag{6.41}$$

When $\bar{\mathbf{f}}$ and $\bar{\mathbf{g}}$ are linear functions, this matrix is equivalent to the discrete linear system observability matrix in Eq. 6.7.

Summary of Nonlinear Observability

Observability of nonlinear systems is determined by a linearized test that is only guaranteed within a neighborhood of x_0. This neighborhood may be arbitrarily small or may extend over the entire operating range of interest. Ray [52] states that observability of nonlinear systems is often determined by the structure of the system and is not dependent on the state in a complex manner. Consequently, linearized observability tests are usually adequate. An additional complexity with nonlinear systems is that the observability matrix can be a function of the input. In these cases, observability must be verified for all admissible inputs since there may exist input values for which the nonlinear system looses observability. For a number of practical cases, however, the system is observable independent of the inputs. These systems are referred to as *observable for any input.* Gauthier *et al.* [20] present a characterization of these systems using single input–single output models.

Example 6.2

Consider the batch reactor model presented in Example 6.1. The nonlinear observability matrix in Eq. 6.38 is constructed as follows.

$$\mathcal{O} = \begin{bmatrix} 0 & 1 \\ -2\frac{\Delta H}{\rho C} k C_{\text{A}} & -\frac{E}{RT^2}\frac{\Delta H}{\rho C} k {C_{\text{A}}}^2 - \frac{UA}{V\rho C} \end{bmatrix}$$

When $C_{\text{A}} \neq 0$, the nonlinear observability matrix is full rank and the system is observable. In this example, the system is observable for any input since the observability matrix is not a function of the input. Note that the same observability matrix is obtained from the linearized model using Eq. 6.39.

6.5 Nonlinear Recursive Filters

The optimal linear filter in Eq. 6.15 produces the minimum variance and maximum likelihood estimate of the state of the linear system in Eq. 6.4. To develop a nonlinear analogy, one must first specify which estimate is desired since they may not be the same for a nonlinear stochastic system. The minimum variance estimate is the conditional mean of the state. The maximum likelihood or Bayesian estimate is the conditional mode or most probable estimate. These estimates are determined from the conditional probability density of the state given the output measurements. The conditional probability density of the state of a linear system with Gaussian noise is Gaussian and completely determined by the mean and covariance. For a nonlinear system, the conditional probability density of the state is not Gaussian even when the state and measurement disturbances, $w(t)$ and $v(t)$, are Gaussian. The determination of the conditional probability requires that Eq. 6.34 be represented as a vector nonlinear stochastic differential equation.

$$\begin{aligned} dx(t) &= \mathbf{f}(x, u, t)dt + d\beta \\ \frac{d\beta}{dt} &= w(t) \end{aligned}$$

The evolution of the conditional probability is determined by the use of stochastic calculus and, in general, requires some approximations in order to obtain an implementable filter.

These techniques are beyond the scope and intent of the discussion here. We refer the interested reader to the following references for more information. A Bayesian approach to stochastic state estimation for both linear and nonlinear systems is discussed in Ho and Lee [23]. A detailed discussion of the probabilistic approach to nonlinear state estimation is presented in Jazwinski [25]. An overview is provided in Bryson and Ho [8]. A brief review of the stochastic calculus concepts necessary to understand these techniques is provided in Marcus [42].

6.5.1 Extended Kalman Filter

A straight-forward approximation to optimal nonlinear state estimation is to linearize the nonlinear model about a given operating point and apply optimal linear state estimation to the linearized system. The extended Kalman filter computes a state estimate at each sampling time by the use of Kalman filtering on a linearized model of the nonlinear system. This technique is justified if there exists a sufficiently large neighborhood in which the linearized model is a good representation of the nonlinear system. If, in addition, the disturbances are well represented by zero mean Gaussian state and measurement noise, the optimal estimate for the linearized system should be a reasonable approximation to the optimal estimate for the nonlinear system. In this case, extended Kalman filtering is expected to provide an accurate estimate of the state of the nonlinear system.

A linearized model of the continuous-time system in Eq. 6.34 about the state x^* can be developed from the Taylor series expansion in Eq. 6.39. Depending on the selection of x^*, several variations of the extended Kalman filter can be developed. The most common approach is the first order filter in which the nonlinear system is linearized about the current state estimate at each sampling time using the first order terms in Eq. 6.39. More complex approaches that attempt to compensate for the inaccuracy caused by linearization, such as iterative and higher order filters, have also been implemented. These extended Kalman filter techniques are presented in the following sections. Further discussion of extended Kalman filtering is contained in Jazwinski [25], Gelb [21], Bryson and Ho [8], Lewis [36], and Stengel [58].

First Order Extended Kalman Filter

For the continuous-time nonlinear system with discrete output measurements in Eq. 6.36, a linearized approximation to the nonlinear model can be obtained by truncating Eq. 6.39 after the first order terms. This approximate model, linearized about the current state, is used to construct a time-varying Kalman filter at each sampling time. In order to ensure that the partial derivatives exist, the system and measurement functions are restricted to continuous functions in x such that $\mathbf{f}, \mathbf{g} \in \mathcal{C}^1$.

The filtered state estimate is determined from the current output measurement in the same manner as the linear recursive filter in Eq. 6.15.

$$\begin{aligned} \hat{x}_{k|k} &= \hat{x}_{k|k-1} + L_k \left(y_k - \mathbf{g}(\hat{x}_{k|k-1}, k\Delta t) \right) \\ \hat{x}_{0|0} &= \bar{x}_0 \end{aligned} \tag{6.42}$$

The Kalman filter gain, L_k, is computed using the discrete, time-varying Kalman filter formulation in Eq. 6.21 in which the linear system matrix $\bar{C}_k$ is replaced by the linearized measurement function G_k.

$$\begin{aligned} L_k &= \hat{P}_{k|k-1} G_k^T \left(G_k \hat{P}_{k|k-1} G_k^T + R \right)^{-1} \\ G_k &= \left. \frac{\partial \mathbf{g}(x,t)}{\partial x} \right|_{x=\hat{x}_{k|k-1},\ t=k\Delta t} \end{aligned} \tag{6.43}$$

The estimated covariance of the state is updated at each sample time due to the contribution of the discrete measurement as follows.

$$\begin{aligned} \hat{P}_{k|k} &= \left(I - L_k G_k \right) \hat{P}_{k|k-1} \\ \hat{P}_{0|0} &= Q_0 \end{aligned} \tag{6.44}$$

Between sampling times, the state estimate is propagated using the nonlinear system model in Eq. 6.36.

$$\begin{aligned} \dot{\hat{x}}\left(\tau \mid k\right) &= \mathbf{f}(\hat{x}\left(\tau \mid k\right), u(\tau), \tau), \qquad \hat{x}\left(k\Delta t \mid k\right) = \hat{x}_{k|k} \\ \hat{x}_{k+1|k} &= \hat{x}\left((k+1)\Delta t \mid k\right) \end{aligned} \tag{6.45}$$

The covariance of the state estimate in Eq. 6.45 is propagated between sampling times using the linear differential equation in Eq. 6.26 in which the linear system matrix $A(t)$ is replaced by the linearized system function $F\left(\tau \mid k\right)$. Since there are no output measurements available between sampling times, the contribution to the covariance from the output measurement is removed.

$$\begin{aligned} \dot{\hat{P}}\left(\tau \mid k\right) &= F\left(\tau \mid k\right) \hat{P}\left(\tau \mid k\right) + \hat{P}\left(\tau \mid k\right) F\left(\tau \mid k\right)^T, \quad \hat{P}\left(k\Delta t \mid k\right) = \hat{P}_{k|k} \\ \hat{P}_{k+1|k} &= \hat{P}\left((k+1)\Delta t \mid k\right) + Q \\ F\left(\tau \mid k\right) &= \left. \frac{\partial \mathbf{f}(x,u,t)}{\partial x} \right|_{x=\hat{x}(\tau|k),\ u=u(k\Delta t),\ t=\tau} \end{aligned} \tag{6.46}$$

This continuous-time linear approximation of the estimated state covariance is used in Eq. 6.42 to compute the Kalman filter gain. A discrete linear approximation to the covariance can also be used in the Kalman filter gain calculation. It is obtained from the time-varying filtering Riccati equation in Eq. 6.21 with the contribution from the output measurement removed.

$$\hat{P}_{k+1|k} = \Phi\left((k+1)\Delta t, k\Delta t\right) \hat{P}_{k|k} \Phi^T\left((k+1)\Delta t, k\Delta t\right) + Q \qquad (6.47)$$

$$\frac{d\Phi\left(t, k\Delta t\right)}{dt} = F\left(t \mid k\right) \Phi\left(t, k\Delta t\right), \qquad \Phi\left(k\Delta t, k\Delta t\right) = I \qquad (6.48)$$

In this expression, $\Phi\left((k+1)\Delta t, k\Delta t\right)$ is the transition matrix of the time-varying linear system obtained by linearization of the nonlinear system. This discrete covariance approximation does not require the solution of a differential equation to determine the covariance, but the state transition matrix must be computed. A further approximation can be made by linearizing about the state $\hat{x}_{k|k}$ and assuming a time-invariant linear system for the transition matrix calculation in Eq. 6.48. Wishner *et al.* [64] compared this method of discrete covariance propagation to the continuous-time method and found agreement within 0.1 percent for the example presented in the reference.

The extended Kalman filter for the discrete nonlinear system in Eq. 6.35 is similar to that presented for continuous-time systems with discrete measurements.

$$\hat{x}_{k|k} = \hat{x}_{k|k-1} + L_k \left(y_k - \bar{\mathbf{g}}(\hat{x}_{k|k-1}, k)\right) \qquad (6.49)$$

$$L_k = \hat{P}_{k|k-1} \bar{G}_k^T \left(\bar{G}_k \hat{P}_{k|k-1} \bar{G}_k^T + R\right)^{-1} \qquad (6.50)$$

$$\hat{P}_{k|k} = \left(I - L_k \bar{G}_k\right) \hat{P}_{k|k-1} \qquad (6.51)$$

$$\hat{x}_{k+1|k} = \bar{\mathbf{f}}(\hat{x}_{k|k}, u_k, k) \qquad (6.52)$$

$$\hat{P}_{k+1|k} = \bar{F}_k \hat{P}_{k|k} \bar{F}_k^T + Q \qquad (6.53)$$

The discrete linear system matrices are formed from the partial derivatives of the discrete nonlinear measurement and system functions.

$$\bar{G}_k = \left. \frac{\partial \bar{\mathbf{g}}(x, k)}{\partial x} \right|_{x = \hat{x}_{k|k-1}} \qquad (6.54)$$

$$\bar{F}_k = \left. \frac{\partial \bar{\mathbf{f}}(x, u, k)}{\partial x} \right|_{x = \hat{x}_{k|k},\ u = u_k} \qquad (6.55)$$

The functions $\bar{\mathbf{f}}$ and $\bar{\mathbf{g}}$ are also restricted to continuous functions in x so that the partial derivatives in Eqs. 6.54 and 6.55 exist.

A first order extended Kalman filter was implemented on an experimental batch reactor to estimate heat generation by DeVallière and Bonvin [16] and on an experimental batch styrene reactor to estimate conversion and chain length distribution by Schuler and Suzhen [55]. Kozub and MacGregor [31] implemented an extended Kalman filter on an semi-batch polymerization reactor to estimate monomer concentration and a reactive impurity disturbance. Dimitratos *et al.* [19] estimated the monomer concentration in a copolymerization reactor. Myers and Luecke [48] discuss an efficient numerical integration technique for the system and covariance equations and used this technique to estimate the specific growth rate in a fed-batch fermenter.

Linearized Extended Kalman Filter

The linearized extended Kalman filter determines the Kalman filter gains from a nominal state and input trajectory [21]. The nominal state, $\tilde{x}(t)$, and input, $\tilde{u}(t)$, are specified *a priori* based on an assumed behavior of the system. The Kalman filter gain for continuous-time systems with discrete measurements is then computed as shown in Eqs. 6.43 through 6.47 in which the measurement and system functions are linearized about the nominal state and input trajectories.

$$G_k = \left.\frac{\partial \mathbf{g}(x,t)}{\partial x}\right|_{x=\tilde{x}(k\Delta t),\ t=k\Delta t} \tag{6.56}$$

$$F\left(\tau \mid k\right) = \left.\frac{\partial \mathbf{f}(x,u,t)}{\partial x}\right|_{x=\tilde{x}(\tau),\ u=\tilde{u}(k\Delta t),\ t=\tau} \tag{6.57}$$

The Kalman filter gain for discrete systems is computed as shown in Eqs. 6.50 through 6.53 in which the discrete measurement and system functions are linearized about the nominal trajectories.

$$\bar{G}_k = \left.\frac{\partial \bar{\mathbf{g}}(x,k)}{\partial x}\right|_{x=\tilde{x}_k} \tag{6.58}$$

$$\bar{F}_k = \left.\frac{\partial \bar{\mathbf{f}}(x,u,k)}{\partial x}\right|_{x=\tilde{x}_k,\ u=\tilde{u}_k} \tag{6.59}$$

The advantage to this approach is that the filter gains can be computed off-line, resulting in a considerable reduction in the real-time computational requirements of the filter. The disadvantage is the lack of feedback from the process in the determination of the filter gain. Since the estimated state is normally a better representation of the true state than the nominal state, the linearized filter is less accurate. Systems that have large process disturbances or are controlled between multiple operating points generally are not well suited to this approach. This filter is applicable for systems with a well-defined state and input trajectory. It is normally used when the improvement in the performance of the first order filter does not justify the real-time computations necessary to implement it.

Because of the approximations made in the propagation of the covariance matrix in extended Kalman filtering, this covariance matrix can become a poor estimate in some applications. In these cases, updating the filter gain at each sampling period may result in computational effort that does not improve the performance of the filter. Safonov and Athans [54] present a linearized extended Kalman filter formulation that uses a constant filter gain and present sufficient conditions for convergence of this filter. This approach consists of a steady-state Kalman filter applied to the nonlinear process. The nominal state trajectory is composed of a single operating point. The robustness of the Kalman filter to model mismatch is sufficient to obtain acceptable performance with this technique in some applications.

Iterative Extended Kalman Filter

In the first order extended Kalman filter approaches already discussed, the nonlinear system equations are linearized about some estimate of the state. If this estimate is not close to the true state, the linearized model may be a poor approximation to the nonlinear system. Iterative schemes have been developed in which a linearized model is determined from an updated state estimate at each iteration. These schemes attempt to reduce the estimation error by improving the approximation to the nonlinear system that is used in the determination of the filtered state.

One iterative scheme that can be employed is to repeat the calculation of $\hat{x}_{k|k}$ in Eq. 6.42 [25]. This method is useful when compensating for nonlinearity in the measurement function. Letting $\hat{x}_{k|k,i}$ represent the ith iterate of the filtered state, the next iteration is determined as follows.

$$\begin{aligned} \hat{x}_{k|k,i+1} &= \hat{x}_{k|k-1} + L_{k,i}\left(y_k - \mathbf{g}(\hat{x}_{k|k,i}, k\Delta t) - G_{k,i}(\hat{x}_{k|k-1} - \hat{x}_{k|k,i})\right) \quad (6.60) \\ \hat{x}_{k|k,0} &= \hat{x}_{k|k-1} \end{aligned}$$

The filter gain and linearized measurement function are recomputed based on the current iterate of the filtered state.

$$L_{k,i} = \hat{P}_{k|k-1}G_{k,i}^T\left(G_{k,i}\hat{P}_{k|k-1}G_{k,i}^T + R\right)^{-1} \tag{6.61}$$

$$G_{k,i} = \left.\frac{\partial \mathbf{g}(x,t)}{\partial x}\right|_{x=\hat{x}_{k|k,i},\ t=k\Delta t} \tag{6.62}$$

Eqs. 6.60 through 6.62 are repeated until there is no significant difference between the iterated filtered states. The estimated state covariance is then updated in the same manner as Eq. 6.44 using the converged gain and linearized measurement function.

$$\hat{P}_{k|k} = \left(I - L_{k,i}G_{k,i}\right)\hat{P}_{k|k-1} \tag{6.63}$$

The state estimate and covariance are propagated between sampling times as shown in Eqs. 6.45 and 6.46 or Eq. 6.47 for continuous-time systems. For discrete systems, Eqs. 6.52 and 6.53 are used. Note that a single iteration results in the first order extended Kalman filter.

Nonlinearity in the system function is not addressed by the preceding method. A second iterative method that takes system function nonlinearity into account can be developed by also updating the estimated state $\hat{x}_{k|k-1}$ in the iteration scheme [64]. The filtered state is iterated in the same manner as the first method in which the filter gain, $L_{k,i}$, and linearized measurement function, $G_{k,i}$, are computed as shown in Eqs. 6.61 and 6.62.

$$\hat{x}_{k|k,i+1} = \hat{z}_{k|k-1,i} + L_{k,i}\left(y_k - \mathbf{g}(\hat{x}_{k|k,i}, k\Delta t) - G_{k,i}(\hat{z}_{k|k-1,i} - \hat{x}_{k|k,i})\right) \tag{6.64}$$

$$\hat{x}_{k|k,0} = \hat{x}_{k|k-1}$$

The current value of the estimated state, $\hat{z}_{k|k-1,i}$, used in the filtered state calculation is computed using the following procedure.

$$\hat{z}_{k|k-1,i} = \hat{z}_{k|k,i} + \Phi\left(k\Delta t, (k-1)\Delta t, i\right)\left(\hat{x}_{k-1|k-1} - \hat{z}_{k-1|k,i}\right) \tag{6.65}$$

$$\hat{z}_{k|k-1,0} = \hat{x}_{k|k-1}$$

The estimated state update requires a smoothed estimate at sample time $k-1$ and a filtered state at sample time k. The smoothed state estimate, $\hat{z}_{k-1|k,i}$, is computed iteratively as follows.

$$\hat{z}_{k-1|k,i+1} = \hat{x}_{k|k} + S_{k,i}\left(\hat{x}_{k|k,i+1} - \hat{z}_{k|k-1,i}\right) \tag{6.66}$$

$$\hat{z}_{k-1|k,0} = \hat{x}_{k-1|k-1}$$

See Jazwinski [25] or Gelb [21] for a review of the nonlinear smoothing techniques used to generate this expression. The filtered state estimate $\hat{z}_{k|k,i}$ used in the estimated state calculation in Eq. 6.65 is determined by propagating the current smoothed estimate forward one sample period using the system equation.

$$\begin{aligned} \dot{\hat{z}}(\tau \mid k,i) &= \mathbf{f}(\hat{z}(\tau \mid k,i), u(\tau), \tau), \qquad \hat{z}((k-1)\Delta t \mid k,i) = \hat{z}_{k-1|k,i} \quad (6.67) \\ \hat{z}_{k|k,i} &= \hat{z}(k\Delta t \mid k,i), \qquad \hat{z}_{k|k,0} = \hat{x}_{k|k-1} \end{aligned}$$

The transition matrix used in the estimated state update in Eq. 6.65, $\Phi(k\Delta t, (k-1)\Delta t, i)$, is determined by linearizing about the current smoothed state estimate $\hat{z}_{k-1|k,i}$.

$$\begin{aligned} \frac{d\Phi(t,(k-1)\Delta t, i)}{dt} &= F(t \mid k,i)\,\Phi(t,(k-1)\Delta t,i), \qquad (6.68) \\ \Phi((k-1)\Delta t,(k-1)\Delta t,i) &= I \\ F(\tau \mid k,i) &= \left.\frac{\partial \mathbf{f}(x,u,t)}{\partial x}\right|_{x=\hat{z}_{k-1|k,i},\ u=u((k-1)\Delta t),\ t=\tau} \end{aligned}$$

The gain matrix used in the smoothed state iteration, $S_{k,i}$, is computed from this transition matrix as follows.

$$S_{k,i} = \hat{P}_{k-1|k-1}\Phi^T(k\Delta t,(k-1)\Delta t,i)\,\hat{P}^{-1}_{k|k-1,i} \qquad (6.69)$$

$$\hat{P}_{k|k-1,i} = \Phi(k\Delta t,(k-1)\Delta t,i)\,\hat{P}_{k-1|k-1}\Phi^T(k\Delta t,(k-1)\Delta t,i) + Q \qquad (6.70)$$

In this iterative method, Eqs. 6.64 through 6.70 are repeated until there is no significant difference between the iterated filtered states. The estimated state covariance is then updated as shown in Eq. 6.63. The state estimate and covariance are propagated between sampling times as in the previous iterative method. Wishner *et al.* [64] state that 95 percent of filter performance is achieved in two iterations for the example presented in the reference. A single iteration of this method results in the first order extended Kalman filter.

This iterative method uses a smoothed estimate from the previous sample time to determine the iterated state estimates. A *globally iterated* extended Kalman filter can be obtained by determining a smoothed estimate of the initial state, $\hat{z}_{0|k}$, and then reprocessing all of the output measurements starting from this smoothed initial state estimate [25]. Since the *a priori* estimate of the initial state is generally no better than an educated guess, this procedure can significantly improve the performance of the filter. However, the increased computational burden may become substantial. This technique was implemented on a semi-batch polymerization reactor by Kozub and MacGregor [31]. It was found to improve the convergence of the estimated states to the true state by providing a better estimate of the initial state of the reactor.

These iterative approaches are used to compensate for the errors introduced by linearizing a nonlinear system. However, multiple iterations of the filtered state calculation require additional computation time which must be taken into account when using these methods. In addition, there is no guarantee that the iterations will converge since these methods are approximations generated by successive linearization.

Second Order Extended Kalman Filter

If we restrict the system and measurement functions to twice differentiable functions in x such that $\mathbf{f}, \mathbf{g} \in \mathcal{C}^2$, the second order Taylor series expansion for the continuous-time nonlinear system in Eq. 6.34 is as follows.

$$\begin{aligned} \dot{x} &= \mathbf{f}(x^*, u, t) + F(x - x^*) + \frac{1}{2}\sum_{j=1}^{n} e^j (x - x^*)^T \mathcal{F}^j (x - x^*) \quad (6.71) \\ y &= \mathbf{g}(x^*, t) + G(x - x^*) + \frac{1}{2}\sum_{j=1}^{m} e^j (x - x^*)^T \mathcal{G}^j (x - x^*) \\ e^j &= [0, \ldots, 0, \underbrace{1}_{j\text{th location}}, 0, \ldots, 0]^T \end{aligned}$$

In this expansion, $\mathcal{F}^j$ and $\mathcal{G}^j$ are the second order Hessian matrices of the jth component of the system and measurement functions and e^j are the standard unit basis vectors. The use of this expansion in the approximate filter design results in the second order extended Kalman filter [3].

The filtered state is determined using the discrete output measurements in the same manner as the first order filter in Eq. 6.42 with the addition of a bias correction term Π.

$$\hat{x}_{k|k} = \hat{x}_{k|k-1} + L_k \left(y_k - \mathbf{g}(\hat{x}_{k|k-1}, k\Delta t) - \Pi_k \right) \quad (6.72)$$

The bias correction term is calculated as follows.

$$\Pi_k = \frac{1}{2}\sum_{j=1}^{m} e^j \operatorname{tr}\left(\mathcal{G}_k^j \hat{P}_{k|k-1}\right) \tag{6.73}$$

The Kalman filter gain, L_k, is computed as shown in Eq. 6.43 for the first order filter with an additional correction term Γ.

$$L_k = \hat{P}_{k|k-1}G_k^T\left(G_k\hat{P}_{k|k-1}G_k^T + R + \Gamma_k\right)^{-1} \tag{6.74}$$

$$\Gamma_k = \frac{1}{2}\sum_{i=1}^{m}\sum_{j=1}^{m} e^i e^{jT} \operatorname{tr}\left(\mathcal{G}_k^i \hat{P}_{k|k-1}\mathcal{G}_k^j\hat{P}_{k|k-1}\right) \tag{6.75}$$

$$G_k = \left.\frac{\partial \mathbf{g}(x,t)}{\partial x}\right|_{x=\hat{x}_{k|k-1},\ t=k\Delta t} \qquad \mathcal{G}_k^j = \left.\frac{\partial^2 \mathbf{g}^j(x,t)}{\partial x^2}\right|_{x=\hat{x}_{k|k-1},\ t=k\Delta t}$$

The covariance of the filtered state estimate, $\hat{P}_{k|k}$, is calculated from $\hat{P}_{k|k-1}$ in the same manner as the first order filter using Eq. 6.44.

Due to the second order terms, the state and covariance are coupled between sampling times by a second bias correction term, μ, and are propagated using the following nonlinear system of differential equations.

$$\dot{\hat{x}}(\tau \mid k) = \mathbf{f}(\hat{x}(\tau \mid k), u(\tau), \tau) + \mu(\tau \mid k), \qquad \hat{x}(k\Delta t \mid k) = \hat{x}_{k|k} \tag{6.76}$$

$$\dot{\hat{P}}(\tau \mid k) = F(\tau \mid k)\hat{P}(\tau \mid k) + \hat{P}(\tau \mid k)F(\tau \mid k)^T, \quad \hat{P}(k\Delta t \mid k) = \hat{P}_{k|k} \tag{6.77}$$

$$\mu(\tau \mid k) = \frac{1}{2}\sum_{j=1}^{n} e^j \operatorname{tr}\left(\mathcal{F}^j(\tau \mid k)\hat{P}(\tau \mid k)\right) \tag{6.78}$$

$$F(\tau \mid k) = \left.\frac{\partial \mathbf{f}(x,u,t)}{\partial x}\right|_{x=\hat{x}(\tau|k),\ u=u(k\Delta t),\ t=\tau}$$

$$\mathcal{F}^j(\tau \mid k) = \left.\frac{\partial^2 \mathbf{f}^j(x,u,t)}{\partial x^2}\right|_{x=\hat{x}(\tau|k),\ u=u(k\Delta t),\ t=\tau}$$

The covariance of the state estimate at the next sample time is determined as follows.

$$\hat{P}_{k+1|k} = \hat{P}((k+1)\Delta t \mid k) + Q \tag{6.79}$$

If Π_k, Γ_k, and $\mu(\tau \mid k)$ are removed, the first order filter is recovered. When $\mathbf{g}$ is a linear function, the only difference between the second and first order filters is the term $\mu(\tau \mid k)$ that couples the covariance and estimated state differential equations. Athans *et al.* [3] state that the major improvement over the first order filter is due to this term for the example presented in the reference. A detailed derivation of this filter is presented in [3].

6.5.2 Statistical Approximation

In the statistical approximation approach, the system and measurement functions are represented by a truncated polynomial series expansion. The coefficients of these expansions are determined by minimizing the error between the function and the truncated polynomial based on the statistics of the state. Consider a scalar system in which the system function error is represented as follows.

$$e = \mathbf{f}(x, u, t) - \left(a_0(u,t) + a_1(u,t)x + a_2(u,t)x^2 + \ldots + a_j(u,t)x^j\right)$$

This approach seeks coefficients of the polynomial expansion that minimize the mean square error between the system function and polynomial, $E\left[e^2\right]$, in which E is the expectation operator. The coefficients are determined by taking the partial derivative of the mean square error with respect to each coefficient, equating these derivatives to zero, and interchanging the derivative and expectation operations. The result is a linear system of equations in the coefficients containing moments and cross-moments of the state and system function. The conditional probability density of the state is required to compute these moments which are generated by the expectation operation. Since the exact probability density is normally not available, an approximation must be used.

If the polynomial expansion is truncated after the linear term, the approach is referred to as *statistical linearization* [21] or *quasilinearization* [58]. The coefficients for the statistically linearized expansion of the system function, $\tilde{\mathbf{f}}(x, u, t) = f + Fx$, are determined as follows.

$$f = E\left[\mathbf{f}(x,u,t)\right], \qquad F = \frac{E\left[\mathbf{f}(x,u,t)x^T\right] - E\left[\mathbf{f}(x,u,t)\right]E[x]^T}{E\left[(x - E[x])^2\right]}$$

An approximation for the measurement function is developed in the same manner as the system function approximation. These expressions are referred to as *describing functions* for the nonlinear system.

For continuous-time systems with discrete measurements, the filter equations using the statistically linearized filter are similar to those of the extended Kalman filter. The filtered state is determined recursively from the current output measurement as shown in Eq. 6.42. The filter gain is computed as shown in Eq. 6.43 and the covariance of the filtered state estimate is updated at each sample time as shown in Eq. 6.44 using the statistically linearized measurement function $\tilde{\mathbf{g}}_k$.

$$
\begin{aligned}
L_k &= \hat{P}_{k|k-1}\tilde{\mathbf{g}}_k^T \left(\tilde{\mathbf{g}}_k \hat{P}_{k|k-1}\tilde{\mathbf{g}}_k^T + R\right)^{-1} && (6.80) \\
\hat{P}_{k|k} &= \left(I - L_k\tilde{\mathbf{g}}_k\right)\hat{P}_{k|k-1} && (6.81)
\end{aligned}
$$

Between sampling times, the state estimate is propagated using the nonlinear system function as shown in Eq. 6.45. The covariance of the state estimate is propagated as shown in Eq. 6.46 using the statistically linearized system function $\tilde{\mathbf{f}}_k$.

$$
\begin{aligned}
\dot{\hat{P}}(\tau \mid k) &= \tilde{\mathbf{f}}_k \hat{P}(\tau \mid k) + \hat{P}(\tau \mid k)\tilde{\mathbf{f}}_k^T, \qquad \hat{P}(k\Delta t \mid k) = \hat{P}_{k|k} && (6.82) \\
\hat{P}_{k+1|k} &= \hat{P}((k+1)\Delta t \mid k) + Q
\end{aligned}
$$

The statistically linearized measurement and system functions used in this approach are determined at each sampling time as follows.

$$
\tilde{\mathbf{g}}_k = \left(E\left[\mathbf{g}(\hat{x}_{k|k-1}, k\Delta t)\hat{x}_{k|k-1}^T\right] - E\left[\mathbf{g}(\hat{x}_{k|k-1}, k)\right] E\left[\hat{x}_{k|k-1}^T\right]\right)\hat{P}_{k|k-1}^{-1} \quad (6.83)
$$

$$
\tilde{\mathbf{f}}_k = \left(E\left[\mathbf{f}(\hat{x}_{k|k}, u_k, k\Delta t)\hat{x}_{k|k}^T\right] - E\left[\mathbf{f}(\hat{x}_{k|k}, u_k, k\Delta t)\right] E\left[\hat{x}_{k|k}^T\right]\right)\hat{P}_{k|k}^{-1} \quad (6.84)
$$

The expectation operations required to determine these statistically linearized functions are computed using an assumed probability density function $p(x)$ in which e^j and e^i are the standard unit vectors.

$$
\begin{aligned}
E\left[f(x,u,t)\right] &= \sum_{j=1}^{n} e^j \int_{-\infty}^{\infty} \cdots \int_{-\infty}^{\infty} f^j(x,u,t)p(x)\, dx^1 \ldots dx^n \\
E\left[f(x,u,t)x^T\right] &= \sum_{j=1}^{n}\sum_{i=1}^{n} e^j e^{i^T} \int_{-\infty}^{\infty} \cdots \int_{-\infty}^{\infty} f^j(x,u,t)x^i p(x)\, dx^1 \ldots dx^n
\end{aligned}
$$

A Gaussian distribution function typically is assumed in this approach. Significant computational effort may be required to perform these integrations. These computational aspects are discussed in Gelb [21] and Stengel [58].

Higher order polynomial expansions of the system and measurement functions can be used in the development of an approximate nonlinear estimator. For example, Mahalanabis and Farooq [41] present a second order formulation for scalar systems. The higher order expansions are easily generated and can improve the performance of the estimator. However, the corresponding filter equations require the determination of higher order moments and cross-moments of the state and model functions. Determination of these higher order moments can greatly increase the computational requirements of the filter.

6.5.3 Summary of Nonlinear Recursive Filters

Extended Kalman filtering is typically the first nonlinear state estimation technique that would be attempted in most situations. It has been found to work well for a number of chemical process control applications and is relatively easy to implement. However, the performance of extended Kalman filtering is directly related to the quality of the linear approximations made in the state and covariance estimates. If the state disturbances are large, the initial state estimate is poor, or the system is highly nonlinear, the linearized equations may not be a good approximation to the nonlinear stochastic process. In these cases, the extended Kalman filter estimates may not be close to the optimal nonlinear estimates and may not converge to the true state. Extended Kalman filtering can produce either biased or divergent estimates under these conditions. It can also be difficult to tune the extended Kalman filter to achieve acceptable performance. Poor choices of the covariance matrices Q, Q_0, and R can also lead to biased or divergent state estimates. Agarwal and Bonvin [1] discuss this issue in the context of state estimation for a batch reactor. Dimitratos *et al.* [19] discuss adaptive schemes that estimate the process noise covariance along with the state to improve the filter tuning. Kozub and MacGregor [31] discuss the importance of including integrating disturbance states in the model to prevent bias in the state estimates due to unmeasured disturbances and model error.

Deciding which extended Kalman filtering technique to implement is based on a trade-off between performance and complexity. The linearized formulation is the easiest to implement but may not work well for many applications. The second order and iterative approaches have been shown to produce improved estimates [64], but they are more difficult to implement and require more computational time. These approaches should normally not be considered as an initial choice unless the first order filter is known not to perform well for the given application or it is necessary to achieve higher performance. For some applications, none of the extended Kalman filtering approaches may perform well and another technique will be required.

Statistical approximation techniques have generally been found to be more accurate than extended Kalman filtering [21]. Another advantage of these techniques is that derivatives of the system and measurement functions are not required. Therefore, a much broader class of nonlinear functions can be considered since there are no continuity restrictions that were necessary for the extended Kalman filter. However, the computational requirements may be greater due to the calculation of the expectation operations. These requirements will depend on the system and measurement functions and the assumed probability distribution. Although it is convenient, the choice of a Gaussian distribution can be a gross approximation that may result in poor performance for some applications. Gelb [21] presents a detailed example of the implementation of this approach.

Example 6.3

We again consider the batch reactor presented in Example 6.1 and compare the performance of the linearized and first order extended Kalman filters in estimating the concentration of component A from the temperature measurements. In this example, perfect temperature measurements are available at a discrete sampling period of 30 seconds and there is no model error. The nominal initial state of the reactor is a component A concentration of 1M and a temperature of 20 °C. The filters are tuned to indicate that disturbances are more likely to occur in the composition due to possible polymerization side reactions not accounted for in the model. The covariance of the initial state is chosen to indicate that the initial temperature estimate is much more reliable than the estimate of the initial composition. The model parameters and filter covariances are listed in Table 6.1.

The steady-state Kalman filter gain used in the linearized filter is determined by linearization of the model equations about the nominal initial state. The continuous-time linear approximation of the estimated state covariance is used in the first order filter. We simulate three different initial temperatures using an estimated initial concentration of 1M in which the actual concentration is 0.9M. Figure 6.1 shows the excellent performance of both filters when the initial batch temperature is at the nominal initial condition of 20 °C. When the initial temperature is lowered to 10 °C, both filters demonstrate slower convergence to the true concentration. However, the degradation in performance of the linearized filter is considerably greater as shown in Figure 6.2. When the initial temperature is increased to 30 °C, there is little effect on the first order filter performance when compared to

Model Parameters		Filter Covariances	
$\frac{\Delta H}{\rho C}$	$-30.0 \ \frac{\text{liter-}\ ^\circ\text{C}}{\text{gmole}}$	Q	$\begin{bmatrix} 10 & 0 \\ 0 & 1 \end{bmatrix}$
$\frac{UA}{V\rho C}$	$1.00 \times 10^{-3}\ \text{sec}^{-1}$		
T_c	$20.0\ ^\circ\text{C}$	R	1
k_0	$1.00 \times 10^{8}\ \frac{\text{liter}}{\text{gmole-sec}}$	Q_0	$\begin{bmatrix} 100 & 0 \\ 0 & 1 \end{bmatrix}$
E_a/R	$7.50 \times 10^{3}\ ^\circ\text{C}$		

Table 6.1: Batch Reactor Model Parameters and Filter Covariances

20 °C. As shown in Figure 6.3, however, the linearized filter estimates oscillate around the true state and do not converge. This example demonstrates the good performance of the first order filter and the limited applicability of the linearized filter over this 20 °C initial temperature range. It also suggests that an iterated or higher order extended Kalman filter may not be necessary for this application.

6.6 Nonlinear Observers

The nonlinear recursive filter techniques presented in the previous section are intended to produce an approximate optimal estimate of the state. Similar to the linear observer discussed in Section 6.3.4, nonlinear observers consider the deterministic dynamic behavior of the reconstruction error. In this section, we briefly discuss observer error linearization and extended Luenberger-type observers. Additional approaches are mentioned in the summary.

6.6.1 Observer Error Linearization

Analogous to feedback linearization for nonlinear control presented in Chapter 5, observer error linearization attempts to find a local coordinate transformation such that the corresponding transformed system is linear in the reconstruction error dynamics. An estimate of the state of the transformed system is then obtained using a linear observer. The advantage to this ap-

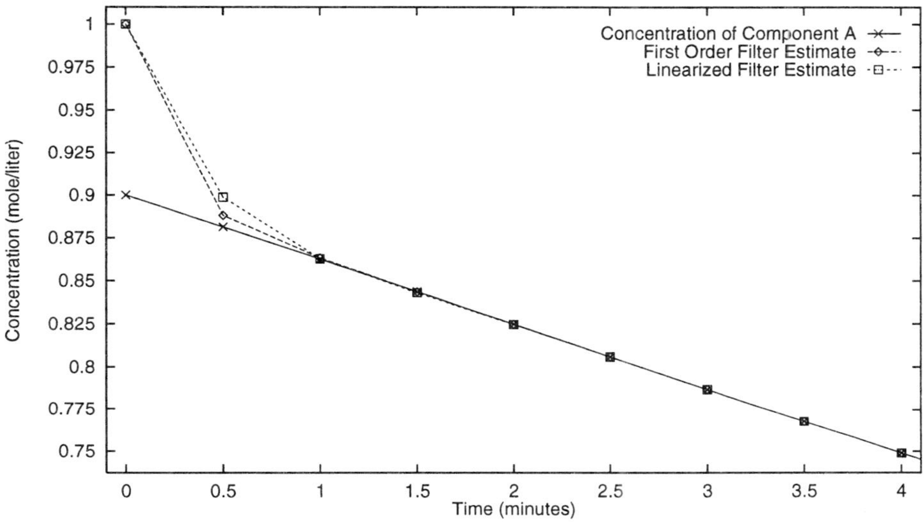

Figure 6.1: First order and linearized extended Kalman filter estimated concentrations at the nominal initial temperature of 20 °C.

proach is that the linearized model is an exact representation of the nonlinear system within a neighborhood of a given nominal state. Therefore, this technique does not introduce the linearization errors present in the recursive filter approaches. The disadvantage is that a coordinate transformation must exist which imposes a restriction on the nonlinear systems that can be considered. This technique was independently introduced using scalar systems by Bestle and Zeitz [6] and Krener and Isidori [32]. It was extended to multivariable systems by Krener and Respondek [33] and Xia and Gao [66].

Consider a nonlinear system described by Eq. 6.34 that is the result of a nonlinear change of coordinates, $x(t) = \mathcal{F}(z(t))$, on the reference system

$$\begin{aligned} \dot{z}(t) &= Az(t) + \gamma(u(t), y(t)) = h(z(t), u(t)) \\ y(t) &= Cz(t) \end{aligned} \tag{6.85}$$

in which the matrices A and C are in *Brunovsky canonical form*. This canonical form results in the following relation for each output y_i in which k_i the corresponding observability index. The observability index is discussed in Section 6.4.2.

$$\dot{z}_2^i = z_1^i, \quad \dot{z}_3^i = z_2^i, \quad \ldots, \quad \dot{z}_{k_i}^i = z_{k_i-1}^i; \quad y_i = z_{k_i}^i \tag{6.86}$$

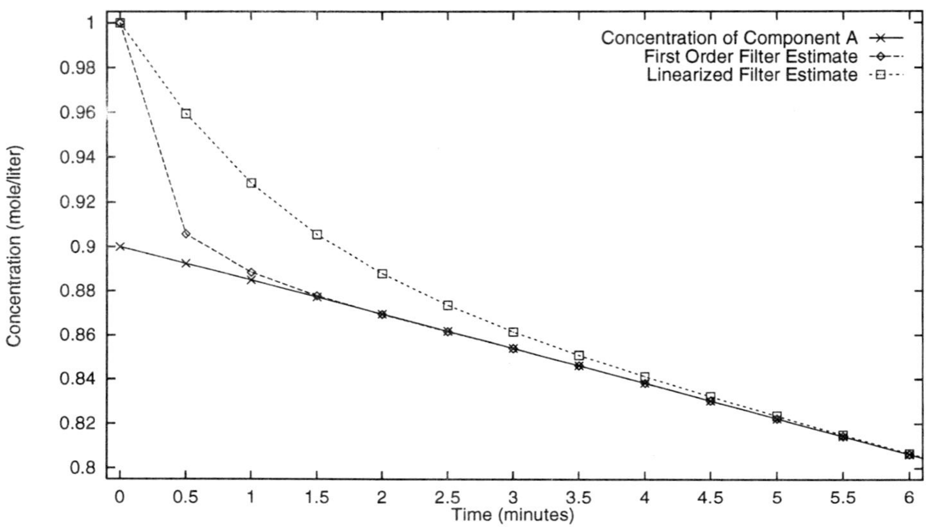

Figure 6.2: First order and linearized extended Kalman filter estimated concentrations at an initial temperature of 10 °C.

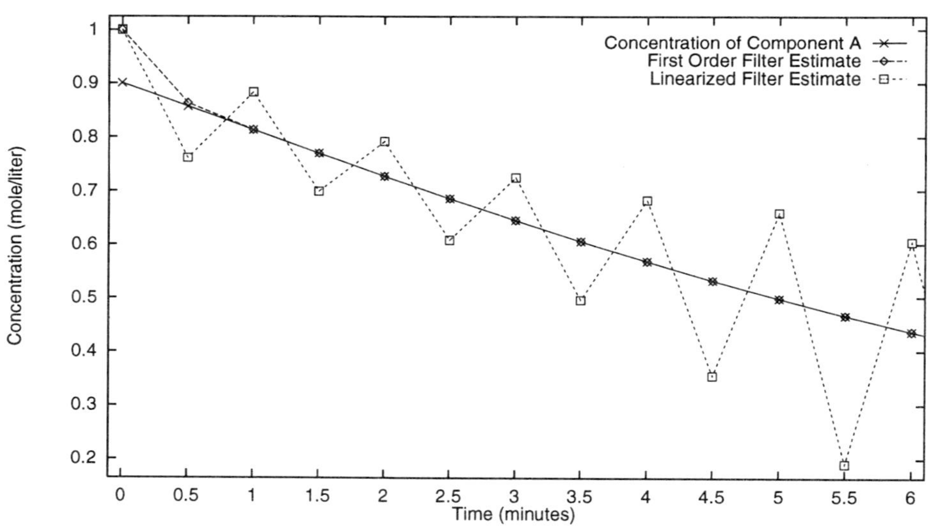

Figure 6.3: First order and linearized extended Kalman filter estimated concentrations at an initial temperature of 30 °C.

The reference system, which is in *observer form*, is linear in the transformed state z with the addition of a nonlinear function of the input and output. An estimate of the state of this system can be obtained using a linear observer as discussed in Section 6.3.4.

$$\dot{\hat{z}}(t) = A\hat{z}(t) + L(y(t) - C\hat{z}(t)) + \gamma(u, y) \tag{6.87}$$

Note that the reconstruction error of the reference system, $z - \hat{z}$, satisfies the linear differential equation in Eq. 6.29. The state estimate for the original nonlinear system is obtained using the nonlinear coordinate transformation.

$$\hat{x}(t) = \mathcal{F}(\hat{z}(t)) \tag{6.88}$$

Let us consider the autonomous system. If a coordinate transformation $\mathcal{F}$ that transforms Eq. 6.85 into Eq. 6.34 exists, the following expressions must hold in which C_j is the $\sum_{i=1}^{j} k_i$ row of C.

$$f(x) = \frac{\partial \mathcal{F}}{\partial z} h(z) \tag{6.89}$$

$$h_j(\mathcal{F}(z)) = C_j z, \quad j = 1, \ldots, p \tag{6.90}$$

In this method, the observer problem consists of determining whether a solution to these equations exists and, if so, finding this solution. Performing the indicated differentiation and rearranging Eq. 6.89 results in a system of differential equations for $i = 1, \ldots, p$ that must be solved.

$$\frac{\partial \mathcal{F}}{\partial z_{j+1}^i} = \frac{\partial f}{\partial x}\frac{\partial \mathcal{F}}{\partial z_j^i} - \frac{\partial}{\partial x}\left(\frac{\partial \mathcal{F}}{\partial z_j^i}\right) f, \qquad j = 1, \ldots, k_i - 1 \tag{6.91}$$

$$\frac{\partial \mathcal{F}}{\partial z}\frac{\partial \gamma}{\partial z_{k_i}^i} = \frac{\partial f}{\partial x}\frac{\partial \mathcal{F}}{\partial z_{k_i}^i} - \frac{\partial}{\partial x}\left(\frac{\partial \mathcal{F}}{\partial z_{k_i}^i}\right) f \tag{6.92}$$

Eq. 6.90 results in a system of linear algebraic equations for $\partial \mathcal{F}/\partial z_1^i$ with $i = 1, \ldots, p$.

$$L_f^{j-1}(dg_k)\frac{\partial \mathcal{F}}{\partial z_1^i} = \delta_{i,k}\delta_{k_i,j} \qquad j = 1, \ldots, k_i;\ k = 1, \ldots, p \tag{6.93}$$

The original nonlinear system in Eq. 6.34 can be expressed in observer form if and only if there is a $\mathcal{F}$ and γ that satisfy Eqs. 6.91 through 6.93. Since the nonlinear system must be observable for a transformation to exist, the relationship in Eq. 6.93 can be expressed using the observability matrix in Eq. 6.38 as follows.

$$\left[\frac{\partial \mathcal{F}}{\partial z_1^1}\ \frac{\partial \mathcal{F}}{\partial z_1^2} \cdots \frac{\partial \mathcal{F}}{\partial z_1^p}\right] = \mathcal{O}^{-1}C \tag{6.94}$$

The necessary and sufficient conditions for the existence of a linearizing transformation along with a detailed computational procedure to determine the transformation is provided in Xia and Gao [66]. For systems with inputs, it can be shown that the conditions are similar to those presented for autonomous systems and must hold for all admissible inputs [66]. In practice, analytical solutions for $\mathcal{F}$ and γ are not obtainable and numerical solutions are used. The formulation presented here considers a continuous output measurement for the estimated state calculation in Eq. 6.87. A corresponding observer for autonomous discrete-time nonlinear systems is discussed in Lee and Nam [35] and Chung and Grizzle [11]. The existence of a continuous-time linearizing transformation for a given nonlinear system does not imply the existence of a discrete linearizing transformation [11]. An excellent presentation of the application of this technique to a continuous-time stirred tank reactor is provided in Kantor [27].

6.6.2 Extended Luenberger Observer

Application of observer error linearization involves the determination of a coordinate transformation. This transformation exists only for a limited class of nonlinear systems and requires the solution of a system of partial differential equations. The observer in Birk and Zeitz [7] does not require the integration of these equations. This formulation is referred to as an *extended Luenberger observer* because the nonlinear observer gain is determined by linearizing these partial differential equations. Consider the following nonlinear observer.

$$\dot{\hat{x}}(t) = \mathbf{f}(\hat{x}, u, t) + L(\hat{x}, u)\,(y(t) - \mathbf{g}(\hat{x}, t)) \tag{6.95}$$

In this approach, an extended linearization of the partial differential equations in Eqs. 6.91 and 6.92 about the estimated state is used to obtain the nonlinear observer gain $L(\hat{x}, u)$. After specifying the characteristic equation or observer poles for the transformed system in Eq. 6.86, the observer gain in Eq. 6.95 becomes

$$L(\hat{x}, u) = \Big[(l_{11} + l_{12}L_f + \ldots + l_{1k_1}L_f^{k_1-1} + L_f^{k_1})\mathbf{s}_1, \ldots, \tag{6.96}$$

$$(l_{p1} + l_{p2} L_f + \ldots + l_{pk_p} L_f^{k_p - 1} + L_f^{k_p}) \mathbf{s}_p \Big] \left[\frac{\partial \mathbf{g}(\hat{x})}{\partial \hat{x}} \right]^{-1}$$

in which l_{ij} are the observer poles for the linearized reconstruction error, k_j are the observability indices, and L_f^j is the Lie derivative operator. The vectors $\mathbf{s}_j$ are determined from the following relationship in which e^j are the standard unit basis vectors and $\partial g_j^* / \partial z_{k_j}^j \neq 0$ can be freely specified such that the calculation of the Lie derivatives in Eq. 6.96 is made as simple as possible.

$$\mathbf{s}_j = \frac{\partial g_j^*(z)}{\partial z_{k_j}^j} \mathcal{O}^{-1} e^j \tag{6.97}$$

A complete derivation of this observer can be found in Birk and Zeitz [7]. The approach is restricted to systems observable for every input with $\mathbf{f}$ and $\mathbf{g}$ smooth analytic functions.

Ciccarella *et al.* [13] present a discrete-time nonlinear Luenberger-type observer that considers single output systems. Let U_k^n represent the series of n past inputs at sample time k and $\Phi(x, U_k^{n-1})$ represent the series of n predicted outputs generated by the discrete nonlinear system in Eq. 6.35 from an initial state x and input sequence U_k^n. We define $\bar{\mathbf{F}}^j(x, U^j)$ as the jth composite of the system function $\bar{\mathbf{f}}$ with input sequence U^j and n is the dimension of the state vector.

$$U_k^n = \begin{bmatrix} u_{k-n+1} \\ u_{k-n+2} \\ \vdots \\ u_k \end{bmatrix}, \quad \Phi(x, U_k^{n-1}) = \begin{bmatrix} \bar{\mathbf{g}}(\bar{\mathbf{F}}^{n-1}(x, U_{k-1}^{n-1})) \\ \bar{\mathbf{g}}(\bar{\mathbf{F}}^{n-2}(x, U_{k-2}^{n-2})) \\ \vdots \\ \bar{\mathbf{g}}(x) \end{bmatrix}$$

The filtered state estimate is computed from the following discrete system in which L is the observer gain and $z_{k-n+j|k} = \bar{\mathbf{F}}^{j-1}(z_{k-n+1|k}, U_{k-n+j-1}^{j-1})$.

$$\begin{aligned} z_{k-n+2|k+1} &= z_{k-n+2|k} + \bar{\mathcal{O}}^{-1}(z_{k-n+1|k}, U_{k-1}^{n-1}) \\ &\quad \times \left[\bar{B} \left(y_{k+1} - \bar{\mathbf{g}}(z_{k+1|k}) \right) + L \left(y_{k-n+1} - \bar{\mathbf{g}}(z_{k-n+1|k}) \right) \right] \end{aligned} \tag{6.98}$$

$$\hat{x}_{k|k} = \bar{\mathbf{F}}^{n-1}(z_{k-n+1|k}, U_{k-1}^{n-1}) \tag{6.99}$$

In this system, $\bar{\mathcal{O}}$ is the discrete nonlinear observability matrix at sample time k computed from the partial derivative of $\Phi(x, U_k^{n-1})$.

$$\bar{\mathcal{O}}(x, U_k^{n-1}) = \frac{\partial \Phi(x, U_k^{n-1})}{\partial x} \tag{6.100}$$

The observer gain is selected such that the eigenvalues of the perturbed linear system for the reconstruction error in z is locally stable.

$$e_{k-n+2|k+1} = (\bar{A} - L\bar{C})e_{k-n+1|k} + O\left(\left\|e_{k-n+1|k}\right\|^2\right) \quad (6.101)$$

This linear system, which is in Brunovsky canonical form, is determined from a nonlinear transformation on z. The approach is restricted to systems observable for every input with $\bar{\mathbf{g}}$ and $\bar{\mathbf{F}}^n(\Phi^{-1}(x, U^{n-1}))$ uniformly Lipschitz continuous functions of the state. Convergence is guaranteed only for a neighborhood of initial state estimates around the true initial state. This neighborhood may be small. A detailed derivation of this observer is contained in [13].

6.6.3 Summary of Nonlinear Observers

In addition to the techniques discussed in this section, there have been a number of other nonlinear observer formulations proposed in the literature. A review of several of these observers for continuous-time systems can be found in Walcott *et al.* [63]. Sliding mode observers for continuous-time systems are discussed in Slotine *et al.* [56]. The conditions under which exponentially convergent nonlinear observers exist for continuous-time systems are presented in Xia and Gao [65]. A Lyapunov function based approach for continuous-time systems is discussed in Tsinias [60]. A Luenberger-type observer that considers single-output nonlinear systems of the form

$$\begin{aligned} \dot{x} &= \mathbf{f}(x) + \mathbf{h}(x)u \quad (6.102) \\ y &= \mathbf{g}(x) \end{aligned}$$

is presented in Gauthier *et al.* [20], Ciccarella *et al.* [12], and Deza *et al.* [18]. Deza *et al.* [18] also consider discrete output measurements. Implementation of this observer on a free radical polymerization CSTR is discussed in Van Dootingh *et al.* [61].

The advantage of nonlinear observers is that the calculation of the state estimate generally involves a gain multiplying the difference between the estimated and measured outputs. After the initial construction of the observer, this calculation can be performed quickly and relatively easily for most of these techniques. The disadvantage is the limitations on the systems that can be considered. Many of these approaches have further restrictions on the system and measurement functions in addition to the observability requirement. The continuous-time formulations are also not appropriate for applications in which the output measurements are available at discrete sampling periods. Another consideration with observer error linearization is the calculations required to construct the observer.

6.7 Nonlinear Moving Horizon Estimator

In this section, the moving horizon state estimator presented previously for linear systems is extended to nonlinear systems. The objective is a deterministic least squares estimator for the discrete nonlinear system in Eq. 6.35 with state and measurement disturbances. In this presentation, we assume that these disturbances are additive in which $v \in \Re^p$ is the measurement disturbance vector and $w \in \Re^n$ is the process or state disturbance vector.

$$\begin{aligned} x_{k+1} &= \bar{\mathbf{f}}(x_k, u_k, k) + w_k \\ y_k &= \bar{\mathbf{g}}(x_k, k) + v_k \end{aligned} \tag{6.103}$$

We also require uniform observability of the nonlinear system as discussed in Section 6.4.2.

6.7.1 Estimated State Disturbance Approach

Because it is not possible to obtain a general closed-form recursive solution to the nonlinear batch state estimation problem, it is difficult to develop a moving horizon estimator equivalent to the batch problem in the nonlinear case. Therefore, a moving horizon estimator similar to that for linear systems is usually constructed.

$$\min_{\{\hat{x}_{k-N|k}, \hat{w}_{k-N|k}, \ldots, \hat{w}_{k-1|k}\}} \sum_{j=k-N}^{k-1} \hat{w}_{j|k}^T Q^{-1} \hat{w}_{j|k} + \sum_{j=k-N}^{k} \hat{v}_{j|k}^T R^{-1} \hat{v}_{j|k} \tag{6.104}$$

$$\text{Subject to:} \quad \begin{aligned} \hat{x}_{j+1|k} &= \bar{\mathbf{f}}(\hat{x}_{j|k}, u_j, j) + \hat{w}_{j|k} \\ \hat{v}_{j|k} &= y_j - \bar{\mathbf{g}}(\hat{x}_{j|k}, j) \end{aligned} \tag{6.105}$$

This nonlinear moving horizon estimator does not penalize the initial state disturbance in the horizon allowing the initial state estimate in the horizon, $\hat{x}_{k-N|k}$, to be chosen freely. The advantage of this approach is that nominal convergence of the estimated state to the true state is achieved [46]. However, this choice ignores information from the output measurements prior to sample time $k-N$. Robertson *et al.* [53] suggest penalizing the initial state disturbance by the inverse of an approximate covariance matrix of the state estimate at time $k-N$. This approach attempts to take prior information into account, but does not have a convergence guarantee. As the horizon length is increased, these two approaches yield similar estimates.

We consider estimated process disturbance and state constraints in which $\bar{\mathbf{h}}$ is the estimated state constraint function.

$$h_{\min} \leq \quad \bar{\mathbf{h}}(\hat{x}_{j|k}) \quad \leq h_{\max}, \qquad j = k-N+1, \ldots, k \tag{6.106}$$

$$w_{\min} \leq \quad \hat{w}_{j|k} \quad \leq w_{\max}, \qquad j = k-N, \ldots, k-1 \tag{6.107}$$

The constraints are chosen to define a convex region in the $\hat{w}$ space containing the origin. For systems forced by an input, the state constraints are shifted by the nominal trajectory due to the input. With the first state estimate unconstrained, these constraints form a feasible set at each sampling time.

Since the process model is nonlinear, this estimator requires the solution to a general nonlinear optimization problem. The techniques required to solve this optimization problem are discussed in Section 6.7.4. The state estimate at sample time $n+j$, in which $n = k-N$, is computed from the optimal solution using the following recursion.

$$\begin{aligned} \hat{x}_{n+j+1|k} &= \bar{\mathbf{f}}(\hat{x}_{n+j|k}, u_{n+j}, n+j) + \hat{w}^*_{n+j} \qquad (6.108) \\ \hat{x}_{n|k} &= \hat{x}^*_{n|k} \end{aligned}$$

A moving horizon estimator for the continuous-time nonlinear system with discrete output measurements presented in Eq. 6.36 is constructed by assuming discrete state and measurement disturbances.

$$\begin{aligned} \dot{x}(\tau) &= \mathbf{f}(x(\tau), u(\tau), \tau), \qquad x(k\Delta t) = x_k \qquad (6.109) \\ x_{k+1} &= x((k+1)\Delta t) + w_k \qquad (6.110) \\ y_k &= \mathbf{g}(x_k, k) + v_k \end{aligned}$$

This model is used for the equality constraint of the moving horizon estimator in Eqs. 6.104 and 6.105. The state estimates are computed from the optimal solution using the recursion in Eqs. 6.109 and 6.110.

6.7.2 Initial State Estimate Approach

A simpler moving horizon estimator can be constructed by not considering estimated state disturbances in the optimization problem and estimating only the initial state in the horizon. This approach is well motivated if the state disturbances are negligible and the measurement is corrupted by zero mean noise [21]. The objective is to determine an initial state estimate that minimizes the difference between the measured and predicted outputs in a least squares sense.

$$\min_{\hat{x}_{k-N-1|k}} \quad \sum_{j=k-N}^{k} \hat{v}_{j|k}^T \hat{v}_{j|k} \tag{6.111}$$

$$\text{Subject to:} \quad \begin{aligned} \hat{x}_{j+1|k} &= \bar{\mathbf{f}}(\hat{x}_{j|k}, u_j, j) \\ \hat{v}_{j|k} &= y_j - \bar{\mathbf{g}}(\hat{x}_{j|k}, j) \end{aligned} \tag{6.112}$$

Nominal convergence of this approach for continuous-time systems is shown by Michalska and Mayne [44]. Nominal convergence for discrete systems follows in the same manner as the moving horizon estimator presented in the previous section. Continuous-time systems with discrete output measurements can also be considered with this approach. In this case, the state disturbance in Eq. 6.110 is removed and this model is used for the equality constraint in Eq. 6.112.

The implementation of this estimator on continuous-time chemical process systems with discrete output measurements is discussed in Jang *et al.* [24], Bequette [5], Kim *et al.* [30], Liebman *et al.* [37], and Ramamurthi *et al.* [51]. The implementation in Jang *et al.* [24], however, is not a moving horizon approach. The observer horizon is completely replaced with N new output measurements before an estimate is made. Kozub and McGregor [31] compare this approach with extended Kalman filtering techniques. The implementation in Ramamurthi *et al.* [51] linearizes the nonlinear model equations about a nominal trajectory within the observer horizon to estimate the initial state. The filtered state estimate is then obtained using the nonlinear model from this initial state. This approach reduces the computational requirements since a quadratic program is solved at each sample time. If the nominal trajectory is close to the estimated trajectory and the system is not highly nonlinear, this implementation can be a reasonable approximation.

The advantage to the initial state estimate approach is a smaller number of decision variables for a given horizon length. This reduction in decision variables results in less computational time to solve the optimization problem. In addition, the horizon length is the only tuning parameter. However, the assumption of negligible process disturbances is not realistic for many applications. Issues such as unmeasured or unmodeled disturbances, process modeling error, and variation in the model parameters cannot be addressed adequately when only the initial state in the horizon is estimated.

6.7.3 Input Observation Error

We have assumed that the input variables of the process are known exactly. However, there is uncertainty in these values in many applications. For example, a control valve may stick and not attain the position indicated by the signal sent to the valve. Errors in the observation of the input values can be taken into account by assuming that there are disturbances to the input. Kim *et al.* [29] and Tjoa and Biegler [59] refer to this technique as an *error-in-variables* method.

A moving horizon estimator for input observation error is obtained by solving a modification of the least squares problem in Eq. 6.104 in which $\mu \in \Re^m$ is the assumed input disturbance.

$$\min_{\{\hat{x}_{k-N|k}, \hat{w}_{j|k}, \hat{\mu}_{j|k}\}} \sum_{j=k-N}^{k-1} \left(\hat{w}_{j|k}^T Q^{-1} \hat{w}_{j|k} + \hat{\mu}_{j|k}^T S^{-1} \hat{\mu}_{j|k} \right) + \sum_{j=k-N}^{k} \hat{v}_{j|k}^T R^{-1} \hat{v}_{j|k} \tag{6.113}$$

$$\text{Subject to:} \quad \begin{aligned} \hat{x}_{j+1|k} &= \bar{\mathbf{f}}(\hat{x}_{j|k}, \hat{u}_{j|k}, j) + \hat{w}_{j|k} \\ \hat{u}_{j|k} &= u_j + \hat{\mu}_{j|k} \\ \hat{v}_{j|k} &= y_j - \bar{\mathbf{g}}(\hat{x}_{j|k}, j) \end{aligned} \tag{6.114}$$

A corresponding moving horizon estimator for continuous-time systems with discrete output measurements is constructed by using the model in Eq. 6.109 for the equality constraint in Eq. 6.114. Removing the estimated state disturbances, as in the initial state estimate approach, results in an estimator that assumes the process disturbances enter the system only through the input.

6.7.4 Solution Techniques

The nonlinear moving horizon estimator problem formulation requires the minimization of a least squares objective function subject to nonlinear differential and algebraic constraints. There are two basic solution strategies that can be employed. The first is referred to as *sequential solution* [30]. In this method, a nonlinear optimization technique is used to determine the estimated state disturbances that are the decision variables of the least squares problem. At each iteration of the optimizer, the objective function is computed by either integrating the nonlinear modeling equations for continuous-time systems or iterating for discrete systems. For gradient based optimizers used with continuous-time systems, the gradient of the objective function can be computed by integration of the sensitivity equations at the same time the model equations are solved [9]. The sequential solution has the advantage of being relatively straightforward to implement and having a small number of decision variables in the optimization problem. However, the computation time required by this method for continuous-time systems may be excessive since the model equations must be integrated at each iteration of the optimizer. This method also does not readily admit algebraic inequality constraints on the states of the system.

The second method is referred to as *simultaneous solution* [37]. In this method, the state variables at each sample time are included as decision variables in an optimization problem that is solved only once. For continuous-time systems, the model equations are discretized in time to convert the differential equation equality constraints into algebraic equality constraints. The discretization is normally performed using orthogonal collocation on finite elements. This strategy easily handles constraints on the states of the system at each sample time. The major drawback is the large number of decision variables and constraint functions that result in a very large order optimization problem that must be solved. The solution normally requires a large-scale, sparse nonlinear optimization technique [37]. A reduced space nonlinear optimization strategy in which the collocation equations are solved inexactly is presented in Logsdon and Biegler [39].

6.7.5 Moving Horizon Algebraic Estimator

Instead of using nonlinear optimization techniques to determine the initial state in the horizon as discussed in Section 6.7.2, Moraal and Grizzle [45] employ Newton's method to solve the system of nonlinear equations generated by the state estimation problem. Let Y_k^N represent the series of N past measured outputs, U_k^N the series of N past inputs, and $\Phi(x, U_k^N)$ the series of N predicted outputs generated by the discrete nonlinear system in

Eq. 6.35 from an initial state x and input sequence U_k^N.

$$Y_k^N = \begin{bmatrix} y_{i+1} \\ y_{i+2} \\ \vdots \\ y_k \end{bmatrix}, \quad U_k^N = \begin{bmatrix} u_{i+1} \\ u_{i+2} \\ \vdots \\ u_k \end{bmatrix}, \quad \Phi(x, U_k^N) = \begin{bmatrix} \bar{\mathbf{g}}(x) \\ \bar{\mathbf{g}}(\bar{\mathbf{F}}^1(x, U_{i+1}^1)) \\ \vdots \\ \bar{\mathbf{g}}(\bar{\mathbf{F}}^{N-1}(x, U_{k-1}^{N-1})) \end{bmatrix}$$

In this expression, $\bar{\mathbf{F}}^j(x, U^j)$ is the jth composite of the system function $\bar{\mathbf{f}}$ with input sequence U^j and $i = k-N$. Similar to the linear algebraic system in Eq. 6.7, a nonlinear system of equations in $\hat{x}_{k-N+1}$ can be constructed using the known inputs and output measurements.

$$Y_k^N - \Phi(\hat{x}_{k-N+1}, U_k^N) \quad = \quad 0 \tag{6.115}$$

The state estimate $\hat{x}_{k-N+1|k}$ is then determined iteratively using Newton's method in which it is assumed that the Jacobian is square.

$$\hat{x}_{k-N+1|k,i+1} \quad = \quad \hat{x}_{k-N+1|k,i} + \left[\frac{\partial \Phi(\hat{x}_{k-N+1|k,i}, U_k^N)}{\partial x}\right]^{-1} \tag{6.116}$$

$$\times \left(Y_k^N - \Phi(\hat{x}_{k-N+1|k,i}, U_k^N)\right)$$

If there are more equations than states, the inverse is replaced by a pseudo inverse. An initial estimate is obtained from the solution at the previous sample time. This approach does not consider constraints.

The observability requirement for this technique is the existence of a bounded N such that the nonlinear system of equations in Eq. 6.115 has a unique solution for all admissible U^N. Sufficient conditions for nominal convergence are provided in [45]. This formulation may fail to converge if not properly initialized and the region of convergence may be small. Modified Newton methods, in which the Jacobian is not explicitly evaluated at each iteration, can be used to reduce the computational requirements. However, these modified methods can effect the rate of convergence. It can be shown that a single iteration of this estimator is equivalent to the extended Kalman filter with $R = \epsilon I$ in the limit as $\epsilon \to 0$. Application of this observer on a continuous stirred tank bioreactor is presented in [45].

6.7.6 Summary of Nonlinear Moving Horizon Estimation

Moving horizon estimation can be applied to a very general class of problems since there are few restrictions on the systems that can be considered.

Nominal convergence of the estimator can also be guaranteed. This approach easily incorporates constraints on the estimates and very general disturbance modeling. It is not restricted to the additive process and measurement disturbances presented in this discussion. Since the nonlinear system is solved exactly, the performance does not suffer from the use of an approximate model. However, this approach requires more computational effort than most of the other nonlinear filters and observers. The additional computational effort can become substantial and prohibit the use of this technique for large-scale systems or systems with short sampling periods.

The solution methods required for an efficient implementation are also more complex. The choice of solution strategy to be employed depends on the application. The sequential approach is normally considered for simple problems due to its relative ease of implementation. This technique is also considered for systems with a large number of states since a prohibitive number of decision variables may be required by the simultaneous approach. However, the sequential approach will normally not be as efficient as the simultaneous approach. The sequential approach may also fail in applications in which there are constraints on the estimated states.

Example 6.4

We demonstrate the performance of the moving horizon estimators presented in Sections 6.7.1 and 6.7.2 on a simple linear integrating process with $A = 1$ and $C = 1$ and no constraints on the estimator. We consider the case of zero mean measurement noise with no process disturbances and the case of random process disturbances with no measurement noise. When only measurement noise is present in the system, the true state is stationary at zero. When only process disturbances are present, the measurement is the true state. The performance of the estimator with state disturbances in Section 6.7.1 is shown in Figures 6.4 and 6.5. For the measurement noise case, this estimator is tuned to reflect that the measurement is unreliable using a horizon length $N = 20$, $Q = 1$, and $R = 100$. The deviations between the estimated output and measured output are not heavily penalized in comparison to the estimated state disturbances which results in state estimates that are very close to the true zero state. For the process disturbance case, the estimator is tuned with $Q = 1$ and $R = 0.1$ to reflect that the measurements are relatively reliable. With a horizon length $N = 20$, the estimated state essentially tracks the true state.

The performance of the initial state estimator in Section 6.7.2 with the same horizon length, $N = 20$, is shown in Figures 6.6 and 6.7. This estimator is essentially identical to the previous approach when only measurement noise is present. However, when process disturbances are present, this estimator does a very poor job of estimating the true state. The estimate of an initial state at the beginning of the horizon cannot take into account the state disturbances that enter the system within the observer horizon.

When the horizon length is shortened to $N = 2$, the performance with process disturbances present improves, as shown in Figure 6.8, but the estimate still lags the true state. With the shorter horizon, however, the performance of this estimator with measurement noise degrades considerably as shown in Figure 6.9. This example clearly shows that initial state estimate approaches are not suitable when significant process disturbances are present.

6.8 Combined State and Parameter Estimation

The assumption of constant model parameters made throughout this discussion may not be realistic in many practical applications. Parameters, such as heat transfer coefficients, can change during the operation of chemical processes. In these cases, a more accurate determination of the state can be made by allowing one or more model parameters to vary. The parameter values are then estimated along with the state. This technique is referred to as combined state and parameter estimation. A brief summary of this topic is presented in this section. Further information on parameter estimation for nonlinear systems is available in Bard [4].

Combined state and parameter estimation typically is performed by augmenting the state of the system to include those parameters to be estimated. The most common assumption made to describe the dynamic behavior of these parameters is an integrated white noise process. Using this assumption, the continuous-time nonlinear model in Eq. 6.34 and the discrete nonlinear model in Eq. 6.35 are augmented as follows in which Θ is the vector of parameters to be estimated.

$$\begin{bmatrix} \dot{x} \\ \dot{\Theta} \end{bmatrix} = \begin{bmatrix} \mathbf{f}(x, u, \Theta, t) \\ 0 \end{bmatrix} ; \qquad \begin{bmatrix} x_{k+1} \\ \Theta_{k+1} \end{bmatrix} = \begin{bmatrix} \bar{\mathbf{f}}(x_k, u_k, \Theta_k, k) \\ \Theta_k \end{bmatrix} \tag{6.117}$$

$$y = \mathbf{g}(x, \Theta, t) ; \qquad y_k = \bar{\mathbf{g}}(x_k, \Theta_k, k)$$

Observability of the augmented system is not implied by observability of the original system and must be verified using the techniques discussed

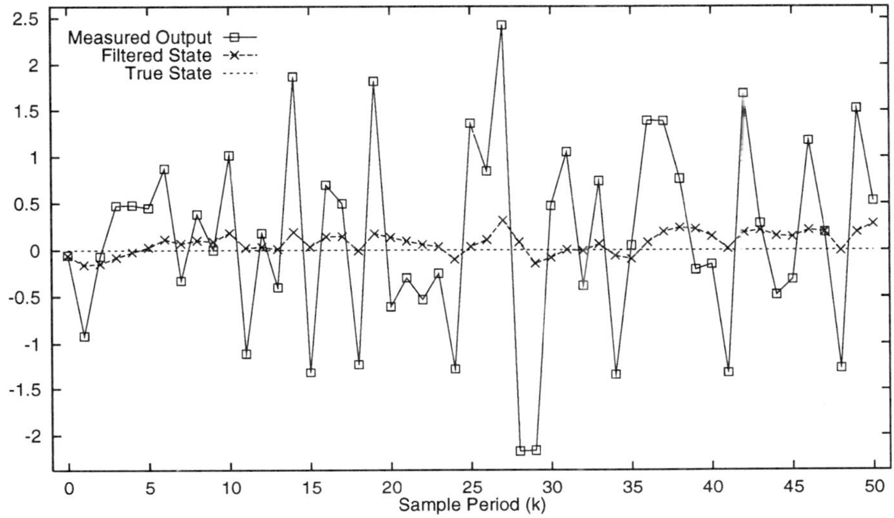

Figure 6.4: Filtered state estimate for the estimated state disturbance moving horizon estimator with measurement noise, $N = 20$.

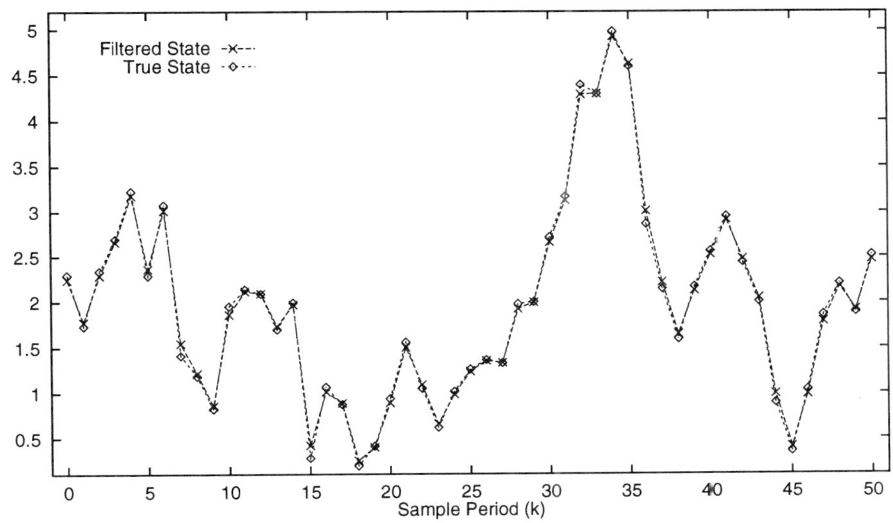

Figure 6.5: Filtered state estimate for the estimated state disturbance moving horizon estimator with process disturbances, $N = 20$.

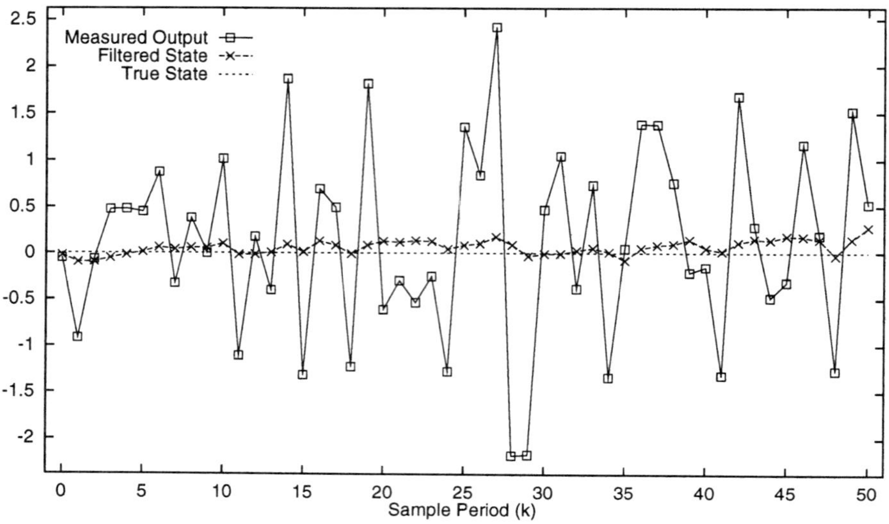

Figure 6.6: Filtered state estimate for the initial state estimate moving horizon estimator with measurement noise, $N = 20$.

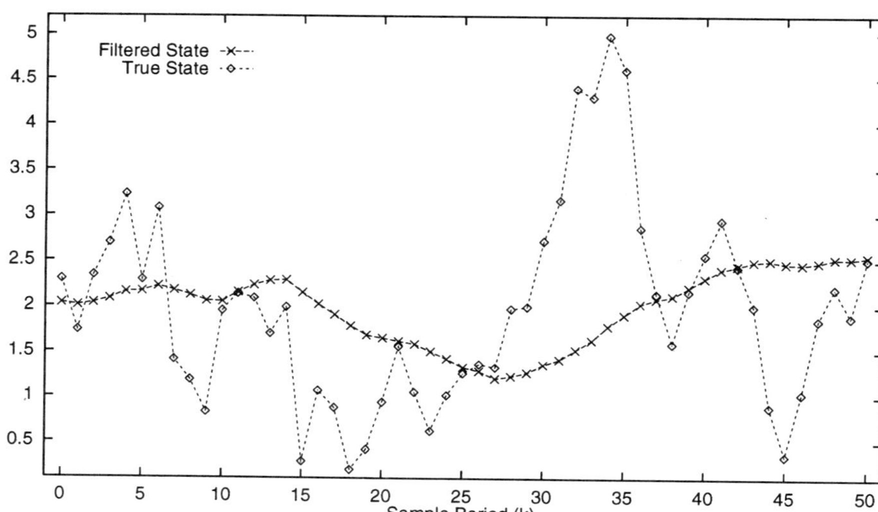

Figure 6.7: Filtered state estimate for the initial state estimate moving horizon estimator with process disturbances, $N = 20$.

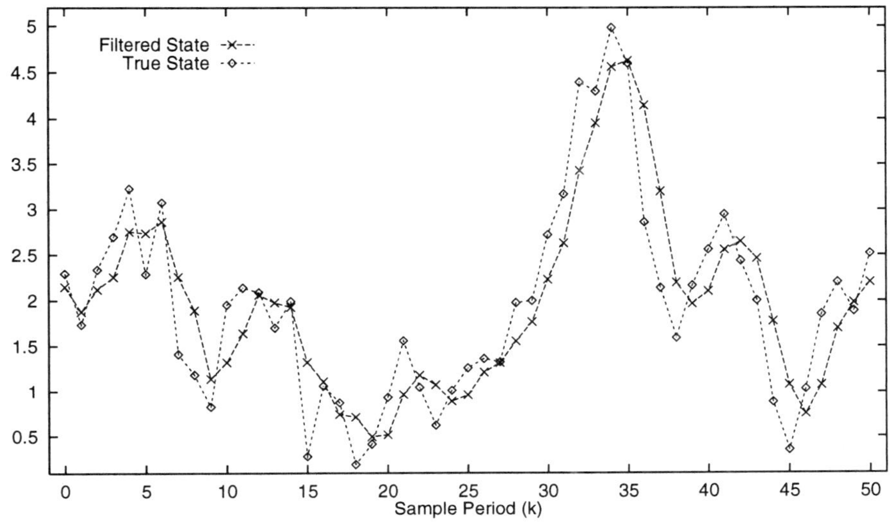

Figure 6.8: Filtered state estimate for the initial state estimate moving horizon estimator with process disturbances, $N = 2$.

in Section 6.4.2. Observability may impose restrictions on the parameters that can be estimated for a given system. Application of the nonlinear state estimation techniques discussed in this chapter on the augmented system is used to provide an estimate of both the state and parameters. In practice, only a small number of the model parameters are estimated.

Parameter estimation using the extended Kalman filter is discussed in Cox [14], Ljung [38], and DeVallière and Bonvin [16]. The extended Schmidt–Kalman filter is discussed in Jazwinski [25]. This filter accounts for the effect of parameter variation on the state, but does not compute an estimate of the parameter values. A discussion of modeling techniques that can improve the estimation of parameter values is presented in DeVallière and Bonvin [17]. Agarwal and Bonvin [1] recommend a decoupled estimator in which the states and parameters are computed separately to reduce the nonlinearity due to parameter-state interactions.

Much of the research in the area of moving horizon estimation has been related to data reconciliation in which parameter estimation is routinely performed. In the moving horizon data reconciliation problem, the model parameters are also included as decision variables in the least squares objective. The optimization problem can be set up to either directly compute

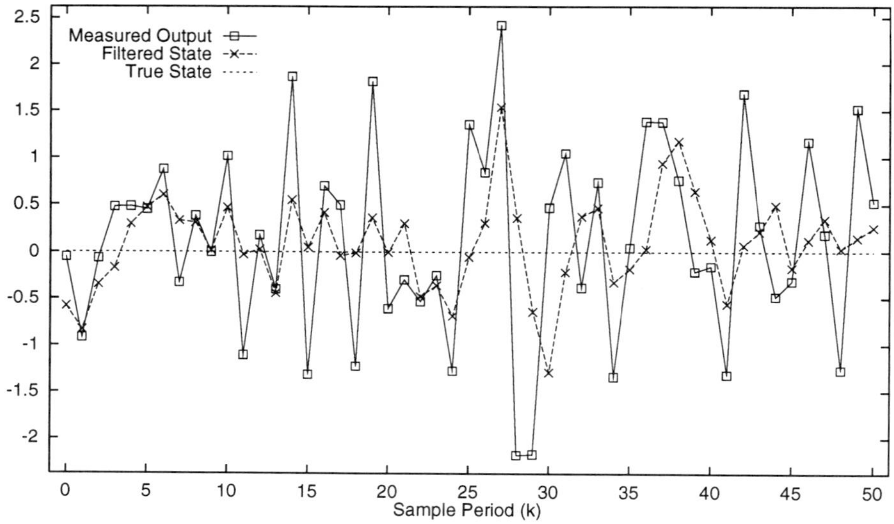

Figure 6.9: Filtered state estimate for the initial state estimate moving horizon estimator with measurement noise, $N = 2$.

the parameter values or compute deviations from nominal values. Combined state and parameter estimation using a moving horizon approach is discussed in Kim *et al.* [30], Liebman *et al.* [37], Tjoa and Biegler [59], and Ramamurthi *et al.* [51].

6.9 Nonlinear State Estimation Summary

The selection of an appropriate nonlinear state estimation technique depends on several factors. A significant consideration is the computational requirement. In order to implement an estimator, it must be able to determine the state estimate within the sampling period of the application. Computationally intensive techniques, such as moving horizon estimation and globally iterated extended Kalman filtering, may not be appropriate for applications with short sampling periods because of the computational time required. Large-scale systems can also restrict the techniques that can be considered since the computational requirement generally increases with the number of states that must be estimated.

Modeling error and unmeasured disturbances are other considerations in the selection of a nonlinear estimator. In essentially every application, there will be a difference between the process and the process model. This difference is referred to as model error or mismatch. There will also be disturbances entering the system that are not accounted for in the model. In most cases, it is not possible to completely eliminate these sources of error that can impose limitations on the performance of the estimator. With a poor or uncertain model, it may not be worthwhile to implement a complex estimation technique since the achievable performance may be no better than simpler approaches. Unfortunately, no general guidelines exist regarding the effect of model error on the selection of a nonlinear state estimation technique.

Simulation studies in which a given technique performs very well for an application may not be a reliable indication of how that technique will perform in practice on similar applications. Slight structural errors in the model, mismatch in the parameter values, and unmeasured disturbances can seriously impact the performance of these techniques. A comprehensive simulation study that includes variation in the model parameters and disturbances to the process and measurements should be carried out before a technique is implemented. The selection of these perturbations is based on engineering judgment of what reasonably can be expected to occur in the process.

The proper nonlinear state estimation technique depends on the application being considered and the computational constraints. For some applications, several attempts at selecting and tuning an estimator may be required before a suitable approach is found. In many cases, however, more than one technique will be appropriate and the choice generally will be based on ease of implementation. The intent here is to present sufficient background information on the selection of nonlinear state estimation techniques for the process control engineer and to provide references for those requiring more detail.

References

[1] M. Agarwal and D. Bonvin. Limitations of the extended Kalman filter for batch reactors. In J. Rinsdorp, editor, *Proceedings of the 1989 IFAC DYCORD Symposium*, pages 287–294, Oxford, 1989. Pergamon Press.

[2] K. J. Åström. *Introduction to Stochastic Control Theory.* Academic Press, San Diego, California, 1970.

[3] M. Athans, R. P. Wishner, and A. Bertolini. Suboptimal state estimation for continuous–time nonlinear systems from discrete noisy measurements. *IEEE Trans. Auto. Cont.*, 13(5):504–514, 1968.

[4] Y. Bard. *Nonlinear Parameter Estimation.* Academic Press, New York, 1974.

[5] B. W. Bequette. Nonlinear predictive control using multi-rate sampling. *Can. J. Chem. Eng.*, 69:136–143, 1991.

[6] D. Bestle and M. Zeitz. Canonical form observer design for non-linear time-variable systems. *Int. J. Control*, 38(2):419–431, 1983.

[7] J. Birk and M. Zeitz. Extended Luenberger observer for non–linear multivariable systems. *Int. J. Control*, 47(6):1823–1836, 1988.

[8] A. E. Bryson and Y. Ho. *Applied Optimal Control.* Hemisphere Publishing, New York, 1975.

[9] M. Caracotsios and W. E. Stewart. Sensitivity analysis of initial value problems with mixed ODEs and algebraic equations. *Comput. Chem. Eng.*, 9(4):359–365, 1985.

[10] D. E. Catlin. *Estimation, Control, and the Discrete Kalman Filter.* Springer-Verlag, New York, 1989.

[11] S. T. Chung and J. W. Grizzle. Observer error linearization for sampled–data systems. *Automatica*, 26:997–1007, 1990.

[12] G. Ciccarella, M. Dalla Mora, and A. Germani. A Luenberger-like observer for nonlinear systems. *Int. J. Control*, 57(3):537–556, 1993.

[13] G. Ciccarella, M. Dalla Mora, and A. Germani. Observers for discrete-time nonlinear systems. *Sys. Cont. Let.*, 20:373–382, 1993.

[14] H. Cox. On the estimation of state variables and parameters for noisy dynamic systems. *IEEE Trans. Auto. Cont.*, 9(1):5–12, 1964.

[15] C. E. De Souza. On stabilizing properties of solutions of the Riccati difference equation. *IEEE Trans. Auto. Cont.*, 34(12):1313–1316, 1989.

[16] P. DeVallière and D. Bonvin. Application of estimation techniques to batch reactors–II. Experimental studies in state and parameter estimation. *Comput. Chem. Eng.*, 13(1):11–20, 1989.

[17] P. DeVallière and D. Bonvin. Application of estimation techniques to batch reactors–III. Modeling refinements which improve the quality of state and parameter estimation. *Comput. Chem. Eng.*, 14(7):799–808, 1990.

[18] F. Deza, E. Busvelle, J. P. Gauthier, and D. Rakotopara. High gain estimation for nonlinear systems. *Sys. Cont. Let.*, 18:295–299, 1992.

[19] J. Dimitratos, C. Georgakis, M. El-Aasser, and A. Klein. Dynamic modeling and state estimation for an emulsion copolymerization reactor. *Comput. Chem. Eng.*, 13(1):21–33, 1989.

[20] J. P. Gauthier, H. Hammouri, and S. Othman. A simple observer for nonlinear systems – Applications to bioreactors. *IEEE Trans. Auto. Cont.*, 37(6):875–880, 1992.

[21] A. Gelb, editor. *Applied Optimal Estimation.* The M.I.T. Press, Cambridge, Massachusetts, 1974.

[22] R. Hermann and A. J. Krener. Nonlinear controllability and observability. *IEEE Trans. Auto. Cont.*, 22(5):728–740, 1977.

[23] Y. C. Ho and R. C. K. Lee. A Bayesian approach to problems in stochastic estimation and control. *IEEE Trans. Auto. Cont.*, 9(5):333–339, 1964.

[24] S.-S. Jang, B. Joseph, and H. Mukai. Comparison of two approaches to on-line prarameter and state estimation of nonlinear systems. *Ind. Eng. Chem. Proc. Des. Dev.*, 25:809–814, 1986.

[25] A. H. Jazwinski. *Stochastic Proceses and Filtering Theory.* Academic Press, New York, 1970.

[26] T. Kailath. *Linear Systems.* Prentice-Hall, Englewood Cliffs, New Jersey, 1980.

[27] J. C. Kantor. A finite dimensional nonlinear observer for an exothermic stirred-tank reactor. *Chem. Eng. Sci.*, 44(7):1503–1510, 1989.

[28] S. S. Keerthi and E. G. Gilbert. Optimal infinite-horizon feedback laws for a general class of constrained discrete-time systems: Stability and moving-horizon approximations. *J. Optim. Theory Appl.*, 57(2):265–293, 1988.

[29] I. Kim, M. Liebman, and T. Edgar. Robust error-in-variables estimation using nonlinear programming techniques. *AIChE J.*, 36(7):985–993, 1990.

[30] I. Kim, M. Liebman, and T. Edgar. A sequential error-in-variables method for nonlinear dynamic systems. *Comput. Chem. Eng.*, 15(9):663–670, 1991.

[31] D. J. Kozub and J. F. MacGregor. State estimation for semi–batch polymerization reactors. *Chem. Eng. Sci.*, 47(5):1047–1062, 1992.

[32] A. J. Krener and A. Isidori. Linearization by ouput injection and nonlinear observers. *Sys. Cont. Let.*, 3(1):47–51, 1983.

[33] A. J. Krener and W. Respondek. Nonlinear observers with linearizable error dynamics. *SIAM J. Cont. Opt.*, 23(2):197–216, 1985.

[34] H. Kwakernaak and R. Sivan. *Linear Optimal Control Systems.* John Wiley and Sons, New York, 1972.

[35] W. Lee and K. Nam. Observer design for autonomous discrete–time nonlinear systems. *Sys. Cont. Let.*, 17(1):49–58, 1991.

[36] F. L. Lewis. *Optimal Estimation.* John Wiley and Sons, New York, 1986.

[37] M. Liebman, T. Edgar, and L. Lasdon. Efficient data reconciliation and estimation for dynamic processes using nonlinear programming techniques. *Comput. Chem. Eng.*, 16(10/11):963–986, 1992.

[38] L. Ljung. Asymptotic behaviour of the extended Kalman filter as a parameter estimator for linear systems. *IEEE Trans. Auto. Cont.*, 24(1):36–50, 1979.

[39] J. Logsdon and L. T. Biegler. A relaxed reduced space SQP strategy for dynamic optimization problems. *Comput. Chem. Eng.*, 17(4):367–372, 1993.

[40] D. G. Luenberger. An introduction to observers. *IEEE Trans. Auto. Cont.*, 16(2):596–602, 1971.

[41] A. K. Mahalanabis and M. Farooq. A second–order method for state estimation of non–linear dynamical systems. *Int. J. Control*, 14(4):631–639, 1971.

[42] S. I. Marcus. Stochastic integrals and stochastic calculus. In M. G. Singh, editor, *Systems and Control Encyclopedia: Theory, Technology, Applications*, pages 4631–4638, Oxford, 1987. Pergamon Press.

[43] E. S. Meadows, K. R. Muske, and J. B. Rawlings. Constrained state estimation and discontinuous feedback in model predictive control. In *Proceedings of the 1993 European Control Conference*, pages 2308–2312. European Automatic Control Council, June 1993.

[44] H. Michalska and D. Q. Mayne. Moving horizon observers and observer-based control. *IEEE Trans. Auto. Cont.*, 40(6):995–1006, 1995.

[45] P. E. Moraal and J. W. Grizzle. Observer design for nonlinear systems with discrete-time measurements. *IEEE Trans. Auto. Cont.*, 40(3):395–404, 1995.

[46] K. R. Muske and J. B. Rawlings. Nonlinear moving horizon state estimation. In R. Berber, editor, *NATO Advanced Study Institute on Methods of Model Based Control Proceedings*, Antalya, Turkey, July 1994. Kluwer Academic Publishers.

[47] K. R. Muske, J. B. Rawlings, and J. H. Lee. Receding horizon recursive state estimation. In *Proceedings of the 1993 American Control Conference*, pages 900–904, June 1993.

[48] M. Myers and R. H. Luecke. Process control applications of an extended Kalman filter algorithm. *Comput. Chem. Eng.*, 15(12):853–857, 1991.

[49] H. Nijmeijer. Observability of autonomous discrete time nonlinear systems: A geometric approach. *Int. J. Control*, 36(5):867–874, 1982.

[50] P. Park and T. Kailath. New square-root algorithms for Kalman filtering. *IEEE Trans. Auto. Cont.*, 40(5):895–899, 1995.

[51] Y. Ramamurthi, P. Sistu, and B. Bequette. Control-relevant dynamic data reconciliation and parameter estimation. *Comput. Chem. Eng.*, 17(1):41–59, 1993.

[52] W. H. Ray. *Advanced Process Control*. McGraw-Hill, New York, 1981.

[53] D. Robertson, J. H. Lee, and J. B. Rawlings. A moving horizon based approach for least squares state estimation. Submitted to *AIChE J.*, 1995.

[54] M. G. Safonov and M. Athans. Robustness and computational aspects of nonlinear stochastic estimators and regulators. *IEEE Trans. Auto. Cont.*, 23(4):717–725, 1978.

[55] H. Schuler and Z. Suzhen. Real-time estimation of the chain length distribution in a polymerization reactor. *Chem. Eng. Sci.*, 40(10):1891–1904, 1985.

[56] J. J. E. Slotine, J. K. Hedrick, and E. A. Misawa. On sliding observers for nonlinear systems. *J. Dynam. Sys. Meas. Cont.*, 109(3):245–252, 1987.

[57] E. D. Sontag. A concept of local observability. *Sys. Cont. Let.*, 5:41–47, 1984.

[58] R. F. Stengel. *Stochastic Optimal Control: Theory and Application.* John Wiley and Sons, New York, 1986.

[59] I.-B. Tjoa and L. T. Biegler. Reduced sucessive quadratic programming strategy for errors–in–variables estimation. *Comput. Chem. Eng.*, 16(6):523–533, 1992.

[60] J. Tsinias. Further results on the observer design problem. *Sys. Cont. Let.*, 14:411–418, 1990.

[61] M. Van Dootingh, F. Viel, D. Rakotopara, J. P. Gauthier, and P. Hobbes. Nonlinear deterministic observer for state estimation: Application to a continuous free radical polymerization reactor. *Comput. Chem. Eng.*, 16(8):777–791, 1992.

[62] M. Vidyasagar. *Nonlinear Systems Analysis.* Prentice-Hall, Inc., Englewood Cliffs, New Jersey, 2nd edition, 1993.

[63] B. Walcott, M. Corless, and S. Żak. Comparative study of non–linear state–observation techniques. *Int. J. Control*, 45(6):2109–2132, 1987.

[64] R. P. Wishner, J. A. Tabaczynski, and M. Athans. A comparison of three nonlinear filters. *Automatica*, 5:478–495, 1969.

[65] X.-H. Xia and W.-B. Gao. On exponential observers for nonlinear systems. *Sys. Cont. Let.*, 11:319–325, 1988.

[66] X.-H. Xia and W.-B. Gao. Nonlinear observer design by observer error linearization. *SIAM J. Cont. Opt.*, 27(1):199–216, 1989.

Chapter 7

Artificial Neural Networks for Nonlinear Process Identification and Control

H. TED SU
Honeywell Technology Center
Honeywell Inc., Minneapolis, Minnesota
Email: tedsu@htc.honeywell.com

THOMAS J. MC AVOY
Department of Chemical Engineering & Institute for Systems Research
University of Maryland, College Park, Maryland
Email: mcavoy@eng.umd.edu

7.1 Introduction to Neural Networks

7.1.1 What Are Neural Networks?

The brains of living creatures provide the control functions necessary for daily existence. These brains consist of a massively parallel network of cells, neurons, that carry out elementary calculations. A neuron has a multitude of synaptic inputs, and it produces a single output which traverses down its axon to become an input for other neurons. There is a vast number of control tasks that the brain must coordinate in real time, including blood pressure, breathing, limb movement, etc. It is interesting how efficient the brain is at its various tasks given the fact that the basic neurons it uses have been estimated [3] to be 5 to 6 orders of magnitude slower than the silicon chips used in modern digital computers. The brain achieves its speed through its massively parallel architecture. Another key aspect of the brain is the fact that it learns by example. According to Kohonen [35], a typical human brain has 10^{11} neurons and 10^{15} synaptic interconnections. These numbers are so large that he concluded that it would be impossible to pre-program the system *a priori*, even with genetic information. While some information is certainly present at birth, the strength of most of the synaptic interconnections is learned from experience. There is a fundamental difference between the way that the brain processes information and a digital computer processes information. Table 7.1 gives a comparison.

Artificial neural networks represent an attempt, albeit primitive, to reproduce the functionalities listed in Table 7.1 in hardware. At the present time, however, most neural networks involve software simulations of the hardware. In looking at neural networks, one should realize that what has been achieved to date is orders of magnitude simpler than what goes on in the brain. We are at the very beginning of trying to model how the brain works.

At this point one might ask why process control engineers would be interested in modeling the brain at all. The answer lies in the nature of process plants. These plants are typically complex, expensive, and unique. The uniqueness stems from such items as different plant ages, feedstocks, piping layouts, etc. For example a company may have 2 or 3 ethylene plants, but each would be different from the others. This unique nature makes first principle modeling costly, relative to the potential profit that can be captured by employing the model. What companies seek is a generic modeling approach that: 1) can make use of readily available process data; 2) is easy to apply; and 3) can be used across a broad spectrum of process applications. Artificial neural networks possess these three characteristics and as a result their use for modeling and control has drawn the attention of engineers in the process industries.

7.1.2 Types of Neural Networks

There is a large number of neural network algorithms that researchers have studied. These algorithms include: multilayer perceptrons (Backpropagation Networks) [71], radial basis functions (RBF) networks [39], cerebellar

Digital Computers (SERIAL)	Neurocomputers (MASSIVELY PARALLEL)
• Make yes/no decisions, using mathematical and logical functions	• Make weighted decisions on the basis of noisy, incomplete, and contradictory data
• Handle data in a rapidly structured sequence of program steps	• Independently formulate methods of processing data
• Find precise answers to a problem given enough time	• Find good, quick, but approximate answers to highly complex problems
• Sort through large data bases to find exact matches, with search time dependent on the size of the data base	• Sort through large data bases to find close matches, with search time independent of the size of the data base
• Store information so that specific data can be retrieved	• Store information so that retrieving any piece of information automatically calls up all related facts

Table 7.1: Comparison of Digital and Neurocomputers

model articulation controller (CMAC) networks [2], Hopfield networks [28], Boltzmann machines [1], Kohonen self organizing networks [34], and adaptive resonance networks [23], to name a few. Of all these networks, the first two, backpropagation and RBF networks have been applied the most in process control applications.

A multilayer perceptron is shown in Fig. 7.1. This type of neural network consists of a collection of neurons that are organized in layers, usually three, with feedforward connections from one layer to another. The input layer stores input values while the output layer calculates a weighted sum of the outputs from the hidden layer, O_j. The hidden layer carries out two calculations. First it calculates a weighted sum of its inputs as:

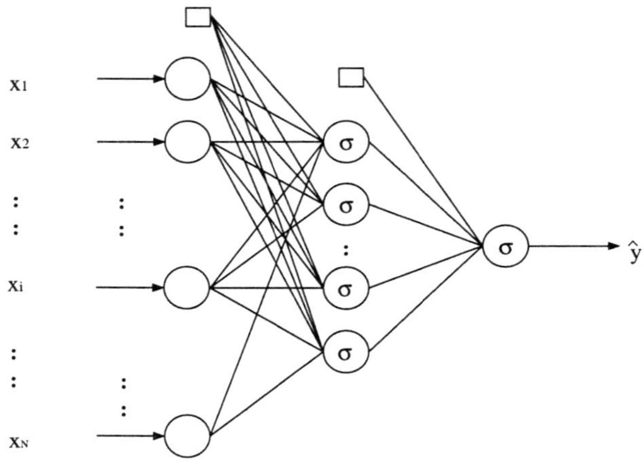

Figure 7.1: Multilayer perceptron

$$S_j = \sum_{i=1}^{N} w_{i,j} X_i + w_{N+1,j} \tag{7.1}$$

Second, the output of the hidden neurons is calculated as the sigma function of S_j:

$$O_j = \sigma(S_j) = \frac{1}{1 + e^{-S_j}} \tag{7.2}$$

In the learning phase, discussed in the following section, the weights are adjusted to minimize the sum of squared errors between predicted and actual outputs. It has been shown [29] that large classes of continuous functions can be approximated by multilayer perceptrons.

Radial basis function networks have an architecture that closely resembles that of multilayer perceptrons. Fig. 7.2 shows a typical RBF network. The first layer, shown simply as X, again acts to store inputs. All the weights from the first layer to the hidden layer have a unity weight. The output layer again calculates a weighted sum of the outputs from the hidden layer, O_j. The hidden layer outputs are given by:

$$O_j = \exp\left(-\frac{||X - X_j||^2}{s_j^2}\right) \tag{7.3}$$

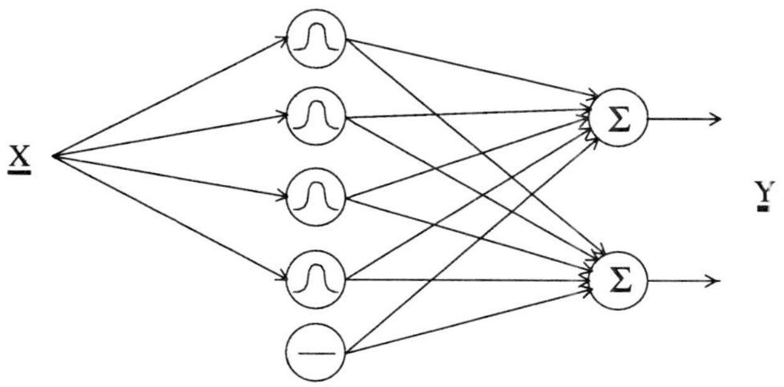

Figure 7.2: Radial Basis Function neural network

The output, O_j, is calculated from a gaussian function, but X_j and s_j are simply parameters and not a mean and standard deviation. The nature of the gaussian function is such that significant values of O_j exist only when X is close to X_j. This aspect of RBF networks distinguishes them from the multilayer perceptrons which use sigmoid functions in their hidden layer. The sigmoid functions produce outputs over a wide range of X values. The localized learning capability of RBF's has been proposed as an advantage of these networks since they don't extrapolate into regions where training data did not exist. RBF's have also been shown [62] to be able to approximate a wide range of continuous functions.

Recently Stephanopoulos and Han [78] proposed that RBF's are superior to backpropagation nets since they solve a regularization problem. Their assertion was based on Poggio and Girosi's results [62]. However, these results are for the limiting case of a large data base where a perfect function fit is theoretically possible. For practical data bases, which are limited in size, it is unclear whether RBF's hold any advantage over backpropagation networks.

7.1.3 Training Neural Networks

This section presents a discussion of how multilayer perceptrons and radial basis function networks are trained. The calculation of the weights in either Fig. 7.1 or Fig. 7.2 is a problem in optimization. For multilayer perceptrons, the first technique that was proposed was called backpropagation, and this name is often used for these networks. With backpropagation, the network weights are adjusted for each exemplar in the data base. The change in weights is proportional to the gradient of the square of the error between the predicted and the actual output. The resulting formulas for the weight changes for the rth data set are:

hidden to output weights

$$\Delta w_{j,k}^{(r)} = \eta\sigma'(S_k)\left(Y_k^{(r)} - O_k^{(r)}\right)O_j \tag{7.4}$$

input to hidden weights

$$\Delta w_{i,j}^{(r)} = \eta\sigma'(S_j)\left((\sigma'(S_k))(Y_k^{(r)} - O_k^{(r)})w_{j,k}^{(r-1)}\right)X_i^{(r)} \tag{7.5}$$

where

$$\sigma'(S_k) = \sigma(S_k)(1 - \sigma(S_k)) \tag{7.6}$$

and η is a tuning parameter that has to be specified. After presentation of the first input-output pair, one proceeds with the next pair and so on. The weights are changed with each presentation.

Since neural networks are typically over-parameterized, an important training issue that arises involves when to stop the training. A simplified version of the statistical technique of cross validation [79], called test set validation, is usually employed. The data set that is used is split into two parts, one for training and one for testing. Periodically, one stops training the network and calculates the error that the network with its current parameters produces on the testing data. Training is terminated when a minimum in the test set error is observed. By using this train-test approach, the fact that a network has too many parameters does not result in a problem and accurate models can be achieved. If such an approach were not used and there were too many network weights, then the network would tend to memorize the training data. Such a network would be very poor when asked to predict outputs for new data.

While the backpropagation algorithm is of historical importance, as a method to calculate network weights it is typically very slow to converge. The reason is due to the fact that it is a purely gradient based approach, and such approaches are known to slow up as one nears a stationary point. Although attempts have been made to improve the algorithm by using past values of weight changes in addition to current gradient information, called momentum, the fact remains that much better approaches to multilayer perceptron network training are available. Some of the more interesting methods involve the coupling of partial least squares with neural networks [68], the use of conjugate gradient optimization [40], and the use of stiff ordinary differential equation solvers [55]. These methods are significantly faster than backpropagation, and they can give more accurate results as well.

The approach that is taken to train RBF networks depends on which parameters are involved. The parameters in the gaussian functions, X_j and s_j in Eq. (7.3), appear in a nonlinear manner. Rather than train these weights, they are usually determined by other means. For example, K-means clustering can be used to calculate the gaussian centers, X_j [45]. Then heuristics can be used to determine the s_j's to ensure smoothness and continuity of the fitted function [45]. By contrast, the output weights in an RBF network appear linearly and as a result a number of methods, such as least squares, can be used to determine them. The fact that only the linearly appearing output weights are determined via optimization gives RBF networks an advantage over backpropagation networks for on-line adaptation [63]. With backpropagation networks some of the weights that have to be determined appear in a nonlinear manner and this fact makes their determination on-line, and adaptively a more difficult problem. A survey of control applications of neural networks can be found in [5].

7.1.4 Uses of Neural Networks for Control

A number of applications of neural networks to process control problems have been reported and several of these are considered in more detail in the next section. A widely studied application involves nonlinear model predictive control [8, 24, 25, 72, 91, 64]. Piovoso et al. [60], have compared neural networks to other modeling approaches for IMC, global linearization, and generic model control and they found the networks to give excellent performance in the face of severe process/model mismatch. Seborg and co-workers ([63, 64, 65, 52, 75]) have used radial basis function neural networks for nonlinear control and they have applied their approaches to simulated systems as well as an actual pH process. They found the neural network based controllers to be superior to other methods in terms of their ease of design and their robustness. A number of authors [30, 37, 85, 86, 88] have used neural networks to detect process faults. Neural networks have also been used for detecting and correcting sensor errors [36], and for tracking batch processes [13]. Although most of the reported neural network studies have been carried out in academia, several real applications have been reported as well. Texas Eastman [73] has used neural networks for steady state modeling to improve significantly the utilization of a costly intermediate chemical species. Neural network modeling was used to identify operating cost savings in a municipal

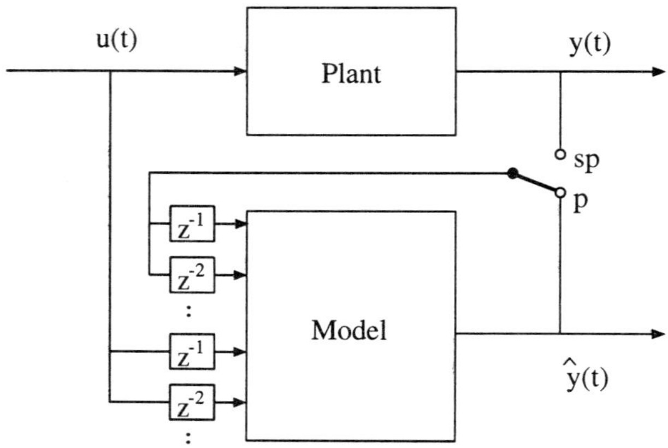

Figure 7.3: Series-parallel or parallel identification methods. z^{-1} is a backward shift operator.

waste water treatment facility [46]. Texaco [26] has reported a feedforward control application on a refinery coker which greatly improved drum switchover. DuPont has used neural networks as soft sensors [61, 74], and for model based control of an acid plant [84]. Similar soft sensor applications on industrial fermentors have been reported by Morris et al. [49]. The soft sensing applications have proven effective because they typically eliminate the time delay associated with getting laboratory assays. This time delay can be on the order of four-eight hours, and eliminating it can produce a significant improvement in control system performance. A number of process companies have applied neural networks, but few have reported their results publicly.

7.2 Dynamic Modeling Using Neural Networks

7.2.1 Series-Parallel vs. Parallel Identification Methods

In general, parameters in a dynamic model, regardless of the form of its mathematical representation, can be estimated by two different approaches: a series-parallel and a parallel identification method [53]. Fig. 7.3 illustrates these two identification approaches. In the case where the pivot connects to "sp", the plant output is fed directly to the model for prediction during the training. This method is referred to as a *series-parallel identification method* since the model is not only in parallel with but also in series with the plant. By comparison, when the pivot connects to "p", the plant output is *not* fed to the model for predicting the plant output. In the latter case, the model is completely in parallel with the plant; thus the method is referred to as a *parallel identification method.*

To further illustrate the difference between the two identification methods, consider the following example. In a sampled-data system where only input and output data are available, a dynamic process is usually described by the following equation:

$$y(t) = \mathcal{F}(y(t-1), \cdots, y(t-K_y), u(t-1), \cdots, u(t-K_u)) + \text{noise} \quad (7.7)$$

The procedure for identifying a model for Eq. (7.7) depends on the assumption of the characteristics of the noise. Conventionally, one takes the following NARX (nonlinear autoregressive with exogenous input) model as a representation of the process shown in Eq. (7.7):

$$\hat{y}(t) = \mathcal{F}_f(y(t-1), \cdots, y(t-K_y), u(t-1), \cdots, u(t-K_u)) \quad (7.8)$$

which corresponds to the series-parallel identification method in Fig. 7.3. In this case, the plant output y is used directly as the input to the model as indicated in the parenthesis of Eq. (7.8). An identification model in the form of Eq. (7.8) is referred to as an equation error model [93, 44].

The alternative approach (the parallel identification method in Fig. 7.3) is to assume that the noise occurs in the output measurement, rather than in the process itself. This approach leads to an alternative estimation method in which the predictions are defined as follows:

$$\hat{y}(t) = \mathcal{F}_r(\hat{y}(t-1), \cdots, \hat{y}(t-K_y), u(t-1), \cdots, u(t-K_u)) \quad (7.9)$$

which is referred to as an output error model [93, 44]. In contrast to Eq. (7.8), the plant output y is not used as the input of the model, but instead, it is replaced with the predicted output $\hat{y}$. Based on the robust estimation school of statistical theory, Eq. (7.8) can yield better forecasts in situations where process noise is present [51], and thus it is also referred to as a *pure robust method* [89]. In the linear case, Wahlberg and Ljung [87] have concluded that the parallel approach tends to filter high frequency noise more than the series-parallel approach. The advantages as well as disadvantages of these

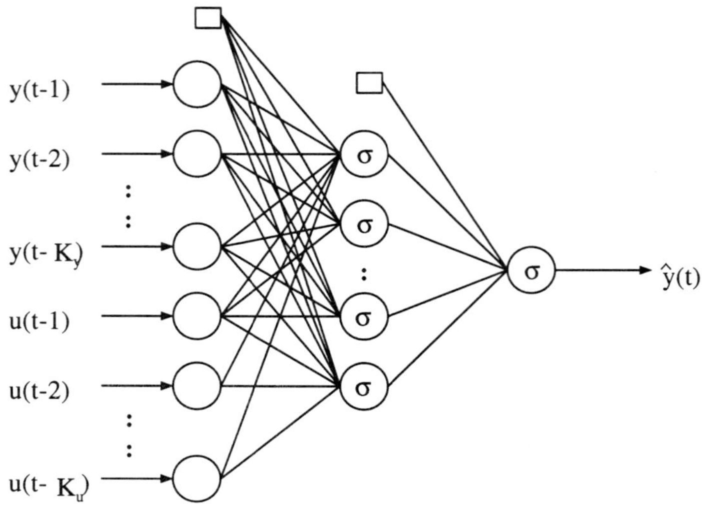

Figure 7.4: An example of a backpropagation feedforward network(FF) dynamic model for SISO system.

two approaches (for linear systems) have been thoroughly discussed in [44]. In the nonlinear case, few results showing the differences between these two approaches have been reported. In the discussion below, backpropagation networks are used for illustration. The identification approaches discussed apply equally well to RBF networks.

NARMAX Modeling

A One-Step Feedforward Modeling Approach

In most of the applications to date, a multilayer feedforward network is employed as a nonlinear auto-regressive with exogenous input model (NARX), in which the network uses a number of past (delayed) plant inputs and outputs to predict the future system output. A NARX model is a subset of the general NARMAX model (MA stands for "Moving Average") in which additional moving average terms are present for modeling the stochastic components of a dynamic process. For example, Bhat and McAvoy [7] have used the series-parallel identification approach for modeling a pH reactor, where the past plant inputs and outputs are fed as the input to the network. Fig. 7.4 illustrates how a backpropagation feedforward network is used to model a single-input-single-output (SISO) system in the series-parallel approach. In the series-parallel approach, the neural network input vector is defined as follows:

$$x(t) \triangleq (y(t-1)^\top, \cdots, y(t-K_y)^\top, u(t-1)^\top, \cdots, u(t-K_u)^\top)^\top \qquad (7.10)$$

There are $N_x = N_y K_y + N_u K_u$ input neurons in total. The dynamic backpropagation network model is actually a special form of the nonlinear function $\mathcal{F}_f$ shown in Eq. (7.8). Suppose that a series of input-output data, $\{u(t)\}$ and $\{y(t)\}$, is given, where $t = -K+1, \cdots, 0, 1, 2, \cdots, N$ and $K = \max\{K_y, K_u\}$. The network training then is used to adjust the weights (V, v, W, and ω) to minimize the error between the network output and the plant output. Usually the error function is defined as:

$$E \triangleq \frac{1}{2}\sum_{t=1}^{N} ||\hat{y}(t) - y(t)||^2 \qquad (7.11)$$

For convenience, let the input and output to the network be expressed in a matrix form. Specifically, the input matrix (containing the past inputs and the past outputs) is:

$$\mathbf{X} \triangleq \begin{bmatrix} y(0)^\top & \ldots & y(-K_y)^\top & u(0)^\top & \ldots & u(-K_u)^\top \\ y(1)^\top & \ldots & y(1-K_y)^\top & u(1)^\top & \ldots & u(1-K_u)^\top \\ \vdots & \ddots & \vdots & \vdots & \ddots & \vdots \\ y(t-1)^\top & \ldots & y(t-K_y)^\top & u(t-1)^\top & \ldots & u(t-K_u)^\top \\ \vdots & \ddots & \vdots & \vdots & \ddots & \vdots \\ y(N-1)^\top & \ldots & y(N-K_y)^\top & u(N-1)^\top & \ldots & u(N-K_u)^\top \end{bmatrix} \qquad (7.12)$$

and the output matrix (containing the target value of the outputs) is:

$$\mathbf{Y} \triangleq \begin{bmatrix} y(1)^\top \\ y(2)^\top \\ \vdots \\ y(t)^\top \\ \vdots \\ y(N)^\top \end{bmatrix} \qquad (7.13)$$

Notice that both of the data matrices are fixed for a given input/output data set. In effect, the FFN modeling is learning the underlying function

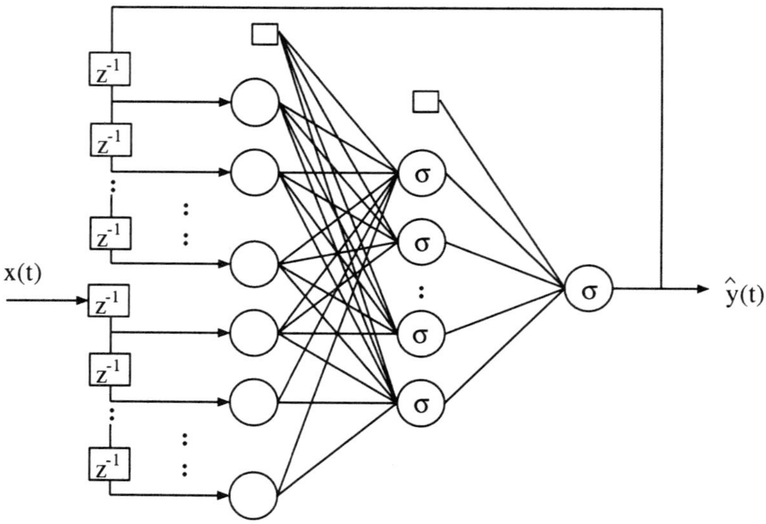

Figure 7.5: An example of a backpropagation recurrent network (RNN) dynamic model for a SISO system.

between the two matrices. In other words, the FFN modeling is similar to the traditional use of neural networks for static modeling.

A Long-Range Recurrent Modeling Approach

Alternatively, the feedforward network can be configured completely in parallel with the plant as shown in Fig. 7.3 with the pivot in the "p" position. In this case, the input to the network is:

$$x(t) \triangleq (\hat{y}(t-1)^\top, \cdots, \hat{y}(t-K_y)^\top, u(t-1)^\top, \cdots, u(t-K_u)^\top)^\top \tag{7.14}$$

Recall that $\hat{y}(t-\tau)$ is a time-delayed (by τ time units) output from the network itself. In that sense, the feedforward network model in the parallel configuration becomes a special form of time-lag recurrent networks. Such a neural network is referred to as an *external recurrent network*. Hereinafter, a recurrent network refers to an external recurrent network unless otherwise indicated. Fig. 7.5 illustrates an external recurrent network resulting from the parallel identification of a feedforward network for an SISO system. An external recurrent network approximates $\mathcal{F}_r$ shown in Eq. (7.9).

Given the same data set as above, the **Y** matrix for a recurrent neural network (RNN) model is identical to that for an FFN model (c.f. Eq. (7.13)).

However, the input matrix changes as the weights change at each iteration (θ) during the training. The input matrix is expressed as follows:

$$
\mathbf{X}|_\theta \equiv \begin{bmatrix} y(0)^\top & \ldots & y(-K_y)^\top & u(0)^\top & \ldots & u(-K_u)^\top \\ \hat{y}(1)^\top & \ldots & y(1-K_y)^\top & u(1)^\top & \ldots & u(1-K_u)^\top \\ \vdots & \ddots & \vdots & \vdots & \ddots & \vdots \\ \hat{y}(t-1)^\top & \ldots & \hat{y}(t-K_y)^\top & u(t-1)^\top & \ldots & u(t-K_u)^\top \\ \vdots & \ddots & \vdots & \vdots & \ddots & \vdots \\ \hat{y}(N-1)^\top & \ldots & \hat{y}(N-K_y)^\top & u(N-1)^\top & \ldots & u(N-K_u)^\top \end{bmatrix}_\theta \tag{7.15}
$$

where $y(t)$ is used for all $t < 1$ when $\hat{y}(t)$ is not available. Initially, one can set the input matrix of a recurrent network as in Eq. (7.12), i.e., $\mathbf{X}|_{\theta=0} = \mathbf{X}$.

7.2.2 Examples

In order to demonstrate the difference between the *one-step* modeling approach and the *long-range* modeling approach, an industrial process is used as an example [83]. The process is a Danish municipal wastewater treatment pilot plant, which involves biological removal of several chemical species. Establishing an analytical model for such a process is extremely difficult. As of 1991, when the plant data were collected (by Technical University of Denmark engineers), no reliable analytical model had been published [58].

A Biological Wastewater Treatment System

The wastewater treatment plant is a complicated process, especially when biological removal of nitrogen, phosphorous and organic material takes place. A pilot scale wastewater treatment facility with such removal was studied by Pedersen et al. [58]. The pilot plant, as shown in Fig. 7.6, is fed with clarified municipal wastewater, from which the large particles, sand, and oils have been removed. The process involves four vessels, in which the biological removal of phosphorous, nitrogen and organic material takes place. The throughput of the plant is on the order of 60 liters per hour. The approximate holding volumes of the pretreatment column (PRE), the two aeration tanks (T1 and T2), and the sedimentation tank (SED) are 200, 800, 800, and 1100 liters, respectively.

The feed to the pretreatment column is a mixture of incoming wastewater and return sludge. The flow rates of the incoming feed and the return sludge

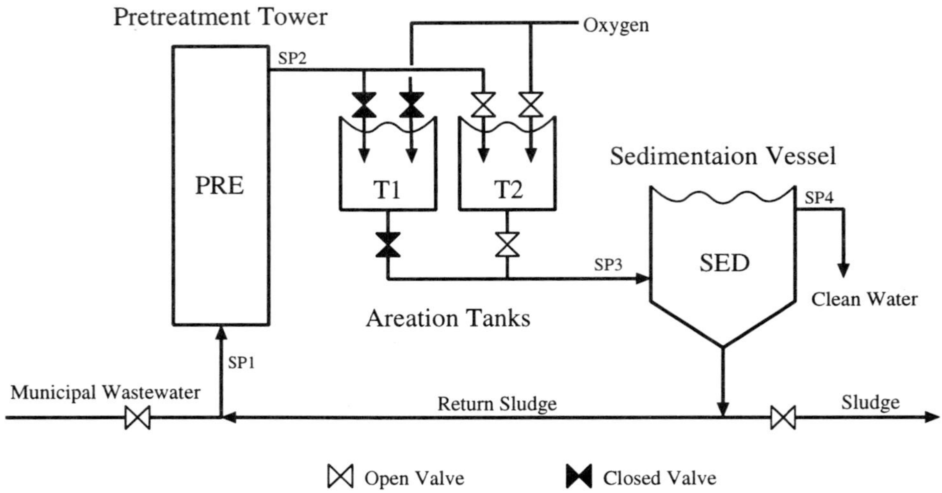

Figure 7.6: Process flow diagram of the wastewater treatment plant.

are kept constant. The pretreatment column is maintained anaerobic and the main biological process is carried out by the phosphorous accumulating bacteria. The two aeration tanks T1 and T2 are operated in an alternating manner which is scheduled prior to the operation. The nitrification and denitrification processes take place in these two tanks alternatively. In the sedimentation vessel, the heavier sludge falls out of suspension. From the top of the vessel the processed effluent is removed from the system. The sludge is returned via the sludge return pump and is mixed with the fluid entering the pretreatment column. The sludge concentration is held approximately constant through intermittent removal of sludge from the system at the point where it exits the sedimentation vessel. Samples are taken from four different locations as indicated in Fig. 7.6, and SP_1 through SP_4 measure the concentrations of NH_4^+, NO_3^{2-}, and PO_4^{3-} in the incoming stream, the pretreatment column, the aeration tanks, and the sedimentation vessel, respectively. The oxygen consumption rates at the aeration tanks are also recorded. A time series of 4 days' data is used to develop a dynamic model for the pilot plant. The data set is obtained by sampling the system every 7 minutes. This sampling results in 900 data points in total. The ranges of all the variables are given by [83]. The data is split into a training set (the first 600 samples) and a testing set (the remaining 300 samples). The outputs of interest are the concentration measurements for the three processing units

	PRE	T12	SED
input	6	8	18
hidden	5	5	6
output	3	3	3

Table 7.2: The number of neurons for each layer of the neural network for the wastewater plant.

(9 variables). The inputs are the concentrations of the incoming stream and the oxygen consumption rates (5 variables).

In this study, three networks are used to model the three processing units, the pretreatment column (PRE), the aeration tanks (T12), and the sedimentation vessel (SED), individually. The pretreatment column is a 3×3 system, the aeration tanks are 3×5, and the sedimentation vessel is 3×3. A first order dynamic model ($K_y = K_u = 1$) is chosen for PRE and T12, and a third order model for SED. The order of the dynamic model is defined by the number of input neurons needed for the networks. The size of the network for each subsystem is summarized in Table 7.2.

After the feedforward network is trained, the one-step ahead prediction by the feedforward network for each system appears to be exceptionally good as illustrated in Fig. 7.7. As a result, one would be tempted to use these models for nonlinear model predictive control [72]. In model predictive control (MPC), the model usually predicts many time steps ahead into the future. Therefore, the feedforward network, which predicts one-step ahead, has to be iterated; i.e., the network has to be chained to itself to go as far as needed into the future. For example, if a predicted output in the future, say $\hat{y}(t+\tau)$, is needed at any time instant t, the predicted output, $\hat{y}(t+1)$ has to replace the actual output $y(t+1)$ measured from the system, and then $\hat{y}(t+2)$ replaces $y(t+2)$ all the way to $t+\tau-1$ since the actual system outputs in the future are not available. However, the chaining will result in a feedforward network being used as an external recurrent network. Due to the difference between y and $\hat{y}$, the prediction error can accumulate during iteration, and large errors can occur.

In order to examine long-range predictive capability, the FFN model is chained to predict the future system output all the way from $t = 1$ to $t = 900$, which includes the training set as well as the testing set. As shown in Fig. 7.8, a feedforward network makes very poor predictions long term. The reason for such poor results made by the feedforward network is simply

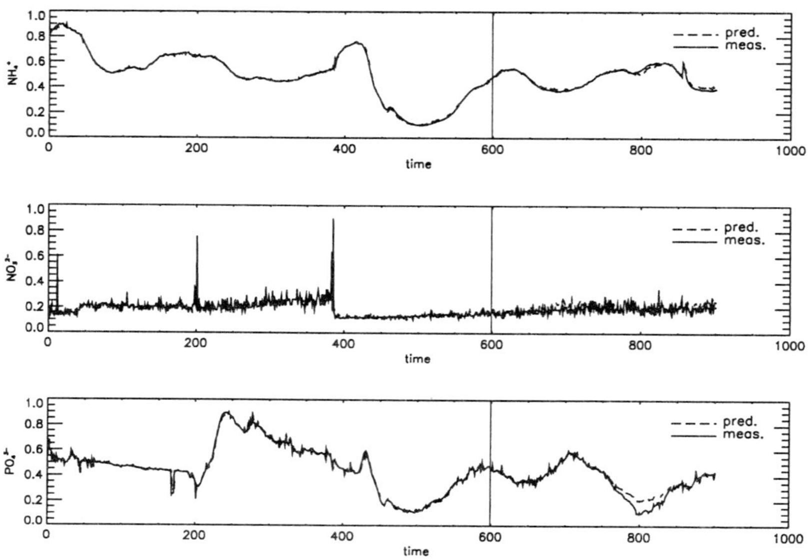

Figure 7.7: One-step ahead prediction of the FFN model for the pretreatment column.

that it is not trained to make long-range predictions. The multiple-step prediction requires iteration or chaining, and feedforward network training does not take such chaining into account.

On the other hand, the external recurrent network is trained to predict the whole trajectory of the system output all the way from $t = 1$ to $t = 600$. For the same PRE system, Fig. 7.9 shows the results given by the trained external recurrent network where the network predicts from $t = 1$ to $t = 900$ (105 hours ahead). For comparison, the root mean squares (RMS) of the prediction error for each subsystem are given in Table 7.3. Clearly, the recurrent network is significantly superior to the feedforward network for long-range predictions. A second industrial example involving data from a catalytic reformer is also discussed in [83].

Results and Discussion

The results presented above have shown that the long-range modeling approach (RNN) has significant advantage over the one-step modeling approach (FFN) for long-range predictions. The prediction horizon (P) is 900 time

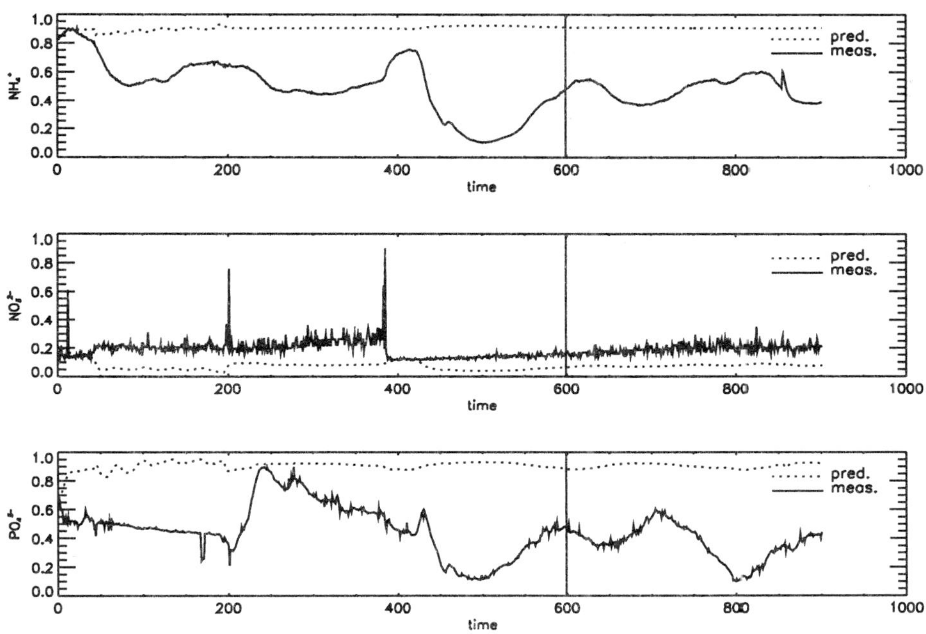

Figure 7.8: Long-rang prediction of the FFN model for the pretreatment column.

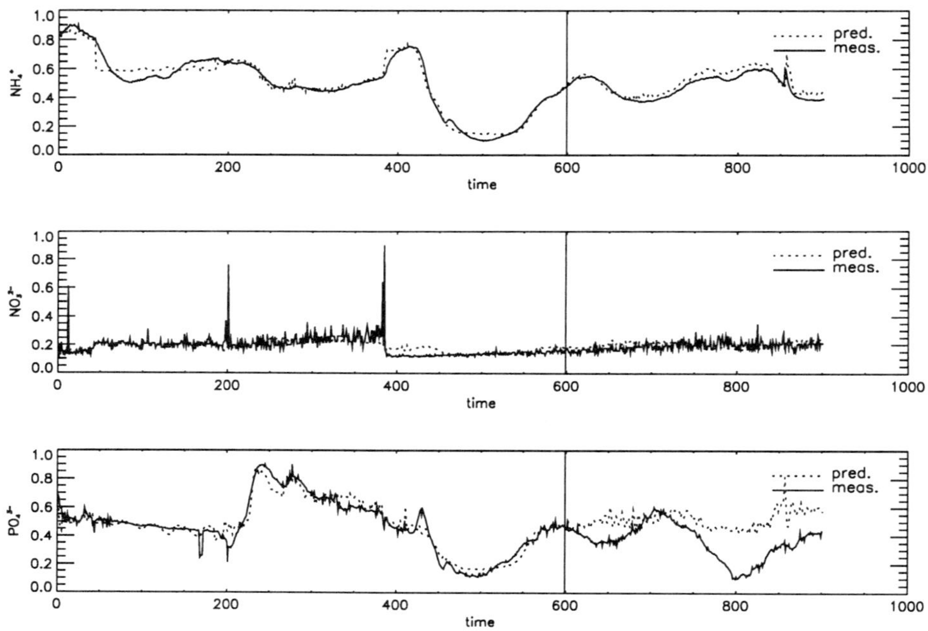

Figure 7.9: Long-rang prediction of the RNN model for the pretreatment column.

unit: 1/100		PRE	T12	SED
one-step ahead FFN training	NH_4^+	3.5	4.3	2.8
	NO_3^{2-}	5.2	3.8	3.9
	PO_4^{3-}	3.6	3.2	1.7
testing	NH_4^+	1.5	3.1	3.8
	NO_3^{2-}	4.1	2.4	1.7
	PO_4^{3-}	4.0	1.8	2.8
long-range FFN training	NH_4^+	45.3	12.5	31.3
	NO_3^{2-}	13.6	8.5	21.7
	PO_4^{3-}	47.0	10.0	18.7
testing	NH_4^+	43.1	10.2	23.6
	NO_3^{2-}	12.0	8.6	11.8
	PO_4^{3-}	54.8	7.4	18.5
long-range RNN training	NH_4^+	4.0	4.5	8.1
	NO_3^{2-}	5.7	5.6	7.0
	PO_4^{3-}	5.1	5.3	4.2
testing	NH_4^+	4.0	9.5	11.4
	NO_3^{2-}	3.8	6.0	6.2
	PO_4^{3-}	18.8	7.0	7.0

Table 7.3: RMS prediction errors for the wastewater plant.

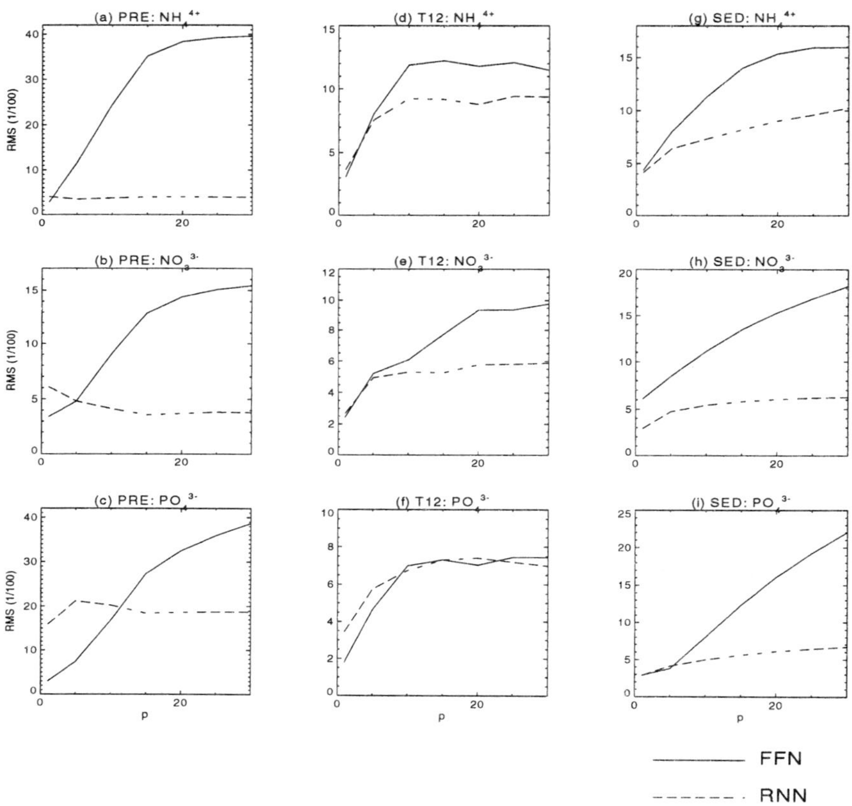

Figure 7.10: RMS prediction errors vs. P for an FFN and an RNN. The RMS errors are calculated based on the scaled variables in all cases.

steps into the future in the wastewater treatment process. In practice, the model does not have to predict so far ahead. Therefore, the network models are compared for small prediction horizons. In particular, the prediction errors are calculated when the prediction horizon is equal to 1, 5, 10, 15, 20, and 30. The prediction error for each output variable is calculated as follows:

$$E_j \triangleq \sum_{t=N+1}^{t=N^*} [\hat{y}_j(t|t-P) - y_j(t)]^2 \tag{7.16}$$

where $\hat{y}(t|\tau)$ stands for the predicted plant output, and $N = 600$, $N^* = 900$. In using $\hat{y}(t|\tau)$, the prediction $\hat{y}(t)$ is made under the condition that the plant output measurements are known at times prior to time τ, and unknown at later times. The RMS of the prediction errors are plotted against the prediction horizon in Fig. 7.10. As can be seen, the prediction errors of a feedforward network increase significantly as the prediction horizon becomes large (see the solid lines), whereas those of a recurrent network are more constant (see the dotted lines). Although in some cases, a feedforward network gives results similar to a recurrent network (e.g. Fig. 7.10-f), the recurrent network gives consistently better results than a feedforward network in all cases when P is greater than 5.

To explain why a recurrent network gives consistently better results than a feedforward network, one can rewrite Eq. (7.11) for the parallel training method using the notation described above:

$$E_{\text{RNN}} = \frac{1}{2} \sum_{t=1}^{N} ||\hat{y}(t|0) - y(t)||^2 \tag{7.17}$$

which implies that the parallel training minimizes the prediction errors for one-step, two-steps, and all the way to N-steps *simultaneously* over the trajectory. Thus it gives better results for long-range predictions. On the other hand, the conventional backpropagation training method uses a criterion given below:

$$E_{\text{FFN}} = \frac{1}{2} \sum_{t=1}^{N} ||\hat{y}(t|t-1) - y(t)||^2 \tag{7.18}$$

which minimizes the prediction error for one-step only. It is trained for a prediction horizon $P = 1$. As a result, the FFN model gives best results for one-step ahead predictions.

It is not necessary to train a recurrent network over the whole trajectory of the training data, i.e. $P = N$, but it can be trained over a smaller trajectory [90]. For illustration, let us consider the following objective,

$$E(t) = \sum_{\tau=t+1}^{t+P} ||\bar{y}(\tau) - \hat{y}(\tau|t)||^2 \tag{7.19}$$

which is equivalent to:

$$\begin{aligned} E(t) = & \sum_{\tau=t+1}^{t+P} ||\bar{y}(\tau) - y(\tau)||^2 \\ & - \sum_{\tau=t+1}^{t+P} [\bar{y}(\tau) - y(\tau)]^\top [y(\tau) - \hat{y}(\tau|t)] + [y(\tau) - \hat{y}(\tau|t)]^\top [\bar{y}(\tau) - y(\tau)] \\ & + \sum_{\tau=t+1}^{t+P} ||\hat{y}(\tau|t) - y(\tau)||^2 \end{aligned} \tag{7.20}$$

where $\bar{y}$ is the desired output, i.e. the setpoint. The ultimate goal is to control the plant output close to the setpoint, i.e. to minimize the first term in Eq. (7.20), by manipulating the input variables. Therefore, one would like to use a neural network model that has a minimal effect on the control performance in achieving this goal. As indicated in Eq. (7.20), the last two terms, where the predicted model output appears, contain the prediction errors of the model, $\hat{y}(\tau|t) - y(\tau)$, over the prediction horizon $(t + 1 \leq \tau \leq t + P)$. Therefore, one can train the model for a trajectory of length P so that the modeling error (i.e. the second and the third terms) is minimal. To do so, one can set up a moving window of length P along the whole training data set, and then train the network over the trajectory within this moving window. With this approach, one reformulates Eq. (7.17) for a RNN as follows:

$$E_{\mathrm{RNN}} = \frac{1}{2} \sum_{t=0}^{N-P} \sum_{\tau=t+1}^{t+P} ||\hat{y}(\tau|t) - y(\tau)||^2 \tag{7.21}$$

Minimizing E_{RNN} in Eq. (7.21) tends to minimize the last two terms in Eq. (7.20), provided that the training data is sufficiently informative. By comparison, reformulating the conventional training method in a similar way gives:

$$E_{\mathrm{FFN}} = \frac{1}{2} \sum_{t=0}^{N-P} \sum_{\tau=t+1}^{t+P} ||\hat{y}(\tau|\tau - 1) - y(\tau)||^2 \tag{7.22}$$

which can be rewritten as follows:

$$E_{\text{FFN}} = \frac{1}{2} \sum_{\theta=1}^{P} \sum_{t=\theta}^{\theta+N-P} ||\hat{y}(t|t-1) - y(t)||^2 \tag{7.23}$$

Typically, N would be much larger than P; therefore, Eq. (7.23) can be approximated by the following equation:

$$E_{\text{FFN}} \approx \quad P \left\{ \frac{1}{2} \sum_{t=1}^{N} ||\hat{y}(t|t-1) - y(t)||^2 \right\} \tag{7.24}$$

When N is much larger than P, the right hand side of Eq. (7.24) contains only a few extra terms than the right hand side of Eq. (7.23), and thus they are very close to one another. Eq. (7.24) is essentially equivalent to Eq. (7.18) with a multiplier of P, i.e., the FFN model still minimizes the one-step ahead prediction error. Additionally, if $P = 1$, Eq. (7.21) reduces to Eq. (7.18). In other words, the FFN method is a special case of the RNN method when the prediction horizon equals one. As will be discussed later, a recurrent network is more appropriate than a feedforward network for model predictive control where accuracy over a trajectory is required.

In an attempt to extend the linear analysis to *nonlinear* neural network models, let us try to explicitly express the noise model for the general nonlinear transfer-function models (c.f. Eq. (7.8) and Eq. (7.9)). For convenience, first, let

$$q^{-n} \triangleq [z^{-1}, z^{-2}, \cdots, z^{-n}]^{\top} \tag{7.25}$$

For an SISO system, Eq. (7.9) (RNN) can be rewritten as follows:

$$\hat{y}(t) = \mathcal{F}(\hat{y}(t)q^{-K_y}, u(t)q^{-K_u}) \tag{7.26}$$

By repetive subsititution of $\hat{y}(t)$ using Eq. (7.26), one obtains a nonlinear *impulse response* model as follows:

$$\hat{y}(t) = \mathcal{F}_{\infty}(u(t)q^{-\infty}) \tag{7.27}$$

where $\mathcal{F}_{\infty}$ denotes a final representation of the function after the substitutions. Hence

$$y(t) = \underbrace{\mathcal{F}_{\infty}(u(t)q^{-\infty})}_{\hat{y}(t)} + \underbrace{e(t)}_{\text{output error}} \tag{7.28}$$

Eq. (7.28) shows that the *noise model* for the RNN approach is a constant. On the other hand, treating Eq. (7.8) for the FFN approach in a similar way results in:

$$y(t) = \mathcal{F}_{\infty}(u(t)q^{-\infty}, \underbrace{e(t)q^{-\infty}) + e(t)}_{\text{equation error}} \tag{7.29}$$

Unlike the RNN method, the *noise model* for the FFN method is a nonlinear *impulse response* model, which of course is not a constant. Intuitively, the different noise models lead to different *relative error* as in the linear case.

Although no general conclusion can be made for the nonlinear neural network case, generally speaking, the observations from the above analysis along with the industrial case study do agree with the theories established for linear systems. Observations for the nonlinear models and the theories for linear models are summarized below:

- In both linear and nonlinear cases, the FFN and RNN modeling approaches result in different noise models. The parallel method (RNN) has a zero-order noise model. The linear zero-order noise model gives a uniform weighting to the *relative* error over the entire frequency region.

- In the linear case, a short prediction horizon leads to emphasis on high frequencies. The series-parallel (FFN) method has the shortest prediction horizon ($P = 1$), and thus the FFN method will result in a high-frequency emphasis in the fit between the model and the plant. In the nonlinear case, the FFN method results in a model which is more accurate for short prediction horizons than for long prediction horizons.

- The RNN method is less sensitive to high frequency noise than the FFN method. The RNN method tends to give more accurate models for long-range predictions than the FFN method in situations where noise is present.

Two other methods of achieving long term predictive accuracy can be mentioned. These methods are closely related to RNN's in that they make use of a trajectory of training data. The neural network training in both of these methods is easier to carry out in practice than training an RNN. The first method involves a Hammerstein model ([80]). With this approach a nonlinear static model is placed in front of a linear dynamic model. The rationale for such an approach lies in the fact that steady state data are typically much more readily available than dynamic data. The steady state

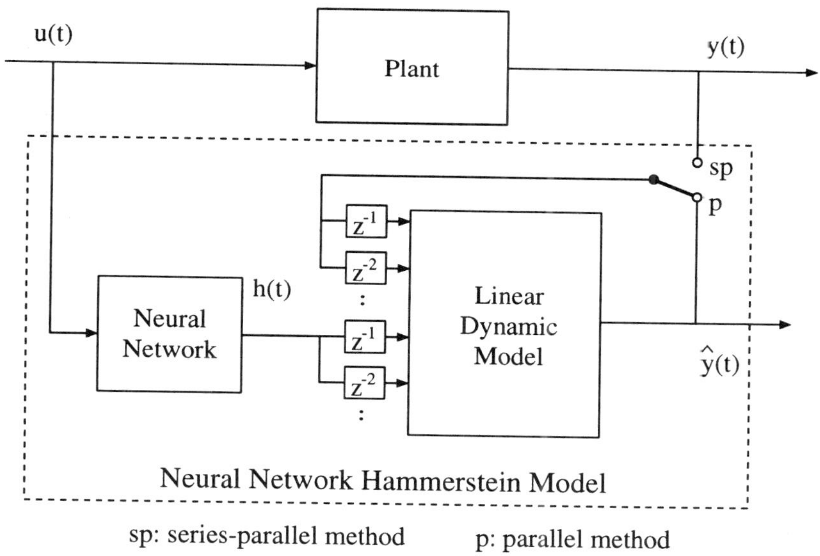

Figure 7.11: The linear dynamic operator in a Hammerstein model can be trained in series-parallel or parallel identification methods. z^{-1} is a backward shift operator.

data are used to develop the static nonlinear model. In Su and Mc Avoy's approach a neural network is used for the static nonlinear model, as shown in Fig. 7.11. For the dynamic model a linear recurrent model is used. Determining the parameters for this linear model is much faster than training a full nonlinear RNN. As shown by Su and Mc Avoy the upstream neural network places a constraint on the final parameters in the linear recurrent model. This constraint essentially forces the total model to satisfy the steady state behavior identified by the neural net preprocessor.

A second method of achieving long term accuracy involves the use of a nonlinear finite impulse response (NFIR) model ([67]). The NFIR model is the nonlinear analog of the linear FIR models used in dynamic matrix control (DMC) ([11]). The NFIR model is given by

$$y(t+1) = \mathcal{N}[u(t), u(t-1), \cdots, u(t-n_u)] \tag{7.30}$$

The trajectory aspect of the NFIR model involves using a large window of past inputs, typically 30 to 40. The sampling time for the window is chosen large enough so that the process can come to steady state at the end of the window after step forcing. The use of the large window helps to insure that the NFIR model achieves long term accuracy in the same way that training over a trajectory (window) insures that a RNN achieves long term accuracy. One of the reasons for the success of DMC is no doubt its use of a training window to achieve long term accuracy. Reference [67] demonstrates the effectiveness of the NFIR approach on data from the catalytic reformer discussed above as well as on data from a large 11x3 industrial process. They used the NNPLS method ([68]) to train the neural net models. Lastly, an approach to dynamic modeling that involves learning the derivative of a time varying function can be mentioned [70, 32]. The method uses a neural net in a Runge Kutta method and it has been used successfully to accurately model very complex surface phenomena, including bifurcation.

7.2.3 Inverse Modeling

In addition to using a neural network as a nonlinear "forward" model for control, one can also find a number of articles that discuss the concept of using an *inverse* neural network model for control [7, 92, 85, 66, 4]. In principle, an inverse neural network model can learn the inverse dynamics under some restrictions (e.g. minimum phase and causality are required). Then the inverse model can be used for control as shown in Fig. 7.12. In practice, especially for discrete-type dynamic models, the inverse model may

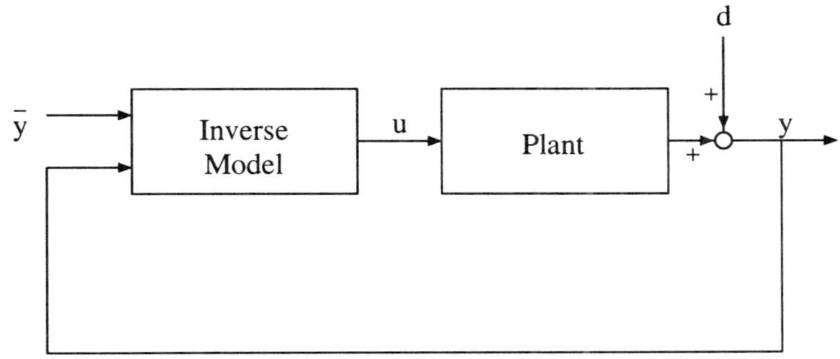

Figure 7.12: An inverse model used for control

not be able to learn the desired inverse dynamics. For example, Psichogios & Ungar [66] trained an inverse model, and then they performed some control case studies using it. They concluded that the neural network controller (inverse model) was unable to handle a disturbance. This conclusion may result from the procedure used to train the network.

The extension of continuous-type model identification to discrete-type identification can give rise to several questions. For illustration, consider that a general causal dynamic system is given as:

$$\frac{dx}{dt} = \mathcal{F}_1(x, u) \tag{7.31}$$

$$y = \mathcal{F}_2(x) \tag{7.32}$$

then one can use a discrete model as follows:

$$y(t+1) = \mathcal{G}_z(y(t), x(t), u(t)) \tag{7.33}$$

to learn the *forward* dynamics of the system from the input/output data. One can rearrange Eq. (7.33) to give an "inverse" model as:

$$u(t) = \mathcal{G}_*^{-1}(y(t+1), y(t), x(t)) \tag{7.34}$$

and then use Eq. (7.34) to learn the *inverse* dynamics from the same input/output data. In other words, given the current states, $x(t)$ and $y(t)$, and the state at the next sampling time, $y(t+1)$, the model is trained to produce the control action, $u(t)$, that drives the system to the target. The inverse model Eq. (7.34) is correct if Eq. (7.33) is an algebraic equation and if $\mathcal{G}_z$ is a monotonic function of $u(t)$. However, a general nonlinear function can possibly contain non-monotonic terms in $u(t)$, and thus treating Eq. (7.34) as a rearrangement of Eq. (7.33) can result in a non-unique solution. For a discussion of the use of inverse models for control, the reader is referred to the article cited above as well as [94, 54, 41]. Additional work is required to decide on the utility of inverse models for process control applications.

7.3 Control via Neural Networks

In this section control systems which use the dynamic neural network models, discussed above, are examined.

7.3.1 Generic Model Control and Global Linearizing Feedback

A very interesting study on using neural networks for reactor control has been published by Piovoso et al [60]. These authors studied the control of a CSTR using three nonlinear approaches: Global Linearizing Feedback (GLF) [27], Internal Model Control (IMC) [15], and Generic Model Control (GMC) [38]. Neural network dynamic models were developed and employed in the GMC and IMC approaches. The GLF results were used as a basis of comparison with results from the other approaches. For the GLF approach a state observer was used. Comparisons were made for perfect knowledge of the reactor model and when significant plant/model mismatch existed. Only the GMC and GLF results are discussed here, and the interested reader is referred to [60] for a discussion of the other results.

Fig. 7.13 gives a schematic of the reactor temperature control scheme studied. The reactor temperature controller is cascaded to a coolant temperature loop, which is a standard approach used in industrial reactor control. It is assumed that reactor composition is not available as an on-line measurement. The particular CSTR studied involved an irreversible first order reaction which had the following dimensionless mass and energy balances [60].

$$\frac{dx_1}{dt} = -x_1 + D_a(1 - x_1)\exp\left(\frac{x_2}{1 + x_2/\gamma}\right) \tag{7.35}$$

$$\frac{dx_2}{dt} = -x_2 + BD_a(1 - x_1)\exp\left(\frac{x_2}{1 + x_2/\gamma}\right) + \beta(y_c - x_2) \tag{7.36}$$

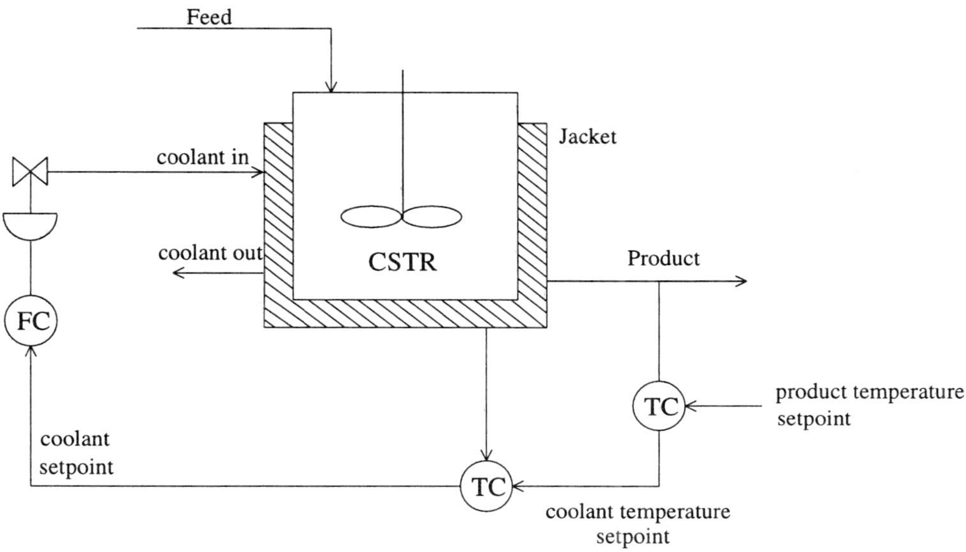

Figure 7.13: CSTR used by Piovoso et al.

State x_1 involved composition and x_2 temperature. The manipulated variable is y_c, a dimensionless cooling jacket temperature. Since only temperatures are assumed measurable, the output equation is given as

$$y = h(x_2) \tag{7.37}$$

The specific parameters used in the simulations are: $D_a = 0.072$, $\gamma = 20.0$, $B = 8.0$, and $\beta = 0.3$. For these parameter values, the reactor exhibits multiple steady state behavior. The control approaches used are discussed next.

Global Linearizing Feedback

As shown in [36], if the output y is equal to x_2, then the relative degree of the system is 1., since x_2 can be written down as an explicit function of y_c. Using the variable transformations

$$z_1 = x_2 \tag{7.38}$$

$$z_2 = x_1 \tag{7.39}$$

then y_c can be determined from the system model as

$$y_c = \frac{1}{\beta}\left[\nu + z_1(1+\beta) - BD_a(1-z_2)\exp\left(\frac{z_1}{1+z_1/\gamma}\right)\right] \tag{7.40}$$

and the resulting equations in z_1 and z_2 are

$$\frac{dz_1}{dt} = \nu \tag{7.41}$$

$$\frac{dz_2}{dt} = -z_2 + D_a(1-z_2)\exp\left(\frac{z_1}{1+z_1/\gamma}\right) \tag{7.42}$$

In order to use GLF on this problem, Eq. (7.40) shows that one must have both z_1 and z_2 available. Since z_2 is assumed to be unavailable, Piovoso et al [36] used an open loop observer to estimate it. They also presented results for the hypothetical case where z_2 was measured. To complete the design ν was chosen as a proportional integral controller giving y_c as

$$y_c = \left[K_p(T^* - z_1) + K_i\int(T^* - z_1)dt + z_1(1+\beta) - \right. \tag{7.43}$$

$$\left. BD_a(1-\bar{z_2})\exp\left(\frac{z_1}{1+z_1/\gamma}\right)\right]/\beta$$

where the overbar indicates that z_2 is calculated from the state observer. Fig. 7.14 shows a schematic diagram of the GLF controller used by Piovoso et al [60]. Two comments can be made about the approach used by these authors. First, they only linearize part of the system, namely the equation associated with the measured state, temperature. Secondly, the choice of a PI controller for ν is arbitrary.

Generic Model Control

The GMC controller design is based on the following model

$$\frac{dx}{dt} = f(x,d,t) + g(x,t)u \tag{7.44}$$

$$y = h(x) \tag{7.45}$$

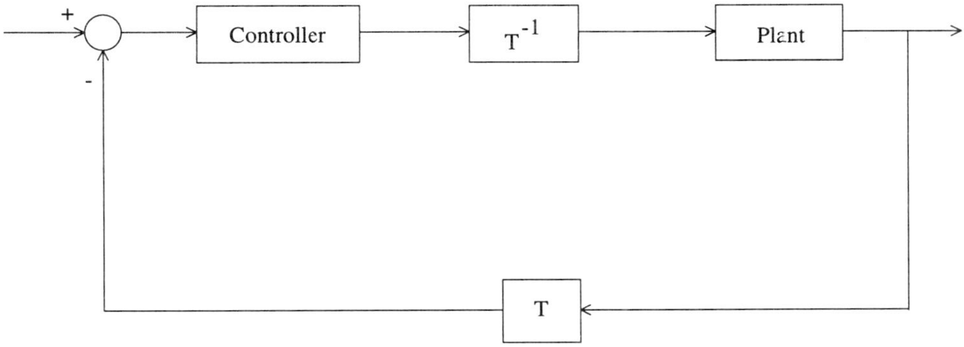

Figure 7.14: Global Linearized Feedback Controller

where f, g, and h are nonlinear functions, x is the state of the system, y is the output, d is the disturbance, and u is the manipulated variable. The derivative of y is assumed to be given by a PI model as

$$\frac{dy}{dt} = K_1(y_{sp} - y) + K_2 \int (y_{sp} - y)dt \tag{7.46}$$

The K_1 and K_2 parameters can be varied to change the controller's response. Eq. (7.46) forces y toward its setpoint, y_{sp}, with zero offset. If Eq. (7.45) is differentiated and together with Eq. (7.46) is substituted into Eq. (7.44), the following control law results

$$u = \left[K_1(y_{sp} - y) + K_2 \int (y_{sp} - y)dt - \frac{dh}{dx} f(x, d, t)\right] \Big/ \left(\frac{dh}{dt} g(x, t)\right) \tag{7.47}$$

For the CSTR studied, the following f, g, and h functions can be determined from the system model, Eq. (7.35) to Eq. (7.37)

$$f(x, t) = -x_2 + BD_a(1 - x_1)\exp\left(\frac{x_2}{1 + x_2/\gamma}\right) - \beta x_2 \tag{7.48}$$

$$g(x, t) = \beta \tag{7.49}$$

$$h(x, t) = x_2 \tag{7.50}$$

β in Eq. (7.49) is assumed to be known. As can be seen from Eq. (7.48), f depends on both x_1 and x_2, but only x_2 is available online. Piovoso et al

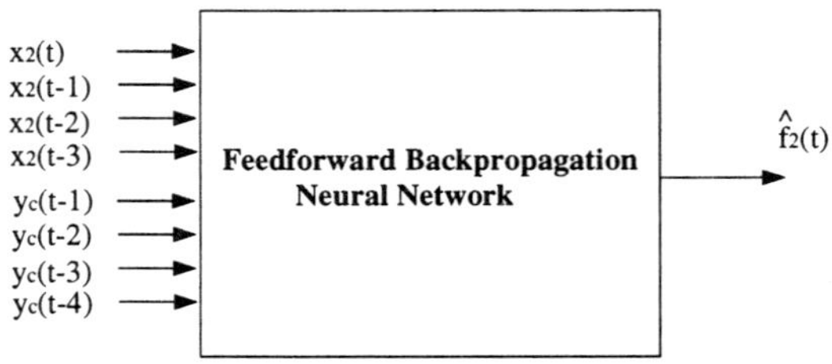

Figure 7.15: BPN topology used by Piovoso et al.

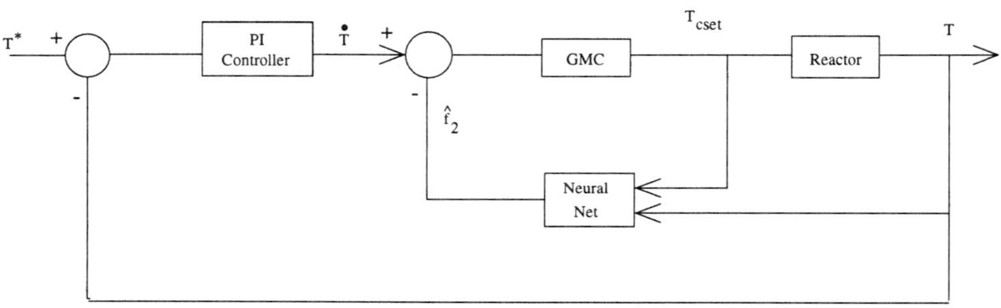

Figure 7.16: GMC architecture used by Piovoso et al.

[60] trained a neural network to estimate f using a time history of x_2 values. Their approach is discussed next.

Piovoso et al [60] used a feedforward backpropagation network in their study. The network was trained using a stiff ODE approach [55]. The network is shown schematically in Fig. 7.15. The inputs to the network consist of the present x_2 measurement and three past values of x_2 and the four previous values of y_c. The network output is the current value of the rate of change of x_2. This approach is interesting since the dependence of f on x_1, which is not measured, is learned implicitly through the use of multiple values of x_2 and y_c. This modeling approach is similar to state estimation using partial state output. The particular network used had 5

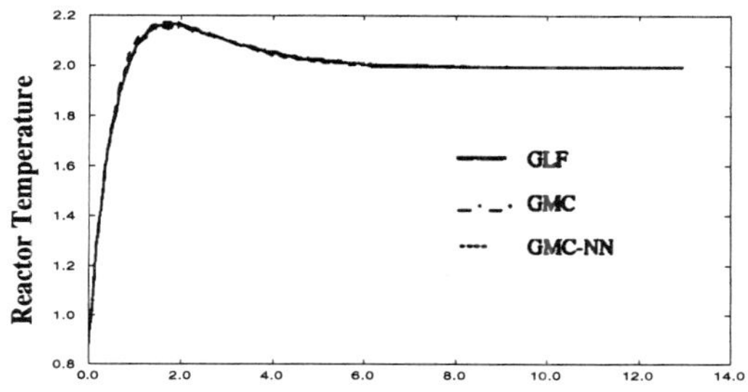

Reactor Temperature Response for a setpoint change, idea case

Figure 7.17: Step response results taken from Piovoso et al.

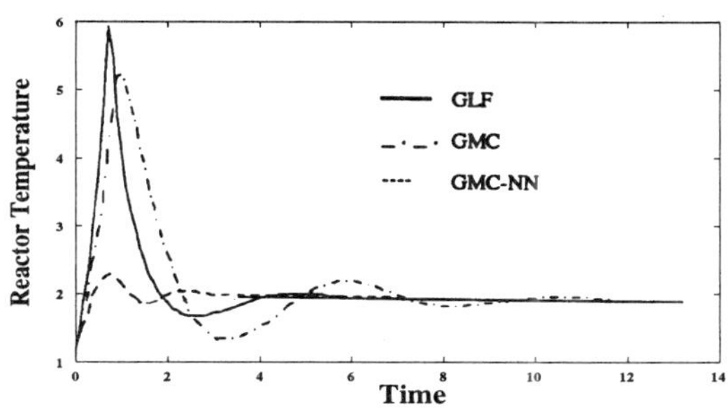

Reactor Temperature response for a setpoint change, model mismatch

Figure 7.18: Set-point change results taken from Piovoso et al.

hidden nodes and the RMS training error was less than 1%. The neural network implementation of GMC is shown schematically in Fig. 7.16.

Piovoso et al [60] presented simulation results for GMC using the neural network dynamic model (GMC-NN), GLF using the state observer, and a hypothetical case of GMC with full state feedback (GMC). It can be noted that this last case is the same as GLF with full state feedback. Fig. 7.17 shows one set of results for a step increase in the reactor setpoint, and no plant/model mismatch. As can be seen all three control schemes give essentially the same response. To simulate plant/model mismatch, the parameters in the simulated CSTR were changed to D_a=0.12, γ=25.0, and B=10. The model CSTR parameters were the same as those presented earlier. Fig. 7.18 gives the results for the same setpoint change as Fig. 7.17. As can be seen both the GLF and GMC schemes give rise to very large overshoots when

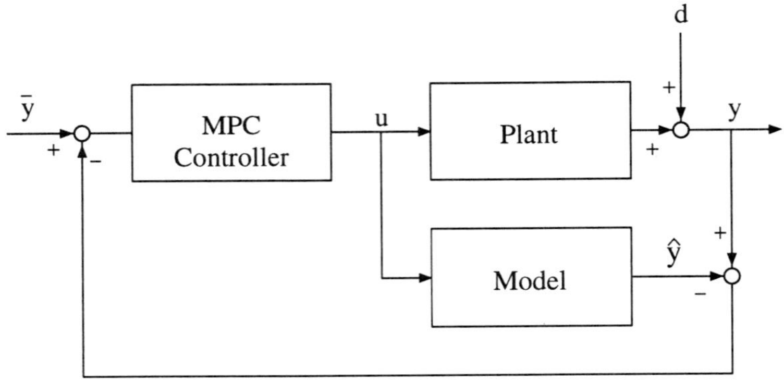

Figure 7.19: The MPC controller in an IMC structure.

model errors exist. By contrast, the overshoot produced by the GMC-NN approach is close to that with no model error. Other results presented in [60] also demonstrate that control systems with neural network models tend to be robust to modeling errors. Determining whether such a conclusion holds in general will require additional study. However, the results presented in [60] are very encouraging and positive in so far as dynamic neural network models are concerned.

Seborg and co-workers ([63, 64, 65, 52, 75]) have carried out a number of studies on both simulated as well as actual systems in which neural networks were used for control. They have reached similar conclusions about the robustness of the neural network based controllers. They also concluded that such controllers were easy to design.

7.3.2 Model Predictive Control

The concept of model predictive control (MPC), in which a model is used as a reference for controlling the plant, has existed for some time, although it has been practiced in an ad hoc fashion. For example, the Smith Predictor, in which a model is used for dead time compensation, has been known since 1957 [77]. In addition, several model predictive control algorithms were proposed in the late 1970's, such as model predictive heuristic control (also known as Model Algorithmic Control, MAC) [69], and Dynamic Matrix Control (DMC) [11, 12]. In 1982, García and Morari [19] proposed a control structure known as Internal Model Control (IMC) to unify the various MPC algorithms. Fig. 7.19 shows a block diagram of model predictive control rep-

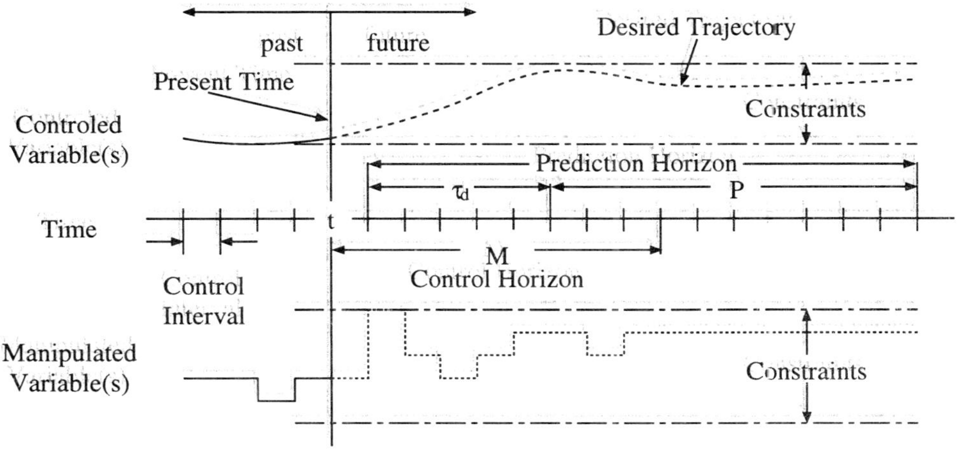

Figure 7.20: Model predictive control approach

resented in the IMC framework. In the IMC structure, the feedback signal contains not only the plant output measurement but also the plant/model mismatch. Therefore, the IMC structure allows the plant/model mismatch to be considered explicitly in the control problem formulation [48]. A thorough review of both theoretical issues and applications of various MPC algorithms can be found in [21]. In addition, Bequette [6] recently reviewed many nonlinear MPC algorithms. Here some examples are discussed. In the following section, all MPC algorithms will be discussed in the framework of IMC.

First the MPC algorithms that use a linear model in the block labeled "Model" in Fig. 7.19 are discussed. For convenience, this class of MPC is referred to as LMPC. From the literature, it appears that the DMC method [12] forms the basis of many other MPC algorithms. Using a *linear* step response model, the original DMC formulates an unconstrained control problem for multi-input-multi-output (MIMO) systems. The basic idea is that a set of M control moves is calculated at every control interval so as to minimize an objective function defining the performance criterion over a prediction horizon (P steps into the future), and the first control move is implemented. Fig. 7.20 illustrates this approach.

In practice, owing to the limitation of physical equipment (e.g. saturation of control valves) and safety considerations, constraints on plant inputs and outputs need to be considered. In 1986, García and Morshedi [20] modified the DMC algorithm so that constraints on the plant input and output

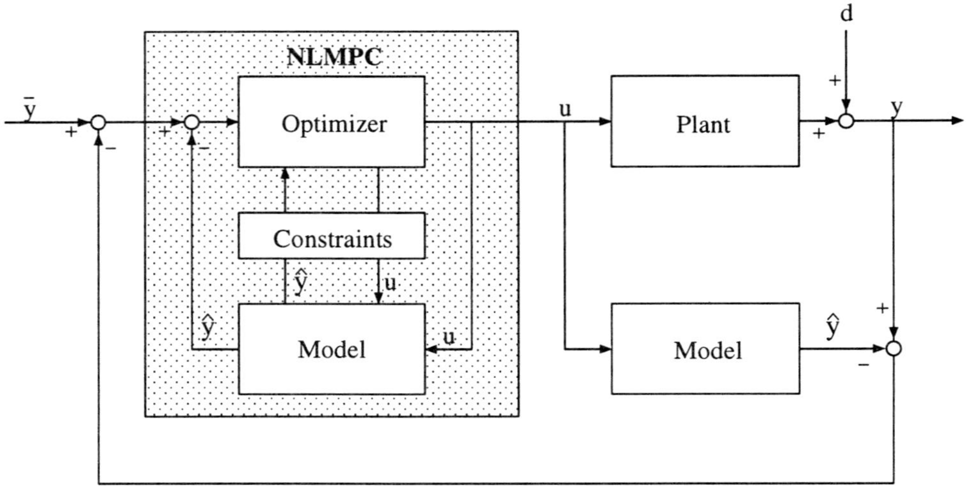

Figure 7.21: Structure of NLMPC controller

variables are explicitly included in the problem formulation. The resulting optimization problem is solved by quadratic programming (QP) techniques, and this control method is therefore referred to as QDMC. In the past few years, the QDMC algorithm has been applied successfully to industrial processes [20].

A number of MPC algorithms using nonlinear models also appear in the literature. For convenience, an MPC that utilizes a nonlinear model is abbreviated as NLMPC. As pointed out by Bequette [6], all NLMPC algorithms are formulated using nonlinear programming (NLP) techniques, in which an objective function is minimized to calculate the future control moves. Fig. 7.21 illustrates a general IMC structure for an NLMPC algorithm. Notice that QDMC (although using a linear model) also fits in this structure where the optimizer is a QP solver. In most of the NLMPC methods, the nonlinear models are represented by a set of nonlinear ordinary differential and algebraic equations. The various NLMPC algorithms differ in the way that the nonlinear model with constraints is solved. In general, three different approaches are used to solve such NLP problems [76]:

- Sequential approach,
- Simultaneous approach, or
- Linearization approach.

In the sequential approach, solution of the ODEs and the calculation of the control moves are performed as two separate tasks. The ODEs are solved by an integrator, and the results are passed to the NLP optimizer [50, 31, 10]. This approach often requires intensive computation time since the optimizer repeatedly solves the ODEs. In the simultaneous approach, the ODEs are solved using collocation methods, such as orthogonal collocation discussed in [17]. In this case, the ODEs are discretized to a set of algebraic equations [43, 57, 14]. This solution technique often results in a large scale optimization problem because additional decision variables are introduced by the collocation methods. Lastly, a number of researchers have developed approaches based on a linearization of the nonlinear model. The optimization of a linearized model is much faster than that of a nonlinear model.

MPC Formulation

Nonlinear model predictive control is formulated using nonlinear programing (NLP) techniques. The approach presented here is an application of neural networks in the nonlinear model predictive control (NLMPC) framework, except that the first principles model (i.e. ODE) is replaced with a discrete-time neural network model. This approach offers significant computation time savings over other NLMPC methods because integration is not needed and additional decision variables are not introduced.

The MPC problem can be stated as follows:

$$\min_{\mathcal{U}(t)} E(t) \triangleq \sum_{k=t+1}^{t+P} \left\{ \| \Theta e(k|t) \|^2 + \| \Lambda \Delta u(k-1|t) \|^2 + \| \Gamma \Delta y(k|t) \|^2 \right\} \tag{7.51}$$

subject to:

$$\begin{aligned} u^- \leq\ & u(k|t) \leq u^+ \\ \Delta u^- \leq\ & \Delta u(k|t) \leq \Delta u^+ \\ y^- - d(t) \leq\ & \hat{y}(k|t) \leq y^+ - d(t) \\ \Delta y^- \leq\ & \Delta y(k|t) \leq \Delta y^+ \\ & u(\tau|t) = u(t+M-1|t),\ \forall \tau > t+M-1 \end{aligned}$$

$$\begin{aligned} \hat{y}(t+1|t) &= \mathcal{N}[y(t), y(t-1), \cdots, u(t|t), u(t-1), \cdots] \\ \hat{y}(t+2|t) &= \mathcal{N}[\hat{y}(t+1|t), y(t), \cdots, u(t+1|t), u(t|t), \cdots] \end{aligned}$$

$$\vdots$$

$$\hat{y}(t{+}P|t) = \mathcal{N}[\hat{y}(t{+}P{-}1|t), \hat{y}(t{+}P{-}2|t), \cdots, u(t{+}P{-}1|t), u(t{+}P{-}2|t), \cdots]$$

where

$$\begin{aligned}
e(k|t) &= \bar{y}(k) - [\hat{y}(k|t) + d(t)] \\
d(t) &= y(t) - \hat{y}(t|t-1) \\
\mathcal{U}(t) &\triangleq [u(t|t)^\top, u(t{+}1|t)^\top, \cdots, u(t{+}M{-}1|t)^\top]^\top \\
\Delta y(\tau|t) &\triangleq \begin{cases} \hat{y}(t{+}1|t) - y(t) & \text{if } \tau = t{+}1 \\ \hat{y}(\tau|t) - \hat{y}(\tau{-}1|t) & \text{otherwise} \end{cases} \\
\Delta u(\tau|t) &\triangleq \begin{cases} u(t|t) - u(t{-}1) & \text{if } \tau = t \\ u(\tau|t) - u(\tau{-}1|t) & \text{otherwise} \end{cases}
\end{aligned}$$

and Θ, Λ, and Γ are diagonal weighting matrices. Here, $(\cdot|t)$ indicates the condition that the plant measurement $y(t)$ is available at time t. For example, $\hat{y}(t{+}1|t)$ stands for the predicted output $\hat{y}(t{+}1)$ when $y(t)$ is available, and $u(t|t)$ stands for the manipulated variable $u(t)$ when $y(t)$ is available. Note that, for convenience, the dead time does not appear in the formulation. It is straightforward to account for the process dead time in the formulation.

Notice that an additional penalty term $\Gamma\Delta y$ is included in the objective function Eq. (7.51). This term (change of the controlled variable) is seldom considered in most MPC formulations. Particularly, it is not considered in the MPC methods where a linearized model is used. However, since the objective function of an MPC formulation can be tailored to meet any specified requirement, one should not limit its form to one in which only one penalty term on Δu is considered. The penalty term on Δy can be useful when the controlled variable cannot be changed drastically. For a non-square MIMO system with more manipulated variables than controlled variables, it can be easier to tune the weight matrix for Δy then for Δu. The case study that follows involves a *multi-input single output* system. It is found that tuning Γ is much more convenient than tuning Λ, and yet both approaches give similar results. For an unconstrained linear MPC problem, the objective function with the additional penalty term on Δy can similarly lead to a closed-form control law as the classical DMC does, except that in DMC the objective function does not have this penalty term [81].

7.3.3 Solution Methods to Nonlinear MPC

FSQP Utilizing Localized Jacobian

The NNMPC algorithm involves a nonlinear programming problem. For solving this NLP problem, several techniques can be used. For example, Edgar and Himmelblau [16] discussed several NLP solution techniques. Among others, the sequential quadratic programming (SQP) method is perhaps the best method. In SQP techniques, a general NLP problem is *locally* approximated by a quadratic problem, and solved by quadratic programming techniques *sequentially*. Superlinear convergence can take place with only first order derivative information of the Lagrangian. Therefore the SQP method is chosen to solve the NNMPC problem. However, as pointed out by some researchers [6], the standard SQP method might not lead to a feasible solution since the objective function as well as all the constraints are linearized. A modified version of the SQP method is proposed by Tits and co-workers [56, 95], in which the feasibility is assured at each iteration. This approach is referred to as FSQP (Feasible SQP). In FSQP, an additional *feasible descent direction* is used. A feasible solution is guaranteed in every single iteration even before an optimal solution is found.

From a practical point of view, it is more important that the solution be feasible than optimal. For example, if an NNMPC controller is implemented on-line, and an optimal solution cannot be obtained within one control interval, one would prefer a sub-optimal but feasible solution than one which is more optimal yet infeasible. In this case, the FSQP algorithm offers greater possibility for on-line application.

In order to have an efficient performance of the FSQP algorithm, one can provide an analytical expression of the gradient for the objective function. To to so, one can use the ordered derivatives to derive the necessary gradient. First, one arranges the variables used in the NNMPC formulation in an *ordered* fashion as shown in Fig. 7.22. Notice that calculating the derivative $\partial E(t)/\partial u(\tau|t)$ is equivalent to calculating the input *delta* of the neural network at time τ in the control horizon, i.e.

$$\frac{\partial^+ E(t)}{\partial u_j(\tau|t)} \equiv \delta^u_{j^*_\tau}(\tau) \tag{7.52}$$

where $\tau = t, t+1, \cdots, t+M-1$, and

$$j^*_\tau = N_y K_y + j \tag{7.53}$$

$$\cdots u(\tau) \rightarrow \left\{ \begin{array}{l} E(t) \\ \underbrace{x^u(\tau) \rightarrow \zeta(\tau) \rightarrow h(\tau) \rightarrow \hat{y}(\tau) \rightarrow}_{\text{network}} \\ x^u(\tau{+}1) \ \cdots \\ x^u(\tau{+}2) \ \cdots \\ \quad \vdots \\ x^u(K^*) \ \cdots \end{array} \right. \left\{ \begin{array}{l} E(t) \\ x^y(\tau{+}1) \ \cdots \\ x^y(\tau{+}2) \ \cdots \\ \quad \vdots \\ x^y(\tau{+}N_y) \ \cdots \end{array} \right.$$

Figure 7.22: The variables in the NNMPC formulation can be arranged in an *ordered* fashion. For example, the variable $u(\tau)$ precedes directly $E(t)$, $x(\tau)$, etc., and $y(\tau)$ precedes directly $\eta(\tau)$. Please refer to the derivation of the Unified Delta Rule.

Although the variable $u(\tau|t)$ has extra "connections" to $E(t)$ and the input variables in the future times (e.g. $x^u(\tau{+}1)$), using the ordered derivatives these extra "connections" do not constitute any inconvenience in terms of the gradient derivation.

Using the *Unified Delta Rule* [83, 82], the calculation of $\delta^u_{j^*_\tau}$ can be easily expressed as follows:

$$\delta^u_{j^*_\tau}(\tau) = \frac{\partial^\circ E(t)}{\partial u_j(\tau|t)} + \sum_{k=1}^{N_h} \delta^h_k(\tau) v_{j^\tau k} + \sum_{\bar{\tau}=\tau+1}^{K^*} \delta^x_{j^u_{\bar{\tau}}}(\bar{\tau}) \tag{7.54}$$

where

$$K^* \triangleq \begin{cases} P & \text{if } \tau = M{-}1 \\ \tau{+}K_u & \text{otherwise} \end{cases} \tag{7.55}$$

and

$$j^u_{\bar{\tau}} = N_y K_y + N_u(\bar{\tau} - \tau{-}1) + j \tag{7.56}$$

Notice that $\partial^\circ E(t)/\partial u_j(\tau|t)$ does not vanish, and that the third term on the right hand side of Eq. (7.54) corresponds to the extra connections from $u(\tau)$ in the future, e.g. $x^u(\tau{+}P)$. Then, the required delta of the network can be easily calculated using the *unified delta rule* as:

$$\delta^x_j(\tau) = \left\{ \sum_{k=1}^{N_h} \delta^h_k(\tau) v_{jk} \right\} \tag{7.57}$$

$$\delta_j^h(\tau) = \left\{ \sum_{k=1}^{N_y} \delta_k^y(\tau) w_{jk} \right\} \sigma_j{}'[h_j(\tau)] \tag{7.58}$$

$$\delta_j^y(\tau) = \left\{ \frac{\partial^\circ E(t)}{\partial \hat{y}_j(\tau)} + \sum_{\bar{\tau}=\tau+1}^{\tau+K_y} \delta_{j_{\bar{\tau}}^y}^x(\bar{\tau}) \right\} \tag{7.59}$$

where

$$j_{\bar{\tau}}^y = N_y(\bar{\tau} - \tau - 1) + j \tag{7.60}$$

Notice that $\delta(\tau) = 0, \ \forall \ \tau > P$.

QP Utilizing Linearized Models

As mentioned earlier, there are a number of solution techniques proposed for solving nonlinear model predictive control problems. Among these techniques, one of the most promising is proposed by García [18]. This technique is an extension of QDMC to nonlinear processes, and thus it is referred to as NLQDMC. In this approach, a nonlinear model is used to evaluate the effects of previous manipulated variable changes and future disturbances. However, the future manipulated variables are calculated based on a linearized version of the nonlinear model. The linearized model is a step response model like DMC. In such calculations, only Quadratic Programming is needed because only a linearized model is involved. Computationally, solving a QP makes this algorithm attractive for on-line implementation. This algorithm however can lead to instabilities when applied to open-loop unstable processes. Recently, Gattu and Zafiriou [22] modified the algorithm so that it can also be applied to open-loop unstable processes. This accomplishment is made possible by incorporating state estimation theory with the original NLQDMC. In this modified NLQDMC, a steady state Kalman filter gain matrix is calculated at each sampling time.

The NLQDMC and its modification can certainly be used for solving the neural network model predictive control problem presented in this chapter. Along with the speed gained by a neural network model over the conventional ODE models, the fast computation time for a QP solution (for solving a linearized model derived from the nonlinear neural network model) can make NNMPC even more attractive as an algorithm for on-line implementation.

Arkun and co-workers [59, 24] proposed an extended DMC approach in which a linearized model is arranged in parallel with the nonlinear model. The linearized model is updated until its output equals that of the nonlinear model. As discussed earlier, in an IMC control scheme, the feedback signal to the controller contains not only the external disturbance to the process but also the plant-model mismatch (c.f. Fig. 7.19 and Fig. 7.21). Therefore, in their method, the discrepancy due to the linearization of the nonlinear model can be attributed to an extra disturbance. This method although it invokes solving the control problem as a QP, requires iteration to handle the additional disturbance due to nonlinearity. Thus it may be more computationally expensive than the NLQDMC approach.

Since the NLQDMC approach linearizes the nonlinear model only once at every control interval, it might produce an inaccurate solution over the prediction horizon. Several researchers have discussed a Newton-type MPC, in which the nonlinear model is repeatedly linearized over the prediction horizon [9, 43, 42]. This repeated linearization can significantly decrease the error of the linearized model. Thus, it may produce a more accurate solution. However, extra computation time will also make this type of approach less attractive for on-line implementation.

7.3.4 Case Study

In this section, a typical chemical process is used to illustrate how a neural network dynamic model can be employed in the Model Predictive Control schemes.

A Polymer Reactor

A continuous copolymerization process is considered, and Fig. 7.23 shows a schematic diagram of the process. Two streams are fed to the reactor: one carries a mixture of two monomers as reactants and the other carries the catalyst. The process also contains a control loop, where the catalyst flow is used to control the reactor temperature (T_I) by a PI (proportional and integral) controller. An on-line measurement of the polymer property is located at the outlet stream. Here, the polymer property is assumed to be the only product quality indicator. The desired polymer property (denoted as y) is maintained by manipulating the feed flow rate of the mixture (u_1), the ratio of the two monomers in the mixture (u_2), and temperature setpoint (u_3). As a result, the simplified process is a 3-input 1-output system. A first principles model developed by Kim and Choi [33] combined with a PI controller is simulated as the plant. Kim & Choi's model is a very complex model, which contains more than 40 highly coupled nonlinear differential

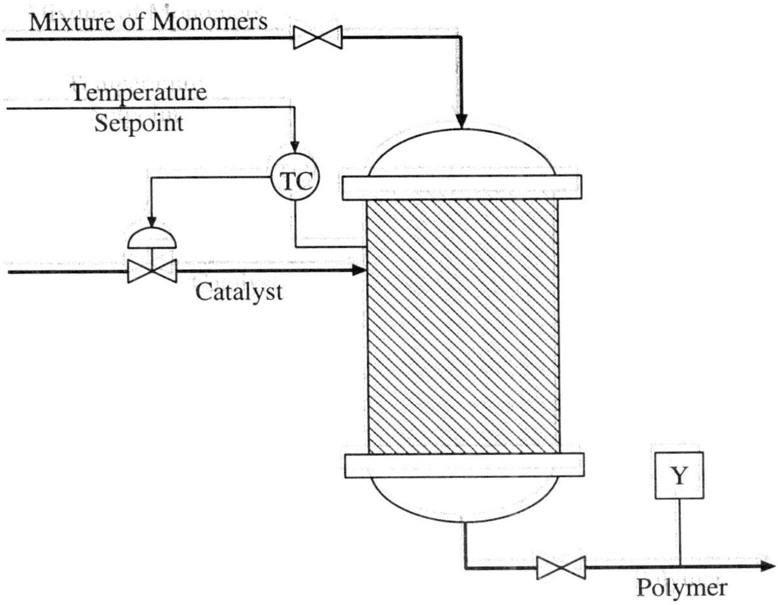

Figure 7.23: A continuous copolymerization process(simplified).

equations describing the fundamental polymerization mechanism. The PI controller, which is properly tuned, is implemented as indicated in Fig. 7.23.

In order to develop a neural network dynamic model, each of the three manipulated variables of the simulated plant is perturbed, independently, by a sequence of pseudo random signals. For a linear system, a pseudo random binary signal (PRBS) is used, but it might not be sufficient to generate informative data for a nonlinear system. Therefore, a pseudo random multi-state signal (PRMS) is used. In particular, the data contains 900 successive data points, in which the first 600 points are used for training and the remaining 300 are used for testing. Each data point is composed of three inputs and one output. A neural network is then trained on the data using the two neural net modeling approaches (FFN and RNN) discussed earlier. During the training phase, the measurement delay is not considered. Since the measurement delay in a real process is not exactly known and can vary over a very wide range, an additional uncertainty of the measurement delay must be accounted for in designing an MPC controller. A nominal measurement delay is denoted as τ_0, and the measurement delay is studied within the range from τ_d=0 to $3\tau_0$. The measurement delay is assumed to be known

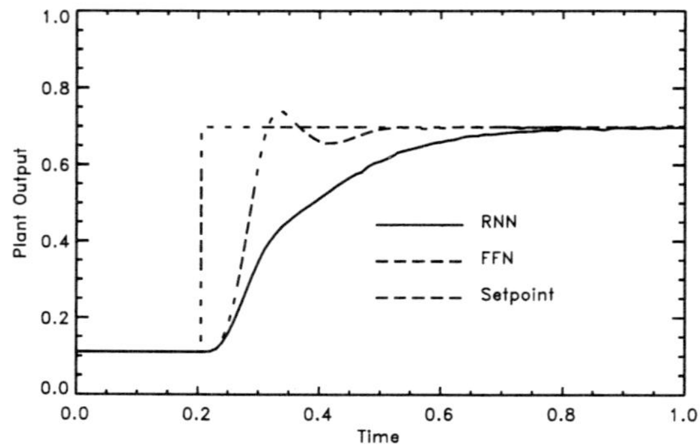

Figure 7.24: At this particular tuning parameter set, the FFN model can outperform the RNN model when the prediction horizon is one-step into the future for setpoint tracking. The tuning parameters are: P=1, M=1, λ=0.5, and γ=0.

within this range. Three cases are studied based upon different assumptions for the measurement delay.

Control Results

In the first case, the process is studied when there is no measurement delay ($\tau_d = 0$). When $\tau_d = 0$, the prediction horizon can be as short as one-step into the future. Recall that the total prediction horizon is $P+\tau_d$. The FFN modeling approach tends to develop a model that is optimal for one-step ahead prediction. It is interesting to see how the FFN model compares to the RNN model when all the tuning NNMPC parameters are fixed. To do so, both of the FFN model and the RNN model used in the NNMPC controller are studied for the case when $P = 1$ and $\tau_d = 0$. As shown in Fig. 7.24, the controller using the FFN model *does* yield better results than the one using the RNN model in the case of setpoint tracking. Similarly, the FFN model yields better results than the RNN model in the disturbance rejection case (see Fig. 7.25). Note that the magnitude of the step change and the pulse disturbance is large compared to the full span of the operating region. Fig. 7.26 and Fig. 7.27 show another example with different tuning parameters for setpoint tracking and disturbance rejection, respectively. In these two cases, the tunning parameters are determined by trial-and-error. The tuning parameters (M and λ) are the best when $P = 1$. As can be seen in these two figures, the performance of an RNN model is better than that of

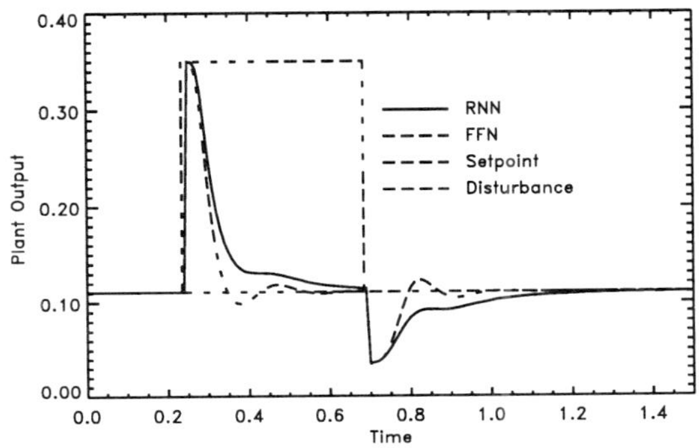

Figure 7.25: At this particular tuning parameter set, the FFN model can outperform the RNN model when the prediction horizon is one-step into the future for disturbance rejection. The tuning parameters are: P=1, M=1, λ=0.5, and γ=0.

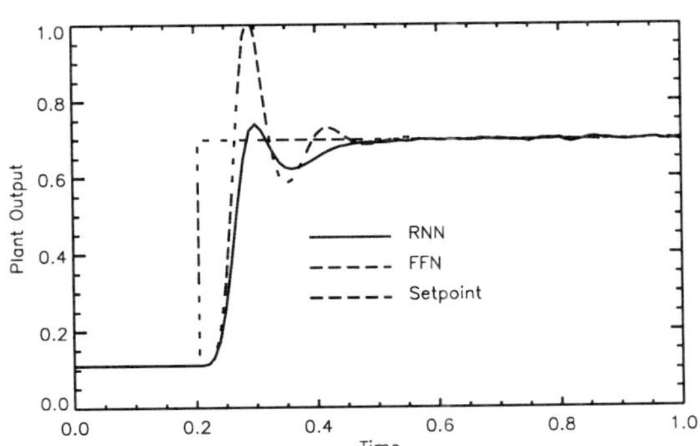

Figure 7.26: It is not always true that the FFN model can outperform the RNN model when the prediction horizon is one-step into the future for set-point tracking. The tuning parameters are: P=1, M=1, λ=0.2, and γ=0.

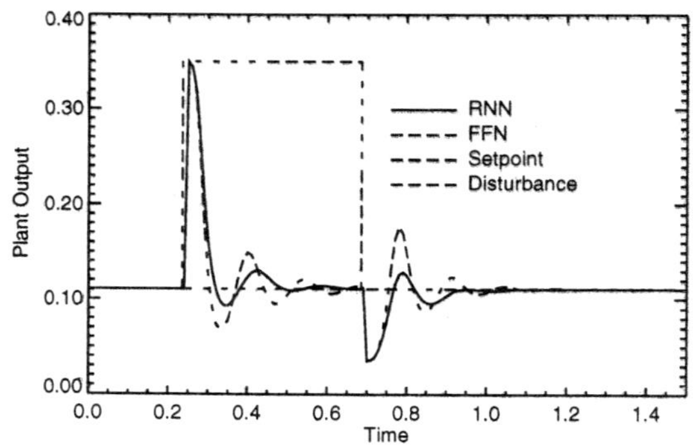

Figure 7.27: It is not always true that the FFN model can outperform the RNN model when the prediction horizon is one-step into the future for disturbance rejection. The tuning parameters are: P=1, M=1, λ=0.2, and γ=0.

the FFN model. The RNN model gives a smaller overshoot in both setpoint tracking and disturbance rejection than the FFN model. Although only the results for two particular sets of tuning parameters are shown here, it has been observed that the FFN model does not necessarily yield better results than the RNN model even when the shortest prediction horizon ($P = 1$ and $\tau_d = 0$) and same tuning parameters are used.

In reality, a measurement delay does exist, and its value is not exactly known. To account for the uncertainty of the measurement delay, one must specify a prediction horizon so that it covers the maximum possible delay. Therefore, in the second case, NNMPC using two different neural network models is studied for a larger prediction horizon. To study the effect of the prediction horizon without the influence of the measurement delay, it is still assumed that the measurement delay is precisely known. For a large prediction horizon, it has been concluded that the RNN model should outperform the FFN model since the the FFN model deteriorates very quickly as the prediction horizon increases [83]. Fig. 7.28 and Fig. 7.29 support this claim for both setpoint tracking and disturbance rejection, respectively, when all other tuning parameters are fixed. In addition, it has been observed that the performance of the FFN model decreases significantly with increasing prediction horizon. For a large prediction horizon, e.g. $P = 30$ or larger, it is extremely difficult to obtain a set of appropriate tuning parameters for the NNMPC controller when the FFN model is used. On the contrary, the NNMPC controller using the RNN model is relatively easy to tune. Conse-

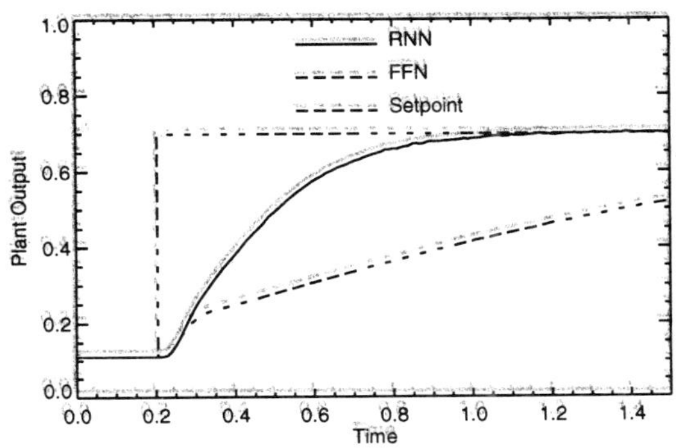

Figure 7.28: As the prediction horizon increase, the RNN tends to outperform the FFN model for setpoint tracking. The tuning parameters are: P=3, M=3, λ=5.0, and γ=0.

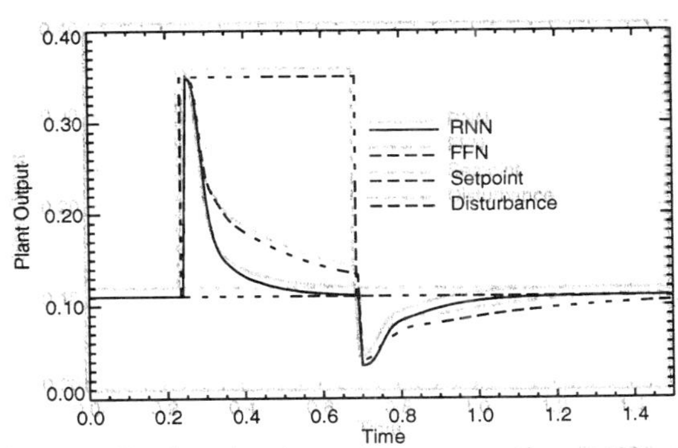

Figure 7.29: As the prediction horizon increase, the RNN tends to outperform the FFN model for disturbance rejection. The tuning parameters are: P=3, M=3, λ=5.0, and γ=0.

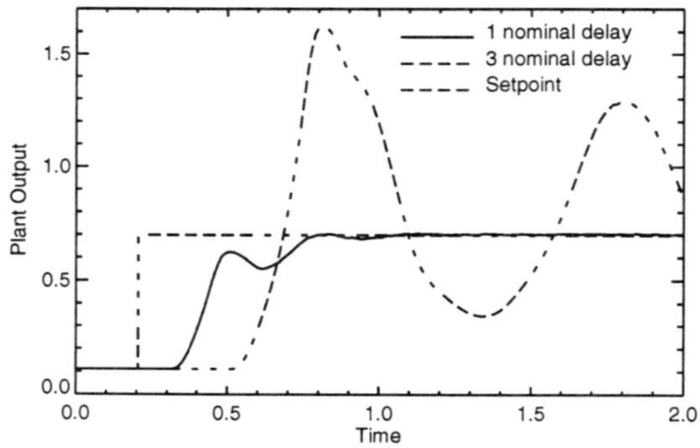

Figure 7.30: With the uncertainty of the measurement delay, the performance of the NNMPC deteriorates. In this case, the tuning parameters are as follows: M=3, P=3,λ=5.0, and γ=2.0. The overall prediction horizon is P+3τ_0.

quently, the RNN model is more appropriate to use than the FFN model in the NNMPC when the prediction horizon is large.

In the third case, the uncertainty of the measurement delay is studied. In order to account for this uncertainty, the prediction horizon must be large enough to cover the uncertainty region of the delay. Therefore, only the RNN model is used in this case. The controller with the RNN model has taken into account a measurement delay whose value is equal to the nominal measurement delay τ_0. The NNMPC controller is first tuned to control a plant where the actual measurement delay is also τ_0. As indicated by the solid line in Fig. 7.30, the closed-loop performance of the controller is quite aggressive. As one can expect, when the plant has a measurement delay other than the nominal τ_0, it is likely that the performance of the NNMPC will deteriorate. When the process has a measurement delay $\tau_d = 3\tau_0$, the closed-loop performance of the system is shown by the dotted line in Fig. 7.30. With the additional uncertainty of the measurement delay, the controller produces a significant overshoot/undershot and oscillation in the process output. Obviously, the controller must be de-tuned in order to inssure reasonable performance for all possible measurement delays. When the controller is de-tuned, it becomes less aggressive, and thus greater stability can be achieved. Although there is currently no general methodology for de-tuning the NLMPC, it is observed that guidelines for tuning the LMPC can be used [47]. In general, the NNMPC is de-tuned when

- The prediction horizon (or P) increases.

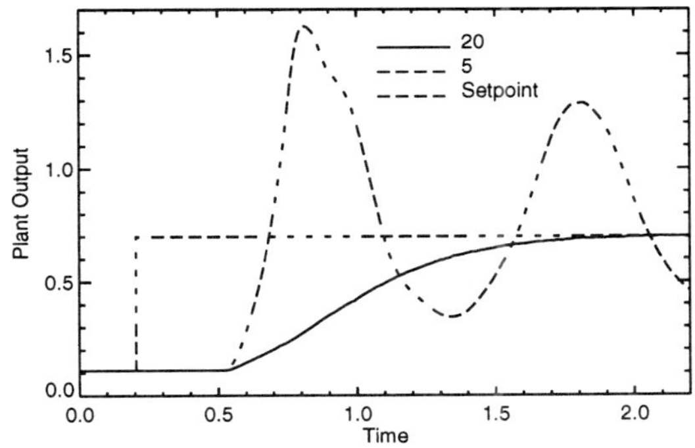

Figure 7.31: When the controller is de-tuned by increasing the λ, greater closed-loop stability can be achieve. The tuning parameters are: M=3, P=3, γ=2.0, and λ= As indicated. The overall prediction horizon is P+$3\tau_0$.

- The control move M horizon decreases.
- The weighting Θ on the error $\mathcal{E}$ decreases.
- The weighting Λ (or λ) on the control move changes $\Delta\mathcal{U}$ increases.
- The weighting Γ (or γ) on the output changes $\Delta\mathcal{Y}$ increases.

Among all these possible tuning methods, an example of the controller de-tuned by increasing the weighting λ on $\Delta\mathcal{U}$ is shown in Fig. 7.31. It can be seen that greater closed-loop stability is achieved by de-tuning the controller.

7.4 Summary and Conclusions

This chapter has given an introduction to neural networks and their use for nonlinear identification and control. Neural networks have been used in a number of process control applications including: nonlinear predictive control, for dynamic modeling in nonlinear control, fault detection, and sensor validation and correction. While most applications to date have been done in universities, a number of real world applications have been reported. This chapter has explored in depth the use of multilayer perceptron networks for dynamic modeling and control. Another important network approach involves radial basis networks. The use of recurrent networks for achieving long term model accuracy has been illustrated on data from a real system, an activated sludge waste water plant. For control of a simulated chemical reactor the robustness of neural network based approaches has been demonstrated.

Neural networks hold great promise as process modeling tools. They are well matched to the modeling requirements of the process industries. Neural networks can be used to develop models, both steady state and dynamic, from data. Thus, they have the potential to be used as cost effective, generic modeling tools. Initial results on their use in control systems have shown neural networks to be very robust to modeling errors. Once a process model is available then it can be used in many ways to improve process operation. As more experience is obtained, it will be possible to get a better assessment of whether neural networks can significantly impact process operation. If so, then their use will certainly grow.

References

[1] D. Ackley, G. Hinton, and T. Sejnowski. A learning algorithm for boltzmann machines. *Cognitive Science*, 9:147–169, 1982.

[2] J. Albus. A new approach to manipulator control: The cerebellar model articulation controller. *Trans. of ASME J. of Dyn. Sys., Meas. and Control*, 97:220–227, 1975.

[3] J. Anderson. *"General Introduction" in NEUROCOMPUTING: Foundations of Research.* MIT Press, 1988. J. A. Anderson and E. Rosenfeld, Editors.

[4] A. Doyle F. Aoyama and Venkatasubramanian V. Control-affine fuzzy neural network approach for nonlinear process control. *J. Process Control*, 1995. (in press).

[5] K. L. Åström and T. J. McAvoy. Intelligent control. *Journal of Process Control*, 2:115–126, 1992.

[6] B. W. Bequette. Nonlinear control of chemical processes: A review. *Industrial & Engineering Chemistry Research*, 30:1391, 1991.

[7] N. Bhat and T. J. McAvoy. Use of neural nets for dynamic modeling and control of chemical process systems. *Computers & Chemical Engineering*, 14(5):573–583, 1990.

[8] N. V. Bhat and T. J. McAvoy. Use of neural nets for dynamic modeling and control of chemical process systems. *Proceedings of American Control Conference*, pages 1336–1341, 1989. Pittsburgh, PA.

[9] D. D. Brengel and W. D. Seider. Multistep nonlinear predictive controller. *Industrial & Engineering Chemistry Research*, 28(12):1812–1822, 1989.

[10] L.H. Chen and B. Joseph. On-line optimization using a two-phase approach: An application study. *Industrial & Engineering Chemistry Research*, 26(9):1924–1930, 1987.

[11] C. R. Cutler and B. L. Ramaker. Dynamic matrix control: A computer control algorithm. *AIChE Annual Meeting*, 1979.

[12] C. R. Cutler and B. L. Ramaker. Dynamic matrix control: A computer control algorithm. *Proceedings of American Control Conference*, 1980.

[13] D. Dong and T. J. McAvoy. Nonlinear principal component analysis using neural networks. *Computers & Chemical Engineering*, 1993. submitted.

[14] J. W. Eaton and J. B. Rawlings. Feedback control chemical processes using on-line optimization techniques. *Chemical Engineering Communication*, 14(4/5):469–497, 1990.

[15] C. G. Economou, M. Morari, and B. O. Palsson. Internal model control: An extension to nonlinear systems. *Industrial & Engineering Chemistry Research*, 25(2):403–411, 1986.

[16] T.F. Edgar and D.M. Himmelblau. *Optimization of Chemical Processes.* Prentice-Hall, 1989.

[17] B.A. Finlayson. *Nonlinear Analysis in Chemical Engineering.* McGraw-Hill, New York, 1980.

[18] C. E. García. Quadratic/dynamic matrix control of nonlinear processes. *AIChE Annual Meeting*, 1984.

[19] C. E. García and M. Morari. Internal model control: 1. a unifying review and some new results. *Industrial & Engineering Chemistry Research*, 21(1):308–323, 1982.

[20] C. E. García and A. M. Morshedi. Quadratic programming solution of dynamic matrix control (QDMC). *Chemical Engineering Communication*, 46:73–87, 1986.

[21] C. E. García, D. M. Prett, and M. Morari. Model predictive control: Theory and practice–a survey. *Automatica*, 25(3):335–348, 1989.

[22] Gangadhar Gattu and Evanghelos Zafiriou. Nonlinear QDMC with state estimation: Formulation and application to an industrial challenge semi batch process. *Proceedings of American Control Conference*, 3:2866–2871, 1991.

[23] S. Grossberg. Adaptive pattern classification and universal recoding: I. parallel development and coding of neural feature detectors. *Biological Cybernetics*, 23:121–134, 1976.

[24] Evelio Hernández and Yaman Arkun. Neural network modeling and an extended DMC algorithm to control nonlinear processes. *Proceedings of American Control Conference*, 3:2454–2459, 1990.

[25] Evelio Hernández and Yaman Arkun. A study of the control relevant properties of backpropagation neural net models of nonlinear dynamic systems. *Computers & Chemical Engineering*, 16(4):227–240, 1992.

[26] G. Hobson. Neural net applications at PSP. *National Petroleum Refiners Association Meeting*, pages CC–90–137, Oct. 1990. Seattle, WA.

[27] K. Hoo and J. Kantor. *Chemical Engineering Communication*, 37:1–10, 1985.

[28] J. J. Hopfield. Neural networks and physical systems with emergent collective computational abilities. *Proceedings of National Academy of Sciences*, 79:2554–2558, 1982.

[29] K. Hornik, M. Stinchcombe, and H. White. Multilayer feedforward networks are universal approximators. *Neural Networks*, 2(5):359–366, 1989.

[30] J. C. Hoskins and D. M. Himmelblau. Artificial neural network models of knowledge representation in chemical engineering. *Computers & Chemical Engineering*, 12:881–890, 1988.

[31] S. Jang, B. Joseph, and H. Mukai. On-line optimization of constrained multivariable chemical processes. *AIChE Journal*, 33(1):26, 1987.

[32] I. Rico-Martinez R. Ecke R. Farber R. Kevrekidis and Lapedes A. Global bifurcations in rayleigh-benard convection. experiments, empirical maps and numerical bifurcation analysis. *Physica D.*, 71:342–262, 1994.

[33] K. J. Kim and K. Y. Choi. Continuous Olefin copolymerization with soluble Ziegler-Natta catalyst. *AIChE Journal*, 37(8):1255–1260, 1991.

[34] T. Kohonen. Self-organizing formation of topologically correct feature maps. *Biological Cybernetics*, 43:59–69, 1982.

[35] T. Kohonen. *Self-Organization and Associative Memory.* Springer-Verlag, third edition, 1989.

[36] M. A. Kramer. Autoassociative neural networks. *Computers & Chemical Engineering*, 16(4):313–328, 1992.

[37] M. A. Kramer and J. Leonard. Limitations of the backpropagation approach to diagnosis and improvement with radial basis functions. *AIChE Annual Meeting*, 1990. Chicago, IL.

[38] P. Lee and G. Sullivan. *Computers & Chemical Engineering*, 12:573–580, 1988.

[39] J. Leonard and M. Kramer. Radial basis function networks for classifying process faults. *IEEE Control Systems Magazine*, 11:31–38, 1991.

[40] J. Leonard and M. A. Kramer. Improvement of the backpropagation algorithm for training neural networks. *Computers & Chemical Engineering*, 14(3):337–341, 1990.

[41] R. Levin, E. Gewirtzman and G. F. Inbar. Neural network architecture for adaptive system modeling and control. *Neural Networks*, 4:185–191, 1991.

[42] W. C. Li and L. T. Biegler. Newton-type controllers for constrained nonlinear processes with uncertainty. *Industrial & Engineering Chemistry Research*, 29:1647–1657, 1990.

[43] W. C. Li and L. W. Biegler. Multi-step Newton type control strategies for constrained nonlinear processes. *Chemical Engineering Research & Design*, 67:562, 1989.

[44] L. Ljung. *System Identification: Theory for the User*. Prentice-Hall, Inc., 1987.

[45] J. MacQueen. Some methods for classification and analysis of multivariate observations. *Proc. of the Fifth Berkley Symposium on Mathematical Statistics and Probability*, 1967. U.C. Berkley Press, CA.

[46] P. A. Minderman, Jr. and T. J. McAvoy. Neural net modeling and control of a municipal waste water process. *Proceedings of American Control Conference*, pages 1480–1484, 1993. San Francisco, CA.

[47] M. Morari, C. E. García, and D. M. Prett. *Model Predictive Control*. in press, 1992.

[48] Manfred Morari and Evanghelos Zafiriou. *Robust Process Control*. Prentice Hall, Englewood Cliffs, 1989.

[49] A. J. Morris, G. Montague, M. Tham, M. Aynsley, C. Massimo, and P. Lant. *Proceedings of Chemical Process Control IV*, page 585, 1991. published by CACHE.

[50] A. M. Morshedi, H. Y. Lin, and R. H. Luecke. Rapid computation of the jacobian matrix for optimization of nonlinear dynamic processes. *Computers & Chemical Engineering*, 10(4):367–376, 1986.

[51] C. F. Mosteller and R. E. Rourke. *Sturdy Statistics: Nonparametric and order statistics*. Addison-Wesley, 1973.

[52] E. Nahas, M. Henson, and D. Seborg. Nonlinear internal model control strategy for neural network models. *Computers & Chemical Engineering*, 16:1039–1057, 1992.

[53] K. S. Narendra and K. Parthasarathy. Identification and control of dynamic system using neural networks. *IEEE Transactions on Neural Networks*, 1(1):4–27, 1990.

[54] D. Nguyen and B. Widrow. Neural networks for self-learning control systems. *IEEE Control Systems Magazine*, pages 18–23, 1990.

[55] A. J. Owens and D. L. Filkin. Efficient training of the back-propagation network by solving a system of stiff ordinary differential equations. *Proceedings of International Conference of Neural Networks*, II:381–386, 1989.

[56] E. R. Panier and André L. Tits. On combining feasibility, descent and superlinear convergence in inequality constrained optimization. *Mathematical Programming*, 59(2):261–276, 1993.

[57] A. A. Patwardhan, J. B. Rawlings, and T. F. Edgar. Nonlinear model predictive control. *Chemical Engineering Communication*, 87:123–141, 1990.

[58] K. M. Pedersen, M. Kümmel, and H. Søeberg. Monitoring and control of biological removal of phosphorus and nitrogen by flow injection analyzers in a municipal pilot scale waste water treatment plant. 1990. Submitted to Analytical Chim. Acta.

[59] T. Peterson, Evelio Hernández, Yaman Arkun, and F. Joe Schork. Nonlinear predictive control of a semi-batch polymerization reactor by an extended DMC. *Proceedings of American Control Conference*, 2:1534–1539, 1989.

[60] M. Piovoso, K. Kosanovich, V. Rokhlenko, and A. Guez. A comparison of three nonlinear controller designs applied to a nonadiabatic first-order exothermic reaction in a cstr. *Proceedings of American Control Conference*, pages 490–494, June 1992. Chicago, IL.

[61] M. Piovoso and A. Owens. Sensor data analysis using artificial neural networks. *Proceedings of Chemical Process Control IV*, page 101, 1991. published by CACHE.

[62] T. Poggio and F. Girosi. Regularization algorithms for learning that are equivalent to mutlilayer networks. *Science*, 247:978–982, 1990.

[63] M. Pottmann and D. Seborg. Identification of non-linear processes using reciprocal multiquadratic functions. *Journal of Process Control*, 2:189–203, 1992.

[64] M. Pottmann and D. Seborg. A nonlinear predictive control strategy based on radial basis function networks. *Proceedings of IFAC DYCORD Symposium*, 1992. College Park, MD.

[65] M. Pottmann and D. Seborg. A radial basis function control strategy and its application to a ph neutralization process. *Proceedings of Second European Control Conf. ECC '93*, 1993. Groningen, Netherlands.

[66] D. C. Psichogios and L. H. Ungar. Direct and indirect model based control using artificial neural networks. *Industrial & Engineering Chemistry Research*, 30(12):2564–2573, 1991.

[67] J. Qin and T. Mc Avoy. Nonlinear FIR modeling via a neural net PLS approach for long term prediction. *Computers & Chemical Engineering*, 18, 1994. in press.

[68] S. Z. Qin and T. J. McAvoy. Nonlinear PLS modeling using neural networks. *Computers & Chemical Engineering*, 16(4):379–392, 1992.

[69] J. Richalet, A. Rault, J. L. Testud, and J. Papon. Model predictive heuristic control: Applications to industrial processes. *Automatica*, 14(5):413, 1978.

[70] R. Rico-Martinez and Kevrekidis I. Continuous-time modeling of nonlinear systems: Aneural network approach. *Proc. 1993 IEE int. Cont. on Neural Networks*, III:1522–1525, 1993.

[71] D. Rumelhardt, G. Hinton, and R. Williams. *Parallel Distributed Processing: Explorations in the Microstructures of Cognition*, volume 1. MIT Press, 1986.

[72] J. Saint-Donat, N. Bhat, and T. J. McAvoy. Neural net based model predictive control. *International Journal of Control*, 54(6):1453–1468, 1991.

[73] G. Samdani. Neural nets. they learn from examples. *Chemical Engineering*, pages 37–45, Aug. 1990.

[74] D. Schnelle and J. Fletcher. Using neural based process modeling in process control. *Proceedings of ISA-90 International Conference*, 1990.

[75] D. Seborg. Experience with nonlinear control and identification strategies. *Proc. Control '94 Conf*, 1994. Coventry, UK.

[76] P. B. Sistu and B. W. Bequette. Nonlinear predictive control of uncertain processes: Application to a CSTR. *AIChE Journal*, 37(11):1171–1723, 1991.

[77] O. J. M. Smith. Closer control of loops with dead time. *Chemical Engineering Progress*, 53(5):217–219, 1957.

[78] G. Stephanopoulos and C. Han. Intelligent systems in process engineering: A review. *Proceedings of PSE'94*, pages 1339–1366, 1994.

[79] M. Stone. Cross-validation: A review. *Math. Operatiosforsch. Statist., Ser. Statistics*, 9:1, 1978.

[80] H. Su and T. Mc Avoy. Integration of multilayer perceptron networks and linear dynamic models: A hammerstein modeling approach. *Industrial & Engineering Chemistry Research*, 32:1927–1936, 1993.

[81] Hongte Ted Su. *Dynamic Modeling and Model Predictive Control Using Generalized Perceptron Networks*. PhD thesis, University of Maryland, College Park, MD., 1992.

[82] Hongte Ted Su and Thomas J. McAvoy. Applications of neural network long-range predictive models for nonlinear model predictive control, 1993. Special MPC Issue. Accepted for publication.

[83] Hongte Ted Su, Thomas J. McAvoy, and Paul J. Werbos. Long-term predictions of chemical processes using recurrent neural networks: A parallel training approach. *Industrial & Engineering Chemistry Research*, 31:1338–1352, 1992.

[84] Kwaku O. Temeng, Phillip Dave Schnelle, Hongte Ted Su, and Thomas J. McAvoy. Neural model predictive control of an industrial packed bed reactor. 1992. AIChE Annual Meeting.

[85] L. H. Ungar, B. A. Powell, and S. N. Kamens. Adaptive networks for fault diagnosis and process control. *Computers & Chemical Engineering*, 14(4/5):561–572, 1990.

[86] V. Venkatasubramanian. Recall and generalization performances of neural networks for process fault diagnosis. *Proceedings of Chemical Process Control IV*, page 647, 1991. published by CACHE.

[87] B. Wahlberg and L. Ljung. Design variables for bias distribution in transfer function estimation. *IEEE Transactions on Automatic Control*, AC-31(2):134–144, 1986.

[88] K. Watanabe, I. Matsuura, M. Abe, M Kubota, and D. M. Himmelblau. Incipient fault diagnosis of chemical processes via artificial neural networks. *AIChE Journal*, 35:1803–1812, 1989.

[89] P. J. Werbos. Neurocontrol and related techniques. In A. J. Maren, editor, *Handbook of Neural Networks*. McGraw Hill, 1990.

[90] R. J. Williams and J. Peng. An efficient gradient-based algorithm for on-line training of recurrent network trajectories. *Neural Computation*, 2(4):490–501, 1990.

[91] M. J. Willis, C. DiMassimo, G. A. Montague, M. T. Tham, and A. J. Morris. On artificial neural networks in process engineering. *IEE part D*, 1991.

[92] B. E. Ydstie. Forecasting and control using adaptive connectionist networks. *Computers & Chemical Engineering*, 14:583–599, 1990.

[93] P. Young. Parameter estimation for continuous-time models: A survey. *Automatica*, 17(1):23–39, 1981.

[94] V. Zeman, R. V. Patel, and K. Khorasani. A neural network based control strategy for flexible-joint manipulators. *Proceedings of Conference Decision & Control*, pages 1759–1764, 1989.

[95] Jian L. Zhou and André L. Tits. Nonmonotone line search for minmax problems. *Journal of Optimization Theory & Application*, 76(3):455–476, 1993.

Index